Environmental Pollution Control Microbiology

Civil and Environmental Engineering

A Series of Reference Books and Textbooks

Editor

Michael D. Meyer

Department of Civil and Environmental Engineering
Georgia Institute of Technology
Atlanta, Georgia

1. **Preliminary Design of Bridges for Architects and Engineers**
   Michele Melaragno
2. **Concrete Formwork Systems**
   Awad S. Hanna
3. **Multilayered Aquifer Systems: Fundamentals and Applications**
   Alexander H.-D. Cheng
4. **Matrix Analysis of Structural Dynamics: Applications and Earthquake Engineering**
   Franklin Y. Cheng
5. **Hazardous Gases Underground: Applications to Tunnel Engineering**
   Barry R. Doyle
6. **Cold-Formed Steel Structures to the AISI Specification**
   Gregory J. Hancock, Thomas M. Murray, and Duane S. Ellifritt
7. **Fundamentals of Infrastructure Engineering: Civil Engineering Systems: Second Edition, Revised and Expanded**
   Patrick H. McDonald
8. **Handbook of Pollution Control and Waste Minimization**
   edited by Abbas Ghassemi
9. **Introduction to Approximate Solution Techniques, Numerical Modeling, and Finite Element Methods**
   Victor N. Kaliakin
10. **Geotechnical Engineering: Principles and Practices of Soil Mechanics and Foundation Engineering**
    V. N. S. Murthy
11. **Estimating Building Costs**
    Calin M. Popescu, Kan Phaobunjong, and Nuntapong Ovararin
12. **Chemical Grouting and Soil Stabilization: Third Edition, Revised and Expanded**
    Reuben H. Karol

13. **Multifunctional Cement-Based Materials**
    Deborah D. L. Chung
14. **Reinforced Soil Engineering: Advances in Research and Practice**
    Hoe I. Ling, Dov Leshchinsky, and Fumio Tatsuoka
15. **Project Scheduling Handbook**
    Jonathan F. Hutchings
16. **Environmental Pollution Control Microbiology**
    Ross E. McKinney

Additional Volumes in Production

Environmental Pollution Control Microbiology

Ross E. McKinney
Professor Emeritus
University of Kansas
Lawrence, Kansas, U.S.A.

MARCEL DEKKER

MARCEL DEKKER, INC. NEW YORK • BASEL

Although great care has been taken to provide accurate and current information, neither the author(s) nor the publisher, nor anyone else associated with this publication, shall be liable for any loss, damage, or liability directly or indirectly caused or alleged to be caused by this book. The material contained herein is not intended to provide specific advice or recommendations for any specific situation.

Trademark notice: Product or corporate names may be trademarks or registered trademarks and are used only for identification and explanation without intent to infringe.

Library of Congress Cataloging-in-Publication Data
A catalog record for this book is available from the Library of Congress.

ISBN: 0-8247-5493-X

This book is printed on acid-free paper.

Headquarters
Marcel Dekker, Inc., 270 Madison Avenue, New York, NY 10016, U.S.A.
tel: 212-696-9000; fax: 212-685-4540

Distribution and Customer Service
Marcel Dekker, Inc., Cimarron Road, Monticello, New York 12701, U.S.A.
tel: 800-228-1160; fax: 845-796-1772

Eastern Hemisphere Distribution
Marcel Dekker AG, Hutgasse 4, Postfach 812, CH-4001 Basel, Switzerland
tel: 41-61-260-6300; fax: 41-61-260-6333

World Wide Web
http://www.dekker.com

The publisher offers discounts on this book when ordered in bulk quantities. For more information, write to Special Sales/Professional Marketing at the headquarters address above.

Current printing (last digit):

10 9 8 7 6 5 4 3 2 1

PRINTED IN THE UNITED STATES OF AMERICA

Preface

My interest in environmental pollution control microbiology began at MIT in 1948 when I realized that very little was known about this vital area of sanitary engineering technology. I became a serious student of environmental pollution control microbiology in 1950 when I began research on the microbiology and the biochemistry of floc-forming bacteria in activated sludge. I must admit that environmental pollution control microbiology has been my area of technical specialization since 1950. It has been a most interesting experience that I have never regretted. I was fortunate to have been a researcher, a teacher, and an engineering consultant on the design and operation of environmental pollution control systems.

In 1960 I reached a milestone when I decided to move from MIT to the University of Kansas. I forced myself to put the knowledge I was teaching at MIT into a book, *Microbiology for Sanitary Engineers*, published in 1962. As the years passed, I could never find the time to write a new version of my book as new information was accumulating at an ever faster rate. I reached a second milestone in 1993 when I decided to retire from the University of Kansas and find the time to write the new version of my book. That was 10 years ago. Time is a precious commodity that slips away before you realize it. My former students have continued to ask how my book is coming along. They seem most anxious to see how I have been able to condense all the mass of information that has been accumulated over the years. I am happy to report that I have finally succeeded in writing my new book.

Environmental Pollution Control Microbiology is not a compilation of all the information in this field. It is the summation of the information that I found to be valuable as a practitioner in the area of environmental pollution control. I hope this information will be of value to beginning students, practicing design engineers, plant operators, regulatory personnel, municipal officials, and industrial plant managers. The need for knowledge and understanding is far greater than the available information. I have tried to present technical information in a manner that is understandable to both technical experts and private citizens. For the experts in our field I commend the technical libraries for detailed information.

The references at the end of each chapter include general bibliographic references as well as specific references. Since many students seem to feel that information over five years old has no real value, I have included many older references to illustrate when important ideas originated. There is no need to rediscover important facts every five years; nor is it necessary for young researchers to disregard previous research in their search for unique ideas. The greatest achievements come from building on earlier knowledge. While much has been accomplished, much more needs to be learned. The future awaits.

I would like to acknowledge the encouragement and assistance of Dr. Michael Switzenbaum of Marquette University, Dr. Terry Baxter of Northern Arizona University, and Dr. Reza Shamskhorzani of U.S.Filter/Jet Tech in helping me to smooth some of the rough spots in my original manuscript. Other eyes are very helpful in seeing things differently. My wife, Margie, deserves all my appreciation for putting up with a decade of writing and rewriting. I know Margie was pleased when we reached the last page.

I owe a special thanks to all my students at MIT and at the University of Kansas. Their thirst for knowledge helped raise me to a level that I could not possibly have achieved alone. We formed a special bond that kept me stimulated, excited, and young in spirit. I also owe a special thanks to all the practicing engineers, equipment manufacturers, and treatment plant operators who let me demonstrate how to convert research results into full scale pollution control plants. We made progress together and had fun at the same time.

Hope springs eternal, each and every spring.

Ross E. McKinney

Contents

Chapter 1

INTRODUCTION

Environmental pollution control microbiology is currently an exciting and challenging area of science and engineering. Environmental pollution control microbiology is concerned with solving a broad spectrum of environmental pollution problems that affect people around the world from a microbiological point of view. On one hand, environmental pollution control microbiology is concerned with protecting people from pathogenic microorganisms; and on the other hand, it is concerned with the application of microbiology to solve a wide range of environmental pollution problems. Emphasis is placed on the environment rather than on individuals. Both science and engineering are involved in environmental pollution control. Scientists determine the environmental problems and evaluate different solutions. Engineers utilize the information supplied by scientists to design the most efficient systems to solve the environmental problems. By working together as a team, scientists and engineers are meeting the challenges of environmental pollution control in our modern world.

The concepts of *Environmental Pollution Control Microbiology* have their roots in conventional microbiology with its concerns for pathogenic microorganisms and public health. The roots are even deeper in chemistry, which forms the basis of all reactions between the chemical compounds that make up our world, and in biochemistry, which focuses on the chemistry of biological systems. The deepest roots are in civil engineering, which provides the basis for all the other areas of engineering contributing to the design and construction of pollution control facilities. Chemical engineering is one of the newest areas of engineering to become involved in solving environmental problems. It joins mechanical engineering and electrical engineering in contributing their special expertise. *Environmental Pollution Control Microbiology* was developed to show how microbiology, chemistry, and engineering are combined to provide real solutions to environmental pollution problems.

MICROBIOLOGY

Microbiology forms one of the cornerstones for environmental pollution control microbiology. Microbiology also has its roots in many different scientific areas. Microbiology is a combination of bacteriology, virology, mycology, phycology, protozoology, zoology, biochemistry, and mathematics. With such a widespread field, microbiology is made up of diverse groups of scientists rather than a single, coherent group. It will take microbiology many more years before it becomes a unified science, if it ever does. Microorganisms form the unifying theme for microbiology. Viruses, bacteria, fungi, algae, protozoa, rotifers, crustaceans and various worms make up the major groups of microorganisms that exist in the environment. Viruses are the most difficult group of organisms to study because they exist in the area between chemistry and microbiology. Viruses have the characteristics of pure chemical compounds and living microorganisms. Their small size and difficulty in culturing makes viruses hard to study; but their negative impact on plants and animals makes them a very important part of microbiology. In recent years environmental microbiologists have found that viruses are having a greater impact on environmental health than in the years past. As bacteria pathogens have been reduced, viruses have moved into the vacancy that has been created. Viruses are all parasitic pathogens without measurable respiration. They are also host specific. The uniqueness of viruses has created a special aura that greatly affects how the public deals with viruses.

Bacteria have attracted the greatest attention in microbiology because they are easy to cultivate and study in pure cultures and because they have had a major influence on the health of people. Most of the early bacteriological studies were directed at medical bacteriology to understand the causes of common diseases affecting people. As progress was made in controlling disease-producing bacteria, bacteriological studies shifted to understanding the biochemistry of bacteria and the quantitative relationships in bacterial growth. The transmission of pathogenic bacteria through food and water has long been the focus of environmental microbiologists. In recent years efforts have shifted from pathogenic bacteria to helpful bacteria, such as those involved in the treatment of municipal wastewater and industrial wastewater. The simplicity of the bacteria biochemistry has proven to be of major importance in pollution control microbiology. Understanding how bacteria metabolize different organic compounds provides the basis for designing and operating new biological wastewater treatment systems. New techniques and new media for growing bacteria have yielded new bacteria to work with. In spite of all the knowledge we have on bacteria, it has been indicated that the majority of bacteria have yet to be found and examined in pure culture. New media and new techniques will be required to find and study these unknown bacteria that currently inhabit our environment.

Although bacteria are related to fungi, mycology, the study of fungi, grew as a separate field. The problems with fungi lie in their complexities when compared with bacteria. Fungi undergo several phases in their life cycle, making them difficult to study. Most of the efforts in mycology have been directed towards taxonomy than towards the biochemistry of fungi. Because of the economic impact of fungi in agriculture, mycology has been oriented primarily towards plants. The discovery of antibiotics and their application in the control of disease producing bacteria helped to stimulate interest in the biochemistry of fungi. Fungi are also important in environmental pollution control microbiology in the treatment of some industrial wastewaters and in composting of solid wastes. Recently, mycologists have been interested in fungi for the mass production of proteins to be used in animal feed. One group of fungi, the yeast, has developed into a separate area of mycology, the same as bacteria. Yeast cells are used in the production of specific foods and beverages. Yeast cells have also been used in some industrial wastewater treatment systems. Because of the economic value of yeast, considerable effort has been made to study and understand the biochemistry of yeast. More is known about the biochemistry of yeast than the other fungi.

Algae are photosynthetic microorganisms that have unique growth characteristics. Since algae depend on light for their source of energy, they are found primarily in water and on moist surfaces exposed to light. The large number of diverse algae in the environment stimulated phycologists, scientists who study algae, to focus their primary attention on taxonomy. A few phycologists were intrigued by the various mechanisms involved in photosynthesis. These phycologists began to examine the biochemistry of algae. Most recently, space research and the future potential of space exploration created renewed interest in algae biochemistry. The fact that algae utilize carbon dioxide and produce oxygen as a metabolic end product indicated that algae might have value in space travel where people will need oxygen and will produce carbon dioxide as one of their major waste products. Algae have also been studied as a potential source of protein for feed supplements. Algae have a special role to play for environmental pollution control microbiologists. Algae can conserve nitrogen and phosphorus in their protoplasm.

Adding algae to the soil can help improve soil characteristics and increase fertility for higher plants. Growth of algae in ponds has proven useful in stimulating fish farming. Since algae and bacteria do not compete against each other, they can be grown together to take advantage of the best characteristics of both groups of microorganisms. The potential for algae in environmental pollution control is just beginning to be recognized.

Microscopic animals are larger and more complex than the microscopic plants. They are easy to see and recognize at relatively low power magnification under the microscope. It is not surprising that the protozoologists are also concerned about taxonomy and the distribution of protozoa in the environment. The difficulty in isolating and growing protozoa in pure culture slowed studies on their biochemistry. Eventually, techniques were developed that permitted the separation and growth of protozoa, free of bacterial contamination. Even then, most of the productive protozoa studies dealt with the interaction of protozoa and bacteria, since protozoa use bacteria as their primary source of nutrients. Like bacteriology, a portion of protozoology has been concerned with pathogenic protozoa. Recently, concerns over *Giardia, Cryptosporidium,* and *Pfisteria* have focused attention on these pathogenic protozoa and their place in environmental concerns. Part of the problem lies in the fact that these pathogenic protozoa are most dangerous to the very young, the elderly, and those individuals with damaged immune systems. The pathogenic protozoa are parasitic pathogens, living off host animals. Fortunately, most protozoa are non-pathogenic and assist in the reduction of pathogenic bacteria in the environment. Protozoa are also major partners in the microbiology of biological wastewater treatment systems.

Higher animals such as rotifers, crustaceans, and worms have come under the scrutiny of zoologists, scientists who study animals. These microscopic animals live on bacteria, algae, and small protozoa and are not grown in sterile environments. They are quite large with some being macroscopic rather than microscopic in size. The higher animals have complex digestive systems and are much more sensitive to toxic materials than the other microorganisms. Since these organisms form an important link in the aquatic food chain, many studies have been made on the environmental conditions that adversely affect their growth. Concern for water pollution has become a major issue for many aquatic zoologists. A number of field studies have been made, using organism counts and species diversity to indicate environmental damage to streams and lakes. Recently, the EPA proposed using *Ceridaphnia* as an indicator of wastewater effluent toxicity. There is no doubt that zoologists will have an increasing role in water pollution control studies.

The very breadth of microbiology poses a challenge to anyone with a major interest in pollution control microbiology. It is not possible for everyone to be an expert in all areas of microbiology. Yet, it is important to have a good understanding of the different areas to provide communication with the technical specialists in the different areas of microbiology. There is nothing wrong in developing a special interest in one area of microbiology, as long as you keep your focus on the total picture and not on the details. It is important to realize that while pure cultures are critical in the study of microbiology, it is equally important to understand the competition that exists between the different organisms. The fastest growing pure culture organism may not be the predominant organism in the real environment. Laboratory conditions often fail to simulate the natural environment and allow entirely different growth patterns to occur. The keys to pollution control microbiology are in examining the different organisms in their natural environment and in understanding the how and why of their growth characteristics in that environment.

CHEMISTRY

Chemistry has a very special place in environmental pollution control microbiology, the same as in the other branches of microbiology. Chemistry is a basic science concerned with the composition of all matter. Chemistry helps to explain how materials react with each other. Experience, experiments, and faith form the basis for chemistry. Experience came first and produced as many wrong answers as right answers. Experience soon produced experiments to find correct answers and to provide a reproducible basis for chemical reactions. Faith is the glue that holds chemistry together. No one sees the atoms that make up matter; but we have faith that the atoms are all there. Experience shows us that the same materials react the same way every time we bring them together for a reaction. Experiments help us learn new reactions and to show others how old reactions occur. The accumulation of knowledge in the literature helps us bridge the gaps between experience and experiments. While parts of chemistry are qualitative, ultimately it must become quantitative. Mathematics is an integral part of chemistry. Mathematics helps us understand how much of chemical A reacts with chemical B to produce chemical C. Mathematics helps us explain how fast chemical reactions occur. We accept the fact atoms are in motion until the temperature reaches absolute zero. Like chemistry, much of mathematics is based on faith.

Chemistry has its specialties: qualitative and quantitative analyses, organic chemistry, inorganic chemistry, physical chemistry, colloidal chemistry, nuclear chemistry, and biochemistry, plus lots of subspecialties. Each specialty is important to the field of chemistry and to environmental pollution control microbiology.

Qualitative analysis and quantitative analysis are essential in environmental pollution. They permit the measurement of different pollutants found in the environment and their magnitude. Organic chemistry is important since all living creatures are composed of a multiplicity of organic compounds. Again we have lots of faith to accept how organic compounds are constructed and how they react. Technology is slowly providing new answers about the structure of complex chemical compounds; but even here faith is important. The same is true of inorganic chemistry. Fortunately, there are not as many different inorganic compounds as there are different organic compounds. The ability of the chemist to create new organic chemicals has led to the production of new products that have greatly changed the world. Not all of the changes have been for the better. Inorganic chemicals form an important part of trace elements that are critical to all biological life. Physical chemistry teaches us about gases, rates of reactions, and energetics of reactions. Colloidal chemistry teaches us about very tiny particles and how those particles react and interact. Bacteria are larger than true colloids; but they react similar to large colloids. Nuclear chemistry is concerned with radioactive elements and compounds and how they react with the rest of the world. Biochemistry is important since it focuses chemistry on biological systems. It is not surprising that biochemistry covers all areas of chemistry, creating as much confusion as light. Although much knowledge has been gained over the years in all areas of chemistry, much is still to be learned. As new knowledge in chemistry is uncovered, that knowledge will be quickly applied to environmental pollution control microbiology.

Environmental chemistry has grown up in recent years as chemists began to recognize that the environment is an important part of applied chemistry. With emphasis on environmental pollution, environmental chemistry has a strong base in chemical analysis and biochemistry, as it affects the air, water and land environments. Environmental chemistry helps make the bridge between environmental pollution control microbiology and all of the major areas of chemistry. As with microbiology, it is not necessary to be a specialist in all areas of chemistry; but one must have knowledge of the major areas of chemistry that affect microbiological systems in the environment.

ENGINEERING

Engineering is the third cornerstone in the foundation for environmental pollution control microbiology. Engineering is concerned with the various structures and systems employed in solving environmental pollution problems. Because of the diversity of environmental problems and the large number of different solutions, there are many different types of engineers involved. Civil engineers were the first group of engineers to recognize environmental pollution problems. Stormwater runoff resulted in civil engineers designing and constructing storm sewers to remove the excess water from city streets and to discharge it in nearby rivers and streams. Domestic sewage was not a problem for civil engineers initially, since sewage was collected in cesspools adjacent to each house. Periodically, the cesspools were cleaned out by being pumped into a tank on the back of a wagon and hauled out into the country for disposal on land as fertilizer. As cities grew, the demand for more fresh water became a serious issue. The number of water wells increased and the water table dropped. The need for a permanent water supply that could meet the current demands, as well as future demands, became one of the primary tasks for civil engineers. As new water supplies were developed, it was possible to pipe water to every customer willing to pay for a connection and the water. The improved municipal water supplies led to the increased use of water closets, bathtubs, and kitchen sinks. An unexpected consequence of the improved water supplies was a corresponding increase in wastewater production. The increased flow of wastewater to the cesspools caused them to overflow into the storm sewers, creating water pollution in the adjacent streets and in the receiving streams. Civil engineers soon had to design sewers to collect both sanitary wastewater and stormwater. It is not surprising that differences of opinion arose as to whether to use separate sewers or to combine both wastewaters into a single sewer. The initial concerns stemmed from the fact that scientists and engineers believed that diseases could be transmitted by air from decomposing sewage. Hydraulic traps were constructed in the house connections to prevent sewage odors from entering the connected houses. It soon became apparent that combined sewers were more economical than separate sewers since both wastewaters were discharged to the same rivers. It was not readily recognized that the discharge of wastewaters into natural waterways would create serious environmental pollution problems for downstream water users as the communities increased in population. It did not take long before water pollution became serious for the large cities in the United States.

Concerns over polluting public water supplies soon led to research on various methods for treating polluted streams. Intermittent sand filters were developed at the Lawrence, Massachusetts, Experiment Station for treating polluted river water to be used for drinking water. It was shown that intermittent sand filters removed the pathogenic bacteria causing typhoid fever that was endemic in the United States. Intermittent sand filters soon gave way to rapid sand filters that could process more water at a faster rate. The application of chlorine to filtered water insured the safety of the water from rapid sand filters by killing the pathogenic bacteria that were not removed in the water treatment plants. The success of intermittent sand filters for treating polluted river water led to their use for treating domestic wastewater prior to discharge in streams and rivers. Since England had greater water pollution problems than the United States in the 1890s, British engineers modified the intermittent sand filter and produced the first rock media trickling filter. The success of trickling filters in reducing pollution from domestic wastewater prior to discharge into streams and rivers led to its use in England, Europe, and the United States. British research on improving wastewater treatment led to the development of the activated sludge process in 1914. Activated sludge proved to be the most efficient wastewater treatment process and is still widely used throughout the world. The success of activated sludge is a primary example of pollution control microbiology at its best. While wastewater treatment removed contaminants from the water, large quantities of sludge remained to be returned to the environment. Since time began, wastewater sludges were always returned back to the environment as fertilizer. Unfortunately, pathogenic microorganisms were transmitted through the wastewater sludge to some crops used for human consumption. Research found that anaerobic treatment could reduce the survival of pathogenic microorganisms in the wastewater sludges; but it was Karl Imhoff's development of a practical design to treat wastewater sludges that resulted in large-scale treatment systems for wastewater sludges. Anaerobic treatment of wastewater sludges developed slowly over the years until researchers became interested in a more detailed examination of the anaerobic microorganisms responsible for sludge digestion. This is one of the more interesting areas of environmental pollution control microbiology that is still evolving. It is not surprising that water treatment plants increased more rapidly than wastewater treatment plants. Clean water was essential for cities to grow and prosper. Wastewater treatment was largely ignored until environmental pollution threatened to stop industrial development. Today, municipal water supplies and municipal wastewater treatment plants are part of the total water environment necessary to sustain future populations.

Cities also produced considerable amounts of solid wastes, which had to be collected and removed at regular intervals. Most city administrators considered solid wastes as a nuisance and did little more than collect the solid wastes. The solid wastes were carried outside the city limits and dumped onto low value land, where less fortunate people scavenged anything of value that remained. Because of limited coal resources for industrial operations, British engineers developed incinerators that burned solid wastes and recovered energy for industrial plants. Since the United States had ample coal for industrial plants, American engineers showed little interest in energy recovery incinerators. Eventually, civil engineers developed sanitary landfills to replace the open dumps for solid wastes. Conversion from open dumps to sanitary landfills progressed slowly since local politicians were reluctant to spend tax money on things that the public could not see and use directly. Ultimately, the environmental movement in the United States raised public concern over sanitary landfills and incinerators for handling solid wastes. Emphasis on solid waste recycling and minimization produced new engineering systems for collecting and separating the various components of solid wastes. A major effort has been directed towards large scale composting operations for yard wastes. It required environmental regulations to force local government to face these important problems. Engineers favored the new environmental regulations since the regulations resulted in new engineering projects.

As America's industries began to grow, they produced more wastes. Industrial plants were just like small cities, producing liquid, gaseous, and solid wastes. It is not surprising that industries followed the lead of cities and dumped their liquid wastes into nearby streams. The solid wastes were dumped on the land; and the gaseous wastes were sent up tall smokestacks into the atmosphere. When cities began to face their environmental pollution problems, a few of the larger industries followed suite. Civil engineers tried to use the same technology for industrial wastes as for municipal wastes. It quickly became apparent that civil engineers designing waste treatment facilities for industries needed more chemistry to understand how to properly process the various industrial wastes. As civil engineers changed their education to handle both municipal and industrial wastes, they evolved into sanitary engineers. The expanding industrial market for sanitary engineers caused chemical engineers to recognize that industrial waste treatment was as much in their area of expertise as it was in the sanitary engineer's area of expertise. Chemical engineers used their knowledge of chemical processes to design efficient industrial wastewater treatment processes, while using the civil engineers to design and construct the industrial treatment plants. The development of more complex mechanical equipment for solid waste collection and incineration brought the mechanical engineers into the environmental picture. The mechanical engineers also became involved in the design of various pieces of mechanical equipment used in water and wastewater treatment plants. The need for improved controls and better instrumentation to operate the various treatment plants allowed electrical engineers to become part of the environmental field. Air pollution created a new need for chemical and mechanical systems to remove the pollutants from the contaminated air before being discharged into the environment. The changing nature of the engineering solutions to our environmental problems produced the environmental engineer. Today, environmental engineers come from many different backgrounds. They may have originally been civil engineers, chemical engineers, mechanical engineers or electrical engineers. Now, all environmental engineers have a common focus on solving environmental pollution problems.

Many environmental pollution problems and solutions deal with microorganisms. In order to solve the microbial pollution problems, environmental engineers require a background in environmental pollution control microbiology. By understanding how the microbes live and grow in the environment, the environmental engineer can develop the best possible solution to biological waste problems. After construction of the desired treatment facilities, the environmental engineer is in position to see that they are properly operated to produce the desired results at the least cost. Engineering solutions to environmental problems are still evolving. There are no set patterns to solve all environmental problems. Each problem has its own unique challenges that require careful evaluation of all the different technical skills. One thing is certain; engineers will be involved in all the solutions.

POLLUTION CONTROL MICROBIOLOGY

This book will take you through a brief discussion of bacteria and their biochemistry. Understanding bacteria provides the basis for understanding the other microorganisms involved in environmental concerns. We will examine bacteria, fungi, algae, protozoa, rotifers, crustaceans, a few worms and viruses from an environmental point of view. Emphasis will be on the quantitative aspects of microbial growth and metabolism. Once we have examined the different organisms, we will look at the different environments, starting with soil. Soil provides a mixed environment for microorganisms and a surface for water to collect. Soil is a complex mixture of different chemical particles. As water moves through the void spaces around the particles, it carries microorganisms and chemical nutrients for their metabolism. The tiny soil particles filter out the large microorganisms and furnish surfaces on which the microorganisms grow. As the water moves deeper into the soil, it forms ground water. It is not surprising that the ground water has the best microbiological characteristics for domestic waster supplies. The deep soil is simply not a suitable environment for microorganisms. On the other hand, surface waters provide a real challenge for treatment when used for domestic water supplies. The water treatment processes making up municipal systems focus on removing microorganisms and other contaminants in the water. The concepts of each treatment unit will be examined from a microbiological point of view. Once the public has used the water, it is collected in sewerage systems and carried to the wastewater treatment plants. Microbial action has already begun when the wastewater enters the public sewerage system and continues throughout the treatment process. The wastewater treatment units process the wastewater by physical, chemical, and biological action to produce an effluent that can be safely returned to the environment. The residual solids removed by the wastewater treatment units must also be properly processed for return back to the environment. Wastewater treatment is not complete until both the liquid stream and the solid stream have been properly processed and safely returned to the environment. Next, we will look at solid wastes and how microbes can be used to stabilize the biodegradable organics. Composting has been used for stabilizing agricultural wastes for centuries. Only recently has composting been used to handle specific fractions of municipal solid wastes. Air microbiology will look at both the inside air and the outside air as a place for microbes to survive. One of the more interesting chapters deals with hazardous wastes. Hazardous wastes were originally defined as solid wastes. It soon became apparent that hazardous wastes were more liquid than solid. Some hazardous wastes are volatile, producing a gaseous phase. Microbial treatment of major types of hazardous wastes will be discussed since hazardous wastes are currently one of the most active environmental problems facing society.

It is hoped that this introduction to environmental pollution control microbiology will stimulate students in environmental studies and in environmental engineering to recognize the role that environmental microbiology plays in solving environmental problems. Although most of the emphasis will always be placed on the pathogenic microorganisms, most of the microorganisms in the environment are non-pathogenic. Learning to use mixtures of microorganisms to control the major environmental systems has become a major challenge for environmental engineers. Environmental pollution control microbiology should help operators of municipal wastewater and industrial waste treatment plants gain a better understanding of how their biological treatment unit's work and what should be done to obtain maximum treatment on a daily basis. Design engineers should gain a better understanding of how microbes provide the desired treatment and what the limitations are for good design. Even regulatory personnel can obtain a better understanding of the limits of their regulations and what concentration of contaminants can be allowed in the environment. Lastly, it is hoped that some members of the general public will gain a better appreciation for the environment in which they exist and what can be done to maintain the quality of that environment for everyone to enjoy.

THINGS TO REMEMBER

1. Microbiology is a combination of bacteriology, virology, mycology, phycology, protozoology, zoology, biochemistry, and mathematics.

2. Viruses are parasitic pathogens that are widely distributed in nature.

3. Bacteria are useful in the biological treatment of wastewater.

4. Fungi are more complex than bacteria, having several life cycle stages.

5. Algae are photosynthetic microorganisms, using sunlight for their energy.

6. Protozoa and bacteria are partners in wastewater treatment systems.

7. Chemistry is the basis for all microbial reactions.

8. Quantitative chemistry is essential for environmental pollution control.

9. Engineers use microbiology and chemistry in the design of biological environmental pollution control plants.

10. Environmental pollution control microbiology covers environmental problems in soil, water, and air.

Chapter 2

BACTERIA

Bacteria are the most important group of microorganisms in the environment. Their basic mission is to convert dead biological matter to stable materials that can be recycled back into new biological matter, keeping the biological cycle functioning without disruption. Only a few bacteria cause diseases in plants and animals; but those few disease-producing bacteria have had a profound impact on human development. It took people centuries before they recognized that bacteria caused many common diseases. Once it became known that bacteria could cause disease, efforts were directed towards developing methods to control the disease-producing bacteria. The first approach to the control of disease-producing bacteria came from research on different chemicals that were toxic to the bacteria being studied. This was primarily the approach taken by medical researchers. A second, experimental research approach examined the environment where the disease-producing bacteria were found. The environmental approach was slower than the medical approach, as it required finding a solution and then making major changes in the environment rather than treating individuals directly. Eventually, the environmental approach led to the development of large-scale environmental systems to reduce human exposure to disease-producing bacteria. The environmental approach proved to be an effective, economical approach for improving the health of the public at large and allowed our populations to increase at a very rapid rate. The increased number of people and their increasing longevity placed new stresses on the environment, creating environmental everywhere. Understanding the biology and the biochemistry of bacteria has helped in the development of biological pollution control systems to minimize the negative impact of pollution. The progress to date has helped make the United States one of the healthiest countries in the world and has shown the way for developing countries to create their own healthy environments. Researchers and engineers are continuously creating new concepts and systems for biological control of environmental pollution.

DEFINITION AND DESCRIPTION

Bacteria are very small, single cell microorganisms that contain all of their essential materials within a simple, continuous sac. To survive and grow, the bacteria must obtain their nutrients from the immediate environment. Since bacteria do not have mouths, their nutrients must be soluble and small enough to pass through the sac-like membrane that surrounds the living cell. The bacteria membrane is considered to be semi-permeable, allowing various materials to pass back and forth across the membrane. It is not surprising that the aqueous environment is the best environment for bacteria. The continuous cell membrane and soluble nutrient requirements resulted in bacteria being classified as plants. Since bacteria do not use *photosynthesis* with chlorophyll, as most plants do, it was hard for some people to accept bacteria as plants. Eventually, it was recognized that bacteria lacked a defined nucleus within the cell, providing a significant difference from the other microorganisms. This difference in basic cell structure allowed microorganisms to be placed into two major groups, *prokaryotes*, cells without a defined nucleus, and *eukaryotes*, cells with a defined nucleus. Bacteria were classified as procaryotes and all the other microorganisms were classified as eucaryotes.

Nuclear material in bacteria is simply suspended within the cell fluid, making it easy for the cell to create additional nuclear material and a second cell. Bacteria grow by a process known as *binary fission.* When the bacteria accumulate sufficient nuclear material, the nuclear material separates into two masses and the bacteria cell begins to create a double layer, internal wall that divides the cell into two similar units. Once the two cells are formed, they split along the internal wall and form two separate cells that continue to grow and repeat the binary fission process. The cell expands as nutrients are processed to form the new cell. The splitting of the cells occurs when the single cell reaches its maximum size. Two cells are produced that are minimum sized. Thus, the bacteria continuously removes soluble nutrients from the surrounding environment, divides into two cells, and repeats the process until the nutrients in the environment are exhausted or toxic end products are generated in the environment. Each new cell has the same chemical characteristics as the old cell. Each new cell competes with the old cell for its share of the soluble nutrients in the immediate environment.

Many bacteria metabolize organic matter to create new cell protoplasm. These bacteria have been designated as *chemotrophic.* The chemotrophic bacteria use part of the metabolized organic matter for energy while converting the remainder of the metabolized organic matter into cell mass. Only a few bacteria use inorganic carbon dioxide as their source of cell carbon. These bacteria are termed *autotrophic.* Most of the autotrophic bacteria also metabolize reduced inorganic compounds as their source of energy. It has long been recognized that many bacteria, using organic matter for cell synthesis, can metabolize both soluble organic compounds and insoluble organic particles. It is natural for the bacteria to metabolize the soluble organics before attacking the suspended organic particles. Hydrolytic enzymes, located on the bacteria surface, are used to convert the suspended organic particles into soluble organic compounds that pass into the bacteria cell for further metabolism. A small portion of the soluble organic compounds will diffuse into the surrounding liquid since the solid particles and the bacteria surfaces do not form a tight seal. The need for direct contact between the bacteria surface and the suspended solids being metabolized is a major factor limiting the rate of metabolism of suspended solids. Hydrolysis of suspended organic solids is critical for the recycling of organic matter in the natural environment. *Hydrolysis* utilizes water molecules to break small organic fragments off the insoluble organic compounds. Hydrolysis is one of the most common metabolic reactions for bacteria.

A few bacteria have special photosynthetic pigments and can obtain energy from sunlight. The photosynthetic pigments in bacteria range in color from red to purple with some green. The green photosynthetic pigments in bacteria are different from the green, chlorophyll pigments found in most plants. The bacteria photosynthetic pigments allow the bacteria to tap the energy from sunlight for cell synthesis, utilizing reduced sulfur compounds. Recently, scientists have reclassified the blue-green algae as bacteria, since the blue-green algae lack a defined nucleus the same as bacteria. The new classification labeled the blue-green photobacteria as *Cyanobacteria.* This change in taxonomy created a little confusion until everyone recognized that blue-green algae really were *procaryotes*, rather than *eucaryotes*.

Further taxonomic problems arose with the research of Carl Woese on methane bacteria in 1977. Carl Woese used RNA analysis and found that the methane bacteria were quite different from the rest of the bacteria. Woese believed that the methane bacteria were not bacteria, but an entirely different microorganism. Woese called the methane bacteria, *Archaebacteria.* It was not long before the group name was changed to *Archaea.* Woese used RNA analysis to divide microorganism classification into three domains: *Archaea*, *Bacteria*, and *Eucarya.* Woese's research on RNA analysis of different bacteria upset the conventional approaches to microbial taxonomy that were based on physical structure of the cells and on biochemical reactions. The development of methods and equipment for rapid replication of bacteria RNA allowed microbiologists to examine RNA from known pure cultures and to build a library of information that could be used to identify unknown bacteria cultures. By the 1990s replicating RNA became a primary method for identifying bacteria. Needless to say, the RNA identification created some chaos in the area of microbial taxonomy. Recently, Radhey Gupta added to the confusion by pointing out that the *Archaea* have the same cell structure as Gram-positive bacteria, raising the question as to whether the *Archaea* are really separate from bacteria. It was noted by Gupta that the 16S rRNA analyses has resulted in the creation of 23 phyla that are entirely different from the bacteria classification based on cell structure and biochemical reactions. Ludwig and Schleifer in 1999 indicated that forthcoming editions of *Bergey's Manual of Systematic Bacteriology* would base their phylogenic relationships between the different bacteria on the 16S rRNA tree. Changes are definitely in store for bacterial taxonomy. It will be interesting to see what the next major change will be. Change is the one thing that is certain to occur in future years as research yields additional data.

Ingraham, Maaloe, and Neidhardt reported that 20% of the dry weight of *E. coli* was RNA. Approximately 81% of the RNA was ribosomal RNA, identified as rRNA. The rRNA is composed of 3 molecular species, 23S, 16S, and 5S. Most of the RNA analyses have been made on 16S rRNA, since it was the easiest to replicate initially. About 15% of the RNA is transfer RNA, identified as tRNA. The remaining 4% of the RNA is messenger RNA, identified as mRNA. RNA analyses have been undergoing rapid changes in recent years, reducing the time required for bacteria identification. While RNA analyses seem to work quite well with pure cultures of bacteria, there have been problems with mixtures of bacteria in the natural environment. Fragments of RNA from dead cells and other organic material in nature have created mixed results. While the RNA approach to bacteria identification appears to have a sound fundamental basis, more research will be required before it can be widely used in environmental pollution control microbiology. The rapid development of automated equipment to process field samples will continue to produce improved RNA analyses and more accurate microbial identification.

Bacteria are very small organisms, 0.2 to 3.0 microns (μ). Individual bacteria can be seen with the aid of an optical microscope. In 1872 F. J. Cohn believed that bacteria were the smallest living organisms. He saw several different shapes of bacteria under his microscope and produced the first classification of bacteria based on their physical characteristics. There were round bacteria, spheres. There were short rods and long rods. Some bacteria were wavy and some were short, screw-like, and some were long, spiral shaped. As microbiologists looked at the various biochemical reactions that the bacteria produced, it became apparent that many bacteria had the same physical appearance with different biochemical characteristics. As a net result, the physical characteristics of bacteria simply became one of many characteristics that were used to evaluate the different bacteria. Unfortunately, the biochemical reactions produced by the bacteria varied from investigator to investigator, creating the impression of greater diversity than actually existed. As a net result, the same bacteria had different names. After many years of effort, bacteriologists standardized the common bacteriological media and the biochemical tests used for identification. Testing the different bacteria under controlled conditions was a slow process. Many of the original names of the different bacteria were consolidated into a single name, when the biochemical tests indicated similarity of organisms. It is easy to see why the 16S rRNA techniques are being touted as the way to greatly simplify bacteria identification.

Currently, it is recognized that there are three basic shapes of bacteria: spheres, rods and spirals. The spherical bacteria are called *cocci.* The rod shaped bacteria are called *baccillus*; and the spiral shaped bacteria are called *spirillum.* As shown in Figure 2-1, the *cocci* can exist as single cells, as diplos (two cells), as squares.

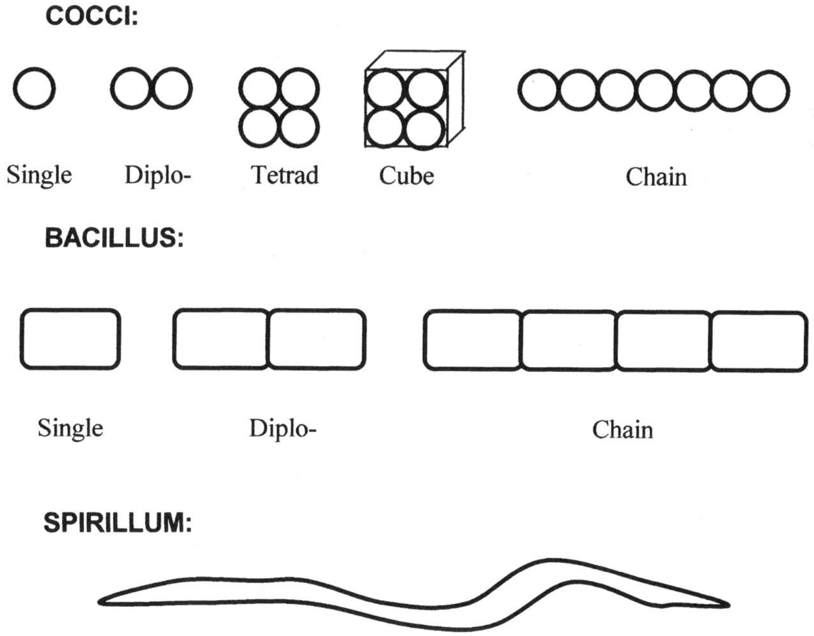

Figure 2-1 TYPICAL SHAPES OF BACTERIA

(four cells), as cubes (eight cells), as chains or as large clumps, having no specific size or shape. The *bacillus* is found as a single cell, diplo cells, and in chains. The *spirillum* exists primarily as single cells or diplos. It is not always easy to identify the shape of the bacteria. Small rod shaped bacteria look very much like spheres. *Spirillum* grows as long thin bacteria that move with an undulating motion. Since the environment has a definite effect on the growth of the bacteria, it is important to isolate the bacteria in pure culture and to grow them in the same media under identical environmental conditions. Some bacteria grow dispersed in one environment and as long filaments in a slightly different environment. For this reason it is essential to describe the environment used when the bacteria form is measured. By understanding the chemical environments where the different bacteria forms occur, we will gain a better understanding in controlling the desired growth form. A deficiency of iron or magnesium has been found to cause filamentous growths in bacteria that normally do not form filaments. Rapidly growing bacteria cultures will show individual cells, as well as, diplo cells. The diplo cells are simply part of the normal growth cycle before the bacteria split into two cells

BACTERIA STRUCTURE

All bacteria have the same general structure. If we examine a single bacterium, moving from the outside to the inside, we would find a slime layer, a cell wall, a cell membrane and the cytosol within the membrane. The slime layer appears to be part of the cell wall that has lost most of its proteins. The net result is the chemical structure of the slime layer is primarily polysaccharide material of relatively low structural strength. Motile bacteria show thin slime layers as the polysaccharide material is easily sheared from the moving cell. Old bacteria that show little motility will have thick slime layers. The cell wall comes next and gives the bacteria shape. Chemically, the cell walls are a mixture of carbohydrates, lipids and proteins. The cell wall also contains the hydrolytic enzymes that exist on the bacteria surface. As the bacteria age, the hydrolytic enzymes break off the cell wall surface and move into the surrounding liquid environment, giving rise to extra-cellular enzymes and increased slime material. The cell membrane is adjacent to the cell wall and is a much denser structure than the cell wall. The cell membrane controls the movement of chemicals into and out of the cell. The cell membrane also contains protein enzymes that allow energy to be transferred from substrates outside the cell to enzymes inside the cell for the synthesis of new cell components. The cytosol makes up the interior of the bacteria. It contains various protein enzymes and the nuclear materials that determine the overall metabolic characteristics of the bacteria. We may even find a number of different storage materials, such as glycogen or polyphosphates, in the cytosol. The organs of motility, the flagella, are also found at the junction of the cell membrane and the cytosol, as are the organs of attachment, the pili. Because of the small size of bacteria, it is necessary to use a microscope to study the bacteria structures. Figure 2-2 is a schematic drawing of a typical rod-shaped bacterium.

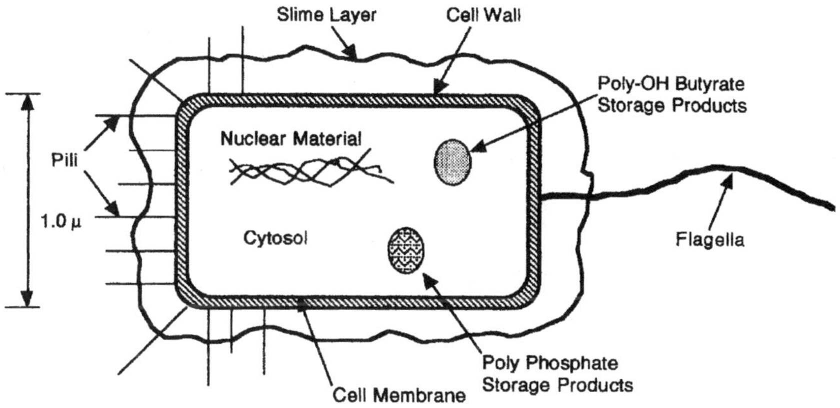

Figure 2-2 SCHEMATIC DIAGRAM OF A TYPICAL BACTERIUM CELL

MICROSCOPES

Optical microscopes are used to examine the overall, physical characteristics of bacteria. The *limit of resolution* of optical microscopes is not adequate for examination of the finer structures in bacteria. It is not possible to observe structures within the bacteria that are smaller than 0.16 microns with optical microscopes. The small size of bacteria and their lack of mass make it difficult to observe individual bacteria. At 100X magnification most bacteria appear as large dots. Even at 430X magnification, it is difficult to recognize the shape of small bacteria. Oil immersion at 970X magnification is required to examine the shape of small bacteria. Oil immersion lenses are specially shaped, requiring a drop of immersion oil between the lens and the sample. Examination of liquid cultures requires a glass cover slip to be placed over the drop of water containing the bacteria originally placed on a glass slide. The drop of oil is placed between the glass cover slip and the oil immersion lens. The focal length of the lens allows only examination of the liquid just under the cover slip. Unfortunately, the density of bacteria is so close to that of water that it is difficult to see the bacteria. The ability to see bacteria in liquid cultures was given a major advance with the advent of the phase contrast microscope. Figure 2-3 shows a rapidly growing bacteria culture

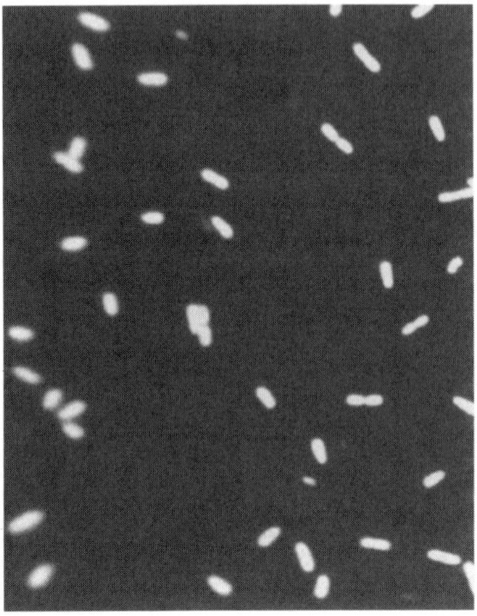

Figure 2-3 BACTERIA AT 970X MAGNIFICATION

using a dark-field to make the background dark, allowing the bacteria to appear bright. The bright bacteria are easy to see against the dark background. Some bacteria are dividing, as evidenced by the diplo- cells. In a rapidly growing culture the size of the bacteria vary from their smallest size, when dividing, to their largest size, after dividing. The photomicrograph has been enlarged to make the bacteria even larger. Note that it is hard to keep all the bacteria in focus at high magnification. A slight difference in depth or in the lens characteristics puts the bacteria at the outer edge out of focus. Phase contrast microscopy changes the light intensity as it passes through materials of slightly different densities. It produces a definite increase in the contrast between the liquid medium and the bacterial cells, making it easier to see the individual bacteria cells. Even the phase contrast microscope does not allow observation of small structures within the bacteria. Phase contrast has become the standard for optical microscopes.

The electron microscope was developed to examine structures that were smaller than 0.16 microns. Electron microscopy depends upon electron absorption to differentiate between structures within the bacteria. Since air absorbs electrons quite readily, it is necessary to place the bacteria in a vacuum prior to showering them with electrons. The bacteria must be completely dehydrated to reduce the absorption of electrons by water inside the cells. A phosphorescent screen is placed after the appropriate enlarging magnets to permit viewing the response of the electrons hitting the screen. The viewing screen resembles a small TV screen. As the electrons are absorbed by the specimen, the screen looks brighter than where the electrons reach the screen unhindered. Initially, it was necessary to coat the bacteria with gold to provide a dense coating to absorb the electrons. Early photomicrographs from electron microscopy looked like shadow pictures of the bacteria. Improvements in electron microscopy have produced improved contrast between different parts of bacteria. Unfortunately, the ability to see internal structures does not provide any knowledge of the significance of the observed structures. Much research remains to be done in evaluating bacteria structures with electron microscopes. Examination of mixtures of bacteria in natural environments

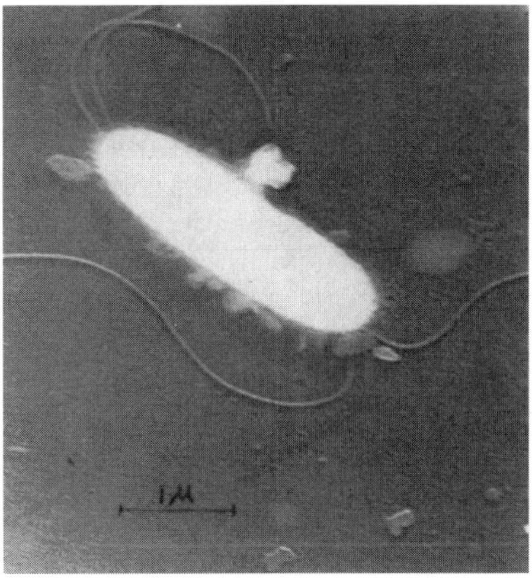

Figure 2-4 ELECTRON PHOTO-MICROGRAPH OF A MOTILE BACTERIUM

requires the use of the scanning electron microscope (SEM). The surfaces of the bacteria are coated with gold to reflect the electron beams. The net result is a picture of the bacteria surfaces with differences in screen density being related to the angle of electron reflection. Figure 2-4 shows a direct transmission electron photomicrograph of a bacterium with bipolar flagella. The gold metal coating makes the bacterium appear bright since the electrons were readily adsorbed by the gold. Current techniques use less metal coatings and lower voltages to produce better images of the bacteria. SEM has gained considerable popularity over direct transmission electron microscopes (TEM) for environmental samples. While electron microscopes produce interesting pictures of the different bacteria and their structures, it is not necessary to use electron microscopy in evaluating water pollution problems.

STAINS AND STAINING

One of the most important techniques for microbiologists to learn is staining. Stains add color to the bacteria, allowing them to be easily seen using the optical microscope. The staining procedure starts with a clean glass slide. A single drop of culture is placed on the glass slide and spread over the desired area of the slide with a sterile wire loop. The culture is allowed to dry in the air. It is then passed over a Bunsen burner flame to fix the bacteria lightly to the glass slide. A drop of the desired stain is placed over the dried smear and spread over the surface of the smear. After 30 seconds to one minute, the stain is washed with water from the surface of the slide. Normally, the slide is washed until the water running off the slide is no longer colored. The slide is air dried and examined under the optical microscope. A simple stain is used for most evaluations of bacteria cultures. Occasionally, the slide is treated with special chemicals and multiple dyes for special stains. Detailed staining techniques can be found in bacteriology laboratory manuals in the library. Students desiring more information on stains and staining should become familiar with related books and laboratory manuals in the library.

CELL WALL

In 1884 a Danish bacteriologist, Christian Gram, found that it was possible to use stains to divide bacteria into "positive" and "negative" groups, depending upon their reactions with his stains. Gram staining has been used over the years for primary differentiation between bacteria groups. Unfortunately, the Gram stain is not as valuable as it was originally conceived. A major problem stems from the fact that Gram-positive bacteria can lose their chemical reactivity and appear Gram-negative. The chemical reactions that affect the Gram stains have long intrigued bacteriologists. Recent advances in analytical chemistry have shown that there are definite differences in the cell wall structures of Gram-positive and Gram-negative bacteria. Gram-positive bacteria have thicker cell walls than Gram-negative bacteria. The basic cell wall materials for both types of bacteria are similar. The major component for cell walls is peptidoglycan. A. L. Koch presented an overview on the growth and form of the bacterial cell wall in 1990. He showed that the cell wall consists of several layers of carbohydrate molecules tied with peptides made from amino acids. The carbohydrates are joined on end, forming long glycan chains. The carbohydrates that predominate in the glycan chains are N-acetyl-glucosamine and N-acetyl-muramic acid. These carbohydrate polymers contain nitrogen and are often mistaken for proteins in chemical analyses. The amino acids in the peptides include alanine and glutamic acid, two of the most commonly produced amino acids. Koch's article is a good place to start for anyone wanting to know more about the cell wall structure of bacteria.

Cell walls are continuously synthesized from the inside surface as the cell expands to form two new cells. It appears that both proteins and lipids are integrated into the cell wall as it is synthesized. The cell wall gives the shape to the bacteria and helps control movement of materials into and out of the cell. The proteins in the cell walls are largely hydrolytic enzymes designed to break down complex proteins and polysaccharides in the liquid around the bacteria into smaller molecules that can move across the cell wall to the cell membrane and into the cell. The lipids are lipo-polysaccharides that provide points on the cell surface for hydrophobic compounds to enter the cell. The hydrophobic compounds in direct contact with the bacteria wall dissolve into the lipo-polysaccharides and move through the cell wall to the cell membrane where they are broken into small molecules by β-oxidation.

As the bacteria age, the synthesis of new cell wall material at the cell membrane surface causes the cell wall to increase in thickness. Young bacteria move rapidly through the liquid where hydraulic shear forces cause the outer layer of the cell wall to break off, giving a uniform cell wall. As the bacteria slow their motility, the cell wall increases its thickness, producing a *capsule* around the bacteria. The proteins in the cell wall tend to break off and move into the surrounding liquid as *extracellular, hydrolytic enzymes*. Some of the lipids are also lost from the cell surface. This leaves the polysaccharides as the primary constituents of the bacteria capsules. The capsular material that accumulates around the bacteria has a low density and is not very chemically reactive. When the bacteria lose their motility, the polysaccharide cell wall material accumulates to a much thicker layer without a specific shape and is called the *slime layer*. The cell wall polysaccharides accumulate as inert organic matter since the bacteria lack the ability to metabolize the cell wall polysaccharides in the aqueous environment. It appears that the enzymes responsible for synthesizing the cell wall polysaccharides exist only in the cell membranes. These polysaccharide enzymes are broken down when the cell dies and undergoes lysis. Without the specific enzymes to hydrolyze the cell wall polysaccharides, the bacteria slime layer accumulates in the liquid environment primarily as inert suspended solids. Some of the polysaccharide materials sheared from the motile bacteria are very tiny particulates that are measured as soluble organic compounds in most environmental analyses. The cell wall polysaccharides can be chemically hydrolyzed to the basic carbohydrates, with the carbohydrate compounds being metabolized by bacteria. There are also some terrestrial fungi that can metabolize the bacteria cell wall materials in the soil environment, but not in an aqueous environment. While the bacteria polysaccharides appear as inert organic compounds in water, they are not completely inert, breaking down very slowly over time.

Normal bacteria stains react with the lipid and the protein portions of the cell walls, but do not react with the polysaccharide portion of the cell walls. The bacteria capsules and accumulated slime materials require special stains to demonstrate their presence in microscopic slides. Alcian Blue 8GN150, a cotton textile dye, has proven to be a reasonable stain for bacteria polysaccharides. When used alone, the Alcian Blue dye colors the polysaccharide materials in the bacteria blue. When Alcian Blue stain is used in combination with a red stain, such as Carbol Fuschin, the bacteria cells appear reddish-blue with light blue capsules. The outer layer of the bacteria polysaccharide retains the blue color while the body of the bacteria is stained red. The visual colors will depend upon the orientation of the bacteria being examined. Some cells may be light blue, and others may be red. Microscopic examination of stained cells will always produce mixtures of colors that require careful evaluation for proper interpretation of the sample being observed.

CELL MEMBRANE

The cell membrane lies next to the inner surface of the cell wall. It is primarily a dense lipoprotein polymer that controls movement of materials into and out of the bacteria. Recently, it has been recognized that the cell membrane is much more important than simply controlling movement of materials. It appears that the primary metabolic reactions for the bacteria occur in the cell membrane. Nutrients are degraded to provide energy for synthesis of cell materials. Cell wall material is synthesized at the membrane surface while RNA, DNA, and proteins are synthesized inside the cell. Generally, the cell membrane is a well-organized, complex of enzymes that allows the most efficient transfer of energy from the substrate being metabolized to the production of all of the cell components. Evaluation of the metabolic reactions in the cell membrane shows why metabolism cannot occur without growth. Studies of bacterial metabolism under non-growth conditions are often studies in which the growth of new cell mass balances the metabolism of existing cell mass, producing a zero cell mass change during the study.

CYTOSOL

The cytosol is the major material inside the cell membrane. It is a colloidal suspension of various materials, primarily proteins. Colloids are tiny particles having very large surface area to mass ratios, making them highly reactive. These protein fragments are important to the success of the bacteria, providing the building blocks for all of the key enzymes that the cell needs. The large size of these colloidal fragments prevents their movement out of the cell until the cell membrane is ruptured.

Since bacteria do not have a defined nucleus, their nuclear material is dispersed throughout the cytosol. The DNA in the bacteria determines their biochemical characteristics and exists as organic strands in the cytosol. The DNA is surrounded by RNA that is responsible for protein synthesis. The dispersed nature of bacterial nuclear material and the permeability of the bacteria cells permit transfer of small nuclear fragments from one cell to another cell, changing the biochemical characteristics of the bacteria. If the changes in biochemical characteristics are permanent, the process is known as *mutation.* If the changes in biochemical characteristics are controlled by the environmental conditions around the bacteria, the process is known as *adaptation.* Both adaptation and mutation are important environmental processes. Adaptation occurs far more often than mutation and is reversible. Once the environmental conditions return to their original state, the bacteria lose their adaptive properties. Considerable research is being conducted on the transfer of genetic materials from one bacteria species to another bacteria species. It is hoped that bacteria with one set of characteristics will develop a second set of characteristics. The results to date have been mixed with a few bacteria developing the new characteristics. Although artificial gene transfer is relatively new, natural gene transfer has been occurring since time began and will continue in the future. Gene transfer among bacteria is a never-ending process.

If the bacteria are in a medium having excess nutrients, the bacteria cannot process all the nutrients into cell components as quickly as the bacteria would like. The excess nutrients that the bacteria remove, but cannot use in the synthesis of new cell mass, are converted into insoluble storage reserves for later metabolism when the available substrate decreases. Excess carbohydrates in the media can be stored as glycogen under the proper environmental conditions. Some bacteria convert excess carbohydrates into extracellular slime that cannot be further metabolized, rather than storing the excess carbohydrates inside the cell. Excess acetic acid can be stored as a polyhydroxybutyric acid polymer. Storage of nutrients inside the cell does not occur in a substrate-limited environment. When the nutrient substrate is limited, the bacteria use all the nutrients as fast as they can for the synthesis of new cell mass.

FLAGELLA AND PILI

Flagella are protein strands that extend from the cell membrane through the cell wall into the liquid. Energy generated at the cell membrane causes the flagella to move and propel the bacteria through the liquid environment. The flagella can be located at one end of the bacteria, at both ends of the cell, or completely covering the bacteria. Flagella are very important for rod shaped bacteria, allowing them to move in search of nutrients when necessary. Flagella are normally smaller than the limit of resolution of optical microscopes and cannot be seen unless chemical precipitates are used to increase the size of the flagella prior to staining. Some bacteria move by flexing their bodies rather than using flagella. *Spirillum* have flagella at the end of the cell that produce a corkscrew type motion. Spherical bacteria are devoid of flagella and lack motility. As bacteria age, they appear to produce many small protein projections that look like flagella. These small protein projections have been called *pili* and appear to help the bacteria attach to various surfaces, including other bacteria. The flagella and pili are best observed with the electron microscope. Figure 2-4 illustrates a bacterium with polar flagella. With more detailed observations of bacteria by new investigators, other bacterial structures may be detected in future years.

CHEMICAL COMPOSITION

Bacteria are approximately 70% to 80% water with 20% to 30% dry matter. The dry matter in bacteria is both organic and inorganic. The percentages of organic matter and inorganic matter depend upon the chemical characteristics of the media in which the bacteria are grown. Media that contain high concentrations of inorganic salts will produce bacteria with higher concentrations of inorganic compounds in the cell mass than media containing low concentrations of inorganic compounds. This relationship is very important in environmental microbiology since the mineral content of water varies widely. The growth media determines both the types and quantities of inorganic compounds in the bacteria. The inorganic fraction in bacteria in normal media ranges from 5% to 15% with an average around 10% dry weight. This means that the organic fraction in bacteria will vary from 85% to 95%, averaging 90% dry weight. Chemical analyses of bacteria are made on the dried residue after centrifuging, washing, and oven drying. The major inorganic ions include the various oxides of sodium, potassium, calcium, magnesium, iron, zinc, copper, nickel, manganese, and cobalt together with chlorides, sulfates, phosphates, and carbonates. The major organic elements include carbon, hydrogen, oxygen, and nitrogen. While phosphorus and sulfur are part of the organic matter in bacteria, they are measured as part of the inorganic matter. There is no way to obtain the absolute chemical composition of the bacteria since every measurement is subject to errors. It is possible to obtain reasonable measurements of the major components of bacteria. Complete balances can only be made by direct determination of every major chemical parameter. Needless to say, determining all major parameters but one allows material balances to always balance from a mathematical point of view. Unfortunately, the errors will not be properly recognized and the results may be less than desired.

Ingraham, Maaloe and Neidhardt reported that the dry matter for *Escherichia coli* contained "approximately 50% carbon, 20% oxygen, 14% nitrogen, 8% hydrogen, 3% phosphorus, 1% sulfur, 2% potassium, 0.05% each of calcium, magnesium, and chlorine, 0.2% iron, and a total of 0.3% trace elements including manganese, cobalt, copper, zinc and molybdenum (Laura, 1960)". Mayberry, Prochazka and Payne found that aerobic bacterial growth on a single organic compound gave 45.6% carbon, 7.6% hydrogen, 12.1% nitrogen and 3.0% ash. Oxygen by difference was 31.7%. Analyses of three different bacteria by Clark gave ranges from 43.6% to 48.0% carbon, 6.64% to 7.14% hydrogen, 26.8% to 35.4% oxygen, 9.9% to 11.8% nitrogen and 7.5% to 11.2% ash. Clark measured all of the chemical parameters separately. These data confirmed that carbon, hydrogen, oxygen and nitrogen are the main components of the organic fraction of bacteria. As previously indicated, the ash or inorganic fraction of the cell mass varies with the media used for growth. Some bacteria require high inorganic salt concentrations for good growth. The salt-loving bacteria, *Halobacteria,* will definitely have higher ash fractions than indicated for normal bacteria. Comparative results for bacteria protoplasm are obtained by examining the carbon, hydrogen, oxygen and nitrogen by themselves. The data showed that carbon was between 47% and 54% with hydrogen between 7.7% and 8.7%, oxygen between 22% and 36% and nitrogen between 12% and 15%. Unfortunately, data on the chemical composition of bacteria are scarce. The data also have definite limitations based on the media used, the method of harvesting, and analytical errors. Recently, E. H. Battley presented data to show that the ash fraction measurement was higher than it should have been. He used the published analysis of *E. coli* to show that the ash error created an error in the organic elements of about 5.5%. Fortunately, the ash weight errors do not change the relative relationships between the major chemical elements. With the inorganic ash having an error of about two times, the true inorganic elements would be about half the ash fraction, giving a range from about 2.5% to about 7.5% instead of the 5% to 15%, previously indicated. Battley's evaluation of the ash data illustrates some of the problems that exist in using data published in the literature. Proper elemental analyses of bacteria cell mass should result in a total mass greater than 100%, if the mineral content is measured from the ash fraction. Most, but not all, of Clark's analyses of bacteria showed a total mass greater than 100%. The error in the ash is sufficiently small that it is easy to attribute the analytical errors to the larger elemental analyses. Overall, the analytical errors have a limited effect on the total chemical analyses. The calculations that follow show how the percentage data are converted into empirical formulas for bacteria cell mass.

****

TYPICAL CALCULATIONS:

1. Using the data of Ingraham, Maaloe and Neidhardt, the following results are presented for C, H, O and N.

C = 0.50, H = 0.08, O = 0.20, N = 0.14 Total C, H, O, N = 0.92

Corrected C = 0.50/0.92 = 0.54 Corrected H = 0.08/0.92 = 0.087

Corrected O = 0.20/0.92 = 0.22 Corrected N = 0.14/0.92 = 0.15

Corrected Total C, H, O, N = 1.00 (no further correction required)

Calculate the general chemical formula for cell protoplasm, using N = 1.0

Number of C atoms = (0.54/12)(14/0.15) = 4.2

Number of H atoms = (0.087/1)(14/0.15) = 8.1

Number of O atoms = (0.22/16)(14/0.15) = 1.3

Chemical formula of cell protoplasm = $C_{4.2}H_{8.1}O_{1.3}N$

Calculate the general chemical formula for cell protoplasm, using C = 1.0

Number of H atoms = (0.087/1)(12/0.54) = 1.9

Number of O atoms = (0.22/16)(12/0.54) = 0.31

Number of N atoms = (0.15/14)(12/0.54) = 0.24

Chemical formula of cell protoplasm = $CH_{1.9}O_{0.31}N_{0.24}$

2. Using the data of Mayberry, Prochazka and Payne, the following results are presented for C, H, O and N.

C = 0.456, H = 0.076, O = 0.317, N = 0.121 Total C, H, O, N = 0.97

Corrected C = 0.456/0.97 = 0.47 Corrected H = 0.076/0.97 = 0.078
Corrected O = 0.317/0.97 = 0.33 Corrected N = 0.121/0.97 = 0.12

Corrected Total C, H, O, N = 1.00 (no further correction required)

Calculate the general chemical formula for cell protoplasm, using N =1.0
Number of C atoms = (0.47/12)(14/0.12) = 4.6
Number of H atoms = (0.078/1)(14/0.12) = 9.1
Number of O atoms = (0.33/16)(14/0.12) = 2.4

Chemical formula of cell protoplasm = $C_{4.6}H_{9.1}O_{2.4}N$

Calculate the general chemical formula for cell protoplasm, using C = 1.0
Number of H atoms = (0.078/1)(12/0.47) = 2.0
Number of O atoms = (0.33/16)(12/0.47) = 0.53
Number of N atoms = (0.12/14)(12/0.47) = 0.22

Chemical formula of cell protoplasm = $CH_{2.0}O_{0.53}N_{0.22}$

3. Using the data of Clark, the following results are presented for C, H, O, and N.

C = 0.458, H = 0.069, O = 0.311, N = 0.108 Total C, H, O, N = 0.946

Corrected C = 0.458/0.946 = 0.48 Corrected H = 0.069/0.946 = 0.073
Corrected O = 0.311/0.946 = 0.33 Corrected N = 0.108/0.946 = 0.11

Corrected Total C, H, O, N = 0.99 (correct largest value by 0.01, C = 0.49 instead of 0.48)

Calculate the general chemical formula for cell protoplasm, using N = 1.0
Number of C atoms = (0.49/12)(14/0.11) = 5.2
Number of H atoms = (0.073/1)(14/0.11) = 9.3
Number of O atoms = (0.33/16)(14/0.11) = 2.6

Chemical formula of cell protoplasm = $C_{5.2}H_{9.3}O_{2.6}N$

Calculate the general chemical formula for cell protoplasm, using C = 1.0

Number of H atoms = (0.073/1)(12/0.49) = 1.8

Number of O atoms = (0.33/16)(12/0.49) = 0.51

Number of N atoms = (0.11/14)(12/0.49) = 0.19

Chemical formula of cell protoplasm = $CH_{1.8}O_{0.51}N_{0.19}$

****

As indicated in the calculations, the C, H, O, N data provide a reasonable evaluation of the organic fraction of the bacteria protoplasm. Environmental microbiologists have tended to favor the formula using N as unity; while bacteriologists have favored the formula using C as unity. Overall, both formulas are identical, making the choice of the specific presentation a personal issue rather than a technical issue. The corrected values of the four primary chemical elements in cell mass are the important values to be used for comparative purposes. Another factor of importance is the use of fractional values of the different elements. There has been a tendency to round off the number of each element to the nearest whole number when using N = 1.0. Rounding off the elements to whole numbers introduces an additional error that can produce misleading results and should not be done. The current accuracy of the different chemical analyses does not justify the use of more than two significant figures when presenting the general chemical formula.

Another way to look at bacterial protoplasm is to examine its major chemical compounds. Most of the bacteria protoplasm is protein, about 60% of the organic compounds. The nucleic acids, RNA and DNA, make up about 25% of the organic compounds. Lipids consist of about 10% of the organic compounds, leaving the remaining 5% as polysaccharides. The proteins are polymers of amino acids formed by splitting water out of two amino acids between the amino group and the carboxyl group and then repeating the process many times over, forming a large complex molecule. RNA and DNA are mixtures of pentose sugars, phosphates, purines, and pyrimidine. The lipids are long chain fatty acids that tend to be combined with proteins into lipoprotein complexes and with the polysaccharides into lipopolysaccharides. The polysaccharides are hexose sugars that are polymerized with the loss of water between two hexose molecules. It is important to realize that many hexose sugars contain amino nitrogen, forming amino polysaccharides. Originally, the organic nitrogen in bacteria was assumed to be associated only with proteins. The development of more sophisticated analytical techniques showed that organic nitrogen is part of the DNA and RNA and the amino polysaccharides. As a net result, organic nitrogen determinations should not be used as a measure of cell proteins. The general analyses of the organic compounds in bacteria indicate that the total N could be about 14%. This is higher than the direct chemical analyses presented by Mayberry, Prokazka and Payne and by Clark.

Further examination of the bacteria shows that the chemical composition of bacteria is not fixed, but changes with various environmental factors. The media in which the bacteria are grown can affect the chemical composition analyses. Since bacteria are about 70% water, the soluble components of the growth media will often accumulate inside the bacteria and affect the chemical analyses. It appears that chemical analyses should be made in nutrient limiting media when the organic substrate metabolism has just been completed. Growth in high organic concentration media leads to anaerobic metabolism and storage of excess nutrients within the cell. While nutrient storage is very common in normal bacteriological media, it is not normal in environmental bacteriological systems that are nutrient limiting. Once the external sources of nutrients are exhausted, the bacteria begin to metabolize proteins within the cell. This protein metabolism results in the release of ammonia nitrogen back into the environment. Continued metabolism without external nutrients results in degradation of the RNA with release of both ammonia nitrogen and phosphates. For these reasons, the best time for chemical analyses of bacteria protoplasm is just before the nutrient substrate has been exhausted. It will be interesting to see what data collected under these conditions will look like and how it will compare with the data previously published. It is easy to see why chemical analyses of bacteria seem to vary over a wide range. One observation that has been noted with chemical analyses of bacteria is the tendency for carbon (C) and hydrogen (H) percentages to remain fairly constant while oxygen (O) and nitrogen (N) vary. As the bacteria release ammonia nitrogen (N), it is replaced by oxygen (O). The percentage of nitrogen (N) in the cell mass decreases with time, while the percentage of oxygen (O) in the cell mass increases.

METABOLISM

Bacteria metabolize nutrients to remain alive and to produce new cells. The production of new cells requires sufficient chemical elements for building all the hundreds of different chemical compounds that comprise the cell. In addition to the chemical elements, the bacteria must have a source of energy to produce the synthesis reactions. Bacteria metabolizing either organic or inorganic chemicals are designated as *chemotrophic.* The organic metabolizing bacteria are known as *heterotrophic.* The inorganic metabolizing bacteria are called *autotrophic.* The bacteria that use light as their source of energy are termed *photosynthetic.* Recently, the blue-green algae had their classification changed to blue-green bacteria, greatly increasing the number of *photobacteria.* Each group of bacteria has their own specific metabolic characteristics that determine their ability to survive in different environments. Although bacteria are biological systems, they must follow normal chemical reactions to survive. The bacteria that currently exist in the environment have adapted to changing chemical characteristics over the eons and represent those bacteria best able to obtain nutrients. The key to understanding bacteria metabolism lies in understanding their biochemistry.

Observation of growing cultures of bacteria shows that they reproduce by binary fission. A single cell expands as it metabolizes nutrients. At a specific stage of growth the DNA creates a duplicate copy of itself. Soon, the cell begins to develop a new cell wall across the middle of the cell. The two copies of DNA move apart, one copy to each side of the new cell wall. Eventually, the cell reaches critical mass and divides into two cells. Each cell is a carbon copy of the original cell. The two new cells can metabolize the same nutrients and divide into two more cells. It has been noted that bacteria tend to divide at the same rate when environmental conditions remain the same. The time required for bacteria to grow and divide into two cells has been termed the *generation time.* Some bacteria can divide in less than an hour at their optimum temperature for growth with excess nutrients. Other bacteria may take a day or two to divide. Under nutrient limiting conditions the rate of bacteria growth is less than their optimum growth rate.

Initially, bacteria were grown in media that contained excess nutrients. Emphasis was on rapid growth in body fluids at body temperature to permit isolation and study of the pathogenic bacteria affecting the health of people and domestic animals. As bacteriologists examined bacteria in soil and streams, they learned that bacteria did not respond the same in the natural environment as in the concentrated nutrient media. The use of dilute nutrient solutions and low temperatures produced slow bacteria growth. The net result was continued use of concentrated nutrient media and elevated temperatures to stimulate rapid bacteria growth even though natural environmental conditions were not duplicated. It took considerable time before applied microbiologists recognized that proper understanding of the bacteria and their relationships to one another required duplication of the real environment with its limiting nutrients and competition, not only from other bacteria but also from other microorganisms. Many initial studies were qualitative; but slowly, the environmental microbiologists recognized that complete understanding of true conditions required quantitative balances both on a chemical basis and an energy basis against time.

ENERGY REACTIONS

Oxidation reactions release heat energy in proportion to the energy change in the chemical reaction. Reacting nutrients are oxidized to a series of chemical end products with the release of heat energy. Because of the importance of oxidation reactions in creating heat energy for modern society, this technology has been studied extensively and evaluated in *thermodynamics*. Engineers have carefully quantified the amount of reactants required and the end products produced. They recognized all the reactants had to be supplied to produce the desired results and the end products produced by the reactions had to be properly returned to the environment. The generation of heat energy is essential for the production of manufactured products in our modern society. Unfortunately, all of the energy from the reactants cannot be converted into work for manufacturing. The efficiency of various manufacturing processes is often related to the amount of energy required to produce the desired products.

The same ideas apply to bacteria that apply to the engineering of new products for our modern society. Only recently have bacteriologists recognized that concept and have attempted to make the kind of material and energy balances that engineers have been making for years. While it is recognized that bacteria are not heat machines, bacteria undergo the same general reactions as chemical systems with energy being obtained from the oxidation of nutrients and cell mass being produced as the major end product of the chemical reactions. The most important idea that has often been ignored is that energy must be gained from a reaction for that reaction to occur. Bacteria do not put energy into a chemical reaction in hopes of getting more energy back. If the reaction does not yield energy to the bacteria, the reaction does not occur. Sometimes it is difficult to properly evaluate all of the specific metabolic reactions that the bacteria use. Many reactions occur on such a micro scale and at such speeds that they cannot be properly identified. In spite of the individual reactions, the overall metabolic reactions must balance on both a chemical element basis and an energy basis.

Glucose has long been used as a simple carbohydrate for bacterial metabolism. Maximum energy is obtained when glucose is oxidized to carbon dioxide and water.

$$C_6H_{12}O_6 + 6O_2 \rightarrow 6CO_2 + 6H_2O + \text{Heat Released} \qquad (2\text{-}1)$$

In this equation the energy in glucose is transformed by the oxidation reaction to carbon dioxide, water and released heat energy. A mol. of glucose always contains the same amount of energy no matter how it was formed. The same is true of carbon dioxide and water. Thus, the amount of energy released from the oxidation reaction will always be the same. The energy released is measured as heat and has been termed the *heat of combustion*. Tables have been prepared for the heat of combustion of many different compounds. Where data are not available, a combustion calorimeter can be used to measure the heat released. The heat of combustion of glucose is approximately 2,813 kJ/mol., 15.6 kJ/g glucose. While bacteria are not heat machines, 2,813 kJ heat energy will be released to the environment when bacteria oxidize a mol. of glucose, 180 g, to carbon dioxide and water. The chemical energy release is the same for the same chemical reaction whether it is done in a calorimeter or by a group of bacteria. The net effect will be a temperature increase in the environment where the reaction occurs. While the general metabolic reaction appears simple, the actual metabolic reaction is much more complex, consisting of a series of interconnected reactions. The biochemical reactions move through a logical sequence of transformations with the release of energy that the bacteria need for synthesis of new cell components. The details of the metabolic reactions are primarily of interest to bacterial physiologists. The environmental microbiologists are concerned with a general understanding of microbial metabolism and the total reaction. Detailed evaluation of the current concepts of metabolism can be found in biochemistry books in the library.

Complex carbohydrates are made up of simple carbohydrates with various chemical linkages. Starch and cellulose are both polymers of glucose. Starch uses a simple chemical linkage system that joins the glucose molecules into long chains as well as in groups of chains. Cellulose uses a more complex linkage than starch, also forming long chains from glucose. The glucose molecules are twisted in cellulose, forming an entirely different chemical compound. The chemical bonds joining the glucose molecules in starch make it more soluble in hot water than cellulose. Solubility of chemical molecules is important in biodegradability. Soluble organic compounds tend to be more easily metabolized than insoluble organic compounds. Starch is more easily metabolized by bacteria than is cellulose. Yet, there are bacteria in nature that can metabolize both of these important polysaccharides. Metabolism of complex organics, such as starch and cellulose, starts with biochemical hydrolysis by water to form glucose units that are soluble and easily metabolized. The initial hydrolysis reaction is illustrated in Equation 2-2.

$$\text{Starch or cellulose} + \text{HOH} + \text{bacteria} \rightarrow \text{glucose} + \text{modified starch or cellulose} \quad (2\text{-}2)$$

The glucose formed from either starch or cellulose is identical and can be easily metabolized. Higher plants contain large quantities of cellulose together with lignin to prevent bacteria from readily metabolizing plant tissue. Removal of the lignin from cellulose allows the cellulose to be readily biodegraded. As already indicated, the slime layer and the cell wall around bacteria are also polysaccharide materials that are not metabolized by bacteria in the aqueous environment.

Fatty acids undergo chemical oxidation the same as carbohydrates with the production of carbon dioxide, water and heat. The main difference between fatty acid metabolism and carbohydrate metabolism is fatty acids release more heat than carbohydrates on a unit weight basis. Palmitic acid ($C_{16}H_{32}O_2$) oxidation produces about 41.3 kJ/g, compared to the 15.6 kJ/g glucose. Amino acids pose an interesting problem because of the amino group. Most amino acids are created from short chain fatty acids. Alanine is a common amino acid that has a chemical structure similar to propionic acid. Alanine has the chemical formula, $C_3H_7O_2N$; while propionic acid has the chemical formula, $C_3H_6O_2$. The heat of combustion of alanine is 1,591 kJ/mol., 18.2 kJ/g. The heat of combustion of propionic acid is 1,526 kJ/mol., 20.6 kJ/g. The difference between the heats of combustion of the two acids lies in the amino nitrogen group that is oxidized to nitrogen gas in the combustion reaction. While amino acids yield more total heat energy than their corresponding fatty acids, the energy yield/g is less. The energy yield for bacteria metabolism is even less than the chemical oxidation reaction, since their oxidation reaction yields carbon dioxide, water, and ammonia instead of carbon dioxide, water, and nitrogen gas. The bacterial oxidation of alanine only yields 1,267 kJ/mol., 14.2 Kcal/g., a significant difference in energy yields. Since bacteria metabolize proteins that contain amino acids, it is important to recognize the differences in energy yields for the bacteria in contrast to energy yields obtained by combustion calorimetry. Proteins are composed of polymers of amino acids. Like the polysaccharides, the proteins are hydrolyzed with water by bacteria to release soluble amino acids that are readily metabolized. Fats are insoluble polymers of glycerol and long chain fatty acids. While fats can be hydrolyzed to glycerol and long chain fatty acids, only glycerol is soluble in water and readily metabolized. The long chain fatty acids are insoluble and are slowly metabolized by bacteria. The long chain fatty acids cannot be hydrolyzed like starch and proteins to form soluble short chain acids that can be easily metabolized. The long chain fatty acids must be biochemically oxidized to produce the soluble short chain acids. For this reason bacterial metabolism of long chain fatty acids occurs in the cell membrane with the bacteria using the energy released during oxidation for synthesis of protoplasm. Oxidation is a high-energy yielding reaction; while hydrolysis is a low energy yielding reaction.

Bacteria can oxidize many different organic compounds. Various bacteria are able to metabolize alcohols, aldehydes, and ketones to carbon dioxide and water. Short chain, soluble organic compounds are more easily metabolized than long chain, insoluble organic compounds. Straight chain compounds are metabolized faster than branched compounds. Amines and amides are two groups of industrially important nitrogen compounds that the bacteria can metabolize. Common soil bacteria metabolize aromatic compounds, such as benzene, toluene, phenol and cresol under the proper environmental conditions. Multiple ring structures tend to be less soluble than single ring compounds and are more difficult for the bacteria to metabolize. Adding chlorine or nitrates to organic compounds makes them more difficult to metabolize. With acclimation bacteria can degrade many substituted organic compounds. The chemical structure of the organic compounds determines how easy or how difficult it is for the bacteria to metabolize the chemicals.

Enzymes mediate bacteria metabolism. Enzymes have been called *organic catalysts* since they increase the rate of reaction without becoming a part of the final products. As already indicated, many different hydrolytic enzymes exist on the surface of bacteria to break complex organic compounds into simple organic compounds for complete metabolism. The energy enzymes are located in the cell membrane with synthesis enzymes located on both the outer surface and inner surface of the cell membrane. Enzymes are proteins with special structures to bring about the desired chemical reactions. Actually, bacteria only bring about a few chemical reactions during metabolism. These basic chemical reactions are simply repeated many times for different chemical compounds. One of the chief characteristics of enzymes is their *specificity*. Enzymes are highly specific, reacting only with one or two compounds. It is the protein structure of the enzyme that makes the enzyme so specific. Physically, the reacting chemical molecule must fit with the protein molecule of the enzyme to allow the desired reaction to take place. The protein part of the enzyme has been designated as the *apo-enzyme*. The reactive part of the enzyme is a highly specialized chemical structure that can be used over and over again in different proteins to bring about the same chemical reaction with different chemical compounds. This reactive part of the enzyme is called the *co-enzyme*. The co-enzyme is assisted by a series of *metallic activators*. The metabolic activators tend to be the reactive points of the co-enzymes. There are a number of heavy metals that act as metallic activators. Iron (Fe) is the most common metallic activator in bacteria enzymes. Other metallic activators include manganese (Mn), cobalt (Co), molybdenum (Mo), copper (Cu), nickel (Ni), and zinc (Zn). Some enzymes may contain more than one metallic activator. Normally, iron will be one of the metallic activators when the enzyme contains more than one metal ion. The alkaline earths: calcium (Ca), magnesium (Mg), and potassium (K), have also been found in some enzymes. The metallic activators are primarily involved with electron transfer reactions. For the most part, the metallic activators are the trace metals that the bacteria find essential for metabolism. Without these trace metals the bacteria would be unable to metabolize the various nutrients. It should be noted that many of the trace metals are considered as toxic heavy metals. This illustrates the fact that chemicals in low concentrations can be useful and toxic at high concentrations. Too often, people make blanket statements about heavy metals being toxic, demanding their complete elimination from the environment. Such action would be disastrous for all biological life.

The bacterial oxidation reactions normally occur in a water environment with oxygen being supplied as dissolved oxygen in the water. These reactions with dissolved oxygen are *aerobic* reactions. Aerobic metabolism is very important in the environment with the production of the most oxidized form of carbon and hydrogen, as well as yielding the most energy for the bacteria. It should be recognized that dissolved oxygen does not react directly with the organic compound being metabolized in most cases. Oxidation occurs indirectly by removing two hydrogen atoms from the molecule to form a carbon-carbon double bond and adding water across the double bond. The two hydrogen atoms are metabolized through a series of enzyme reactions, ultimately reacting with dissolved oxygen to regenerate the water. Although bacterial metabolism of amino acids results in the production of the reduced form of nitrogen, ammonia, there are aerobic bacteria that can oxidize ammonia to nitrites with the production of energy. The initial energy step of nitrification is shown in Equation 2-3.

$$NH_3 + 1.5\ O_2 \rightarrow HNO_2 + HOH + \text{Heat Released} \qquad (2\text{-}3)$$

The energy yield is about 311 kJ/mol., 22.2 kJ/g N. A second group of aerobic bacteria can oxidize the nitrous acid to nitric acid, the most oxidized form of nitrogen, Equation 2-4.

$$HNO_2 + 0.5\ O_2 \rightarrow HNO_3 + \text{Heat Released} \qquad (2\text{-}4)$$

The energy yield for this reaction is only 99.6 kJ/mol, 7.1 kJ/g N. The second group of bacteria does not obtain as much energy as the first group of bacteria, but that does not diminish their importance. The bacteria that oxidize ammonia to nitric acid are known as *nitrifying* bacteria. The group of nitrifying bacteria that oxidize ammonia to nitrous acid are designated as the *Nitroso-* bacteria; and the group of nitrifying bacteria that oxidize the nitrous acid to nitric acid are designated as the *Nitro-* bacteria. In the natural environment at pH values less than 8.0 the ammonia nitrogen will exist as ammonium bicarbonate, ammonium chloride, ammonium sulfate or ammonium phosphate. The production of nitrous acid will create problems for the nitrifying bacteria unless there is adequate alkalinity to neutralize the nitrous acid as fast as it is formed. Sodium or calcium bicarbonates are the major forms of alkalinity in the natural environment to neutralize the nitrous acid. If ammonium bicarbonate is the form of ammonia being oxidized, its alkalinity will be destroyed as well as the sodium or calcium bicarbonate alkalinity required to neutralize the nitrous acid. Equation 2-5 illustrates the neutralization reaction that is essential for nitrification.

$$NH_4HCO_3 + 1.5\ O_2 + NaHCO_3 \rightarrow NaNO_2 + 3\ HOH + 2\ CO_2 \qquad (2\text{-}5)$$

The carbon dioxide produced as an end product will depress the environmental pH slightly unless an excess of alkalinity is available to keep the pH from dropping. Since the nitrous acid has been neutralized, there is no need for additional alkalinity for the oxidation of nitrous acid to nitric acid. Nitrifying bacteria use part of the excess carbon dioxide as their source of carbon for cell synthesis. As a net result, the nitrifying bacteria do not compete with the other bacteria for the carbon required for cell protoplasm; but they will compete for oxygen and for nitrogen.

Another group of aerobic bacteria obtain their energy from the oxidation of reduced sulfur compounds. Hydrogen sulfide, H_2S, is the most reduced form of sulfur. Hydrogen sulfide can be oxidized to sulfuric acid, as shown in Equation 2-6.

$$H_2S + 2O_2 \rightarrow H_2SO_4 + \text{Heat Released} \qquad (2\text{-}6)$$

The sulfur oxidizing bacteria obtain about 848 kJ/mol from the above reaction, 26.5 kJ/g S. While there are bacteria capable of the complete reaction, some sulfur oxidizing bacteria produce only partial metabolism, stopping at pure sulfur or at thiosulfates. The sulfur oxidizing bacteria have primarily been classified as *Thiobacillus.*

Iron oxidizing bacteria are important environmental bacteria, converting soluble ferrous iron, Fe^{2+}, to insoluble ferric iron, Fe^{3+}. Some bacteria convert metallic iron, Fe^0, to ferric iron under aerobic conditions. Energy yield depends upon the end products formed. One of the problems with iron oxidation is that the ferrous iron can be oxidized with dissolved oxygen to ferric iron without the help of bacteria. More data are needed to obtain precise energy yields for the iron bacteria. There is even a group of bacteria that can oxidize hydrogen gas to water. Hydrogen is not very soluble in water, making it difficult to produce sufficient energy for much bacterial growth.

Not all bacteria can use dissolved oxygen for their energy reactions. Many bacteria use oxygen in chemical compounds for their energy metabolism reactions. Yet, it should be realized that these energy reactions are also oxidation reactions. The bacteria that obtain energy without using dissolved oxygen in their metabolism are called *anaerobes.* The anaerobic bacteria do not obtain as much energy per unit of substrate metabolized as the aerobic bacteria. The amount of energy obtained depends on the total metabolic reactions that the different bacteria create. The anaerobic bacteria oxidize their organic compounds by the same initial pathway that the aerobic bacteria use. They both begin oxidation by removing two hydrogen atoms to create the carbon-carbon double bond, Equation 2-7. The enzymes required for removing the two hydrogen ions are essentially the same in aerobic bacteria as in anaerobic bacteria.

$$(\text{-}CH_2\text{-}CH_3) - 2H \rightarrow \text{-}CH{=}CH_2 + \text{Heat Released} \qquad (2\text{-}7)$$

Water is added across the double bond to create a more oxidized compound, Equation 2-8

$$(\text{-CH=CH}_2) + \text{HOH} \rightarrow \text{-CH}_2\text{-CH}_2\text{OH} + \text{Heat Released} \quad (2\text{-}8)$$

The difference between the aerobic bacteria and the anaerobic bacteria lies in how the two hydrogen atoms are processed. While aerobic bacteria always use dissolved oxygen to react with the hydrogen atoms, anaerobic bacteria use a number of different reactions with different energy yields.

Nitrates yield a little less energy for metabolism than dissolved oxygen. Only 83% of the oxidation energy potential in the oxygen in nitrates will be released during the metabolism of nitrates and organic compounds. The cations associated with the nitrates tie up the 17% of the oxidation energy, preventing its use in oxidizing the organic compounds being metabolized. The alkaline earth cations, normally associated with the nitrates, form the corresponding hydroxides that quickly react with carbon dioxide to form the corresponding bicarbonate salts. This is why denitrification generates 3.57 mg alkalinity for every 1.0 mg nitrate-nitrogen reduced to nitrogen gas. Most of the bacteria that use nitrates for their energy metabolism also can use dissolved oxygen. Equation 2-9 is a

$$\text{Organic Matter} + \text{NaNO}_3 \rightarrow \text{NaHCO}_3 + 0.5\text{N}_2 + \text{HOH} + \text{Heat Released} \quad (2\text{-}9)$$

general energy reaction indicating organic metabolism with nitrate reduction. These nitrate-reducing bacteria have been designated as *facultative bacteria* since they can move from aerobic metabolism to anaerobic metabolism and back again, depending upon the availability of dissolved oxygen. Because of the higher energy yields obtained with dissolved oxygen, the facultative bacteria always use dissolved oxygen in their metabolism before using nitrates. It is important to notice that the nitrate reduction reaction results in the production of nitrogen gas, not ammonia, as the reduced nitrogen end product. This reaction is known as *denitrification* and is very important in the loss of nitrogen from both the aquatic environment and the soil environment. If the bacteria use nitrates for their source of cell nitrogen, they must reduce the nitrates to ammonia for synthesis. Since ammonium ions are usually available in the environment, the bacteria will not have to utilize nitrates as their source of cell nitrogen. Nitrates will only be used for cell synthesis when ammonium ions are not available. The bacteria must use additional energy to reduce the nitrates to ammonia. This is energy the bacteria had rather use for additional cell synthesis.

Sulfates can also serve as a source of oxygen for energy metabolism. Sulfate reducing bacteria (SRB) are strict anaerobes that convert sulfates to sulfides while oxidizing organic compounds. The sulfate reducing bacteria are unique in their preference for organic compounds that contain large quantities of hydrogen ions. Long chain fatty acids are the preferred organic compounds for metabolism. Sulfate reducing bacteria can also metabolize short chain organic acids. Lactic acid is more readily metabolized than glucose, although the two compounds have the same ratios of carbon-hydrogen-oxygen. The oxidation reaction for lactic acid by sulfate reducing bacteria is given Equation 2-10.

$$C_3H_6O_3 + 1.5\ Na_2SO_4 \rightarrow 1.5\ NaHS + 1.5\ NaHCO_3 + 1.5\ CO_2 + 1.5\ HOH + \text{Heat Released} \quad (2\text{-}10)$$

The energy yield is only 73.2 kJ/mol lactic acid, 0.79 kJ/g. The energy generation for the sulfate reducing bacteria is quite low compared with the denitrifying bacteria. For this reason, the sulfate reducing bacteria are not competitive when nitrates are in the environment. The sulfate reducing bacteria belong to the *Desulfo-* group of bacteria. They are very important in environmental microbiology since the shift from sulfates to sulfides causes a change from an oxidized environment with the sulfates to a strongly reduced environment with the sulfides. Sulfate reduction also produces an increase in alkalinity and prevents loss of sulfides as a gas by keeping the sulfides as the hydrosulfides in proportion to the pH of the environment. All bacteria need sulfur in their protoplasm; but very few bacteria can use sulfates as their ultimate electron acceptor in metabolism. Two important, sulfur containing amino acids, cystine and methione, are synthesized in bacteria protoplasm.

Oxidized organic compounds can be metabolized for energy, producing lower molecular weight organic compounds. If nitrates or sulfates are not available, the bacteria will metabolize oxidized organic compounds as their electron acceptors. Metabolism attempts to produce the most oxidized state for the end products. Glucose is metabolized to many different organic acids, depending upon the specific bacteria and the environmental conditions in which the bacteria are living. As shown in Equation 2-11, glucose can be metabolized to lactic acid by the lactic acid bacteria.

$$C_6H_{12}O_6 \rightarrow 2C_3H_6O_3 + \text{Heat Release} \quad (2\text{-}11)$$

The energy yield for the production of lactic acid is only 87.8 kJ/mol glucose, 0.49 kJ/g. The low energy yield for this reaction is created by the high energy content of the lactic acid. Essentially, glucose is simply broken into two smaller molecules with some internal rearrangement of atoms. It has long been known that glucose can be converted into many different end products by specific bacteria. Some end products include: formic acid, acetic acid, propionic acid, butyric acid, valeric acid, ethanol, propanol, butanol, acetaldehyde, acetylmethylcarbinol, acetone, 2,3 butylene glycol, carbon dioxide, and hydrogen gas. The bacteria simply produce the end products that yield the maximum amount of energy for the minimum amount of substrate metabolized. Metabolism ceases when the energy reactions reach their lowest level, although the organic end products still contain considerable energy that neither the facultative nor any other anaerobic bacteria can utilize.

Examination of bacteria metabolism shows that it is essentially a series of oxidation-reduction reactions under either aerobic or anaerobic environments. Metabolic reactions proceed until the oxidants are balanced by the reductants. Cell protoplasm is a major part of the reductants with the energy for synthesis obtained from the oxidation reactions. The organic end products of anaerobic metabolism would accumulate in the environment if it were not for the acetogenic bacteria and the methane bacteria. The acetogenic bacteria metabolize the low molecular weight alcohols, aldehydes, ketones, and acids to acetic acid, hydrogen, and carbon dioxide. Since hydrogen is reduced and carbon dioxide is oxidized, the acetogenic bacteria can also combine these two end products to form acetic acid. It is easy to see how the acetogenic bacteria obtained their name. The methane bacteria can take the end products of the acetogenic bacteria and make methane and carbon dioxide as their end products. Since methane is an insoluble saturated hydrocarbon, it diffuses out of the liquid into the atmosphere where it can undergo metabolism by aerobic bacteria at the air-water or air-soil interface. There are two groups of methane bacteria. One group of methane bacteria uses carbon dioxide and hydrogen to produce methane and water as shown below.

$$CO_2 + 4H_2 \rightarrow CH_4 + 2HOH + \text{Heat Released} \qquad (2\text{-}12)$$

This reaction yields 253 kJ/mol methane produced, 15.8 kJ/g methane. This is a very high-energy yield for an anaerobic reaction. It depends upon the large amount of energy contained in the hydrogen gas. The methane produced by this reaction contains considerable energy that can be released by oxidation. This group of methane bacteria competes with the acetogenic bacteria for carbon dioxide and hydrogen. The efficiency of metabolism favors the methane bacteria. Both groups of bacteria exist together in the environment, indicating that neither organism has a real advantage over the other organism. The second group of methane bacteria uses acetic acid to produce methane and carbon dioxide as shown below.

$$CH_3COOH \rightarrow CH_4 + CO_2 + \text{Heat Released} \qquad (2\text{-}13)$$

The energy yield of this reaction is about 41 kJ/mol methane produced, 2.6 kJ/g methane. The acetate metabolizing methane bacteria do not obtain as much energy per g methane produced as the carbon dioxide reducing methane bacteria. Since the two groups of methane bacteria do not compete for the same nutrients, they both exist together in the environment. It has been found the acetate metabolizing methane bacteria are more sensitive to environmental factors than the hydrogen metabolizing methane bacteria. As a net result, the acetate metabolizing methane bacteria are good indicators of the health of the anaerobic environment. Failure to metabolize all the acetate indicates that the acetate utilizing methane bacteria are having difficulties. The primary problems with the methane bacteria are pH, O-R-P, and trace metals. The methane bacteria have been found to metabolize best at pH levels above 6.5. Since many organic compounds are initially metabolized to form organic acids as end products, the pH in the environment will decrease unless there is sufficient alkalinity to neutralize the organic acids as fast as they are formed. Without adequate alkalinity the pH will drop below pH 6 as the organic acids accumulate and the methane bacteria will quickly stop metabolism. The methane bacteria must also have a strongly reduced environment as evidenced by a very low O-R-P. Normally, the sulfate reducing bacteria work with the methane bacteria, creating the highly reduced environment. While the methane bacteria require iron for their enzymes, they also require nickel. Without adequate nickel the methane metabolism slows significantly. Research has indicated that several trace metals are required for optimum methane formation. Zinc, copper, molybdenum, manganese, and selenium have been shown to stimulate growth and metabolism of the methane bacteria. The methane produced as an end product can be collected and used as a source of heat energy. In rural areas of developing nations methane is widely used as a source of fuel for cooking and heating. Methane gas has even been used to power motor vehicles and some of the equipment in wastewater treatment plants. Overall, the methane bacteria play a very important role in the final steps of anaerobic metabolism. They complete the stabilization of organic matter with the release of methane gas to the atmosphere. Although methane gas has been considered as a major factor in the "greenhouse effect" on global warming, there are methane-oxidizing bacteria in the soil to convert the methane back to carbon dioxide and water, completing the stabilization cycle.

The photosynthetic bacteria obtain their energy from light rather than from organic compounds. As previously indicated, the photosynthetic bacteria have different photosensitive pigments than algae and higher plants. There are the purple photobacteria, the green photobacteria, and the blue-green photobacteria. The purple photobacteria and the green photobacteria were the original photosynthetic bacteria. These bacteria use sunlight to obtain hydrogen from hydrogen sulfide and other reduced sulfur compounds for the reduction of carbon dioxide and the production of sulfate, as shown in Equation 2-14.

$$2CO_2 + H_2S + \text{Light Energy} \rightarrow H_2SO_4 + \text{Heat Released} \qquad (2\text{-}14)$$

Carbon from the carbon dioxide is converted into cell mass carbon. This reaction occurs at the water surface of anaerobic environments where hydrogen sulfide is being generated and released to the atmosphere. The photobacteria obtain the hydrogen sulfide and carbon dioxide from the liquid below them and the light from the atmosphere above them. Because of the need for the two different environments, photobacteria grow in a thin layer. The overall energy-synthesis reaction for the metabolism of the hydrogen sulfide also involves the reduction of water and the addition of ammonia nitrogen for protoplasm production.

The blue-green photobacteria are the transition phase between bacteria and algae. They have a mixture of pigments that produce their blue-green color. The basic energy reaction for the *Cyanobacteria* results in water being used as their source of hydrogen, Equation 2-15.

$$CO_2 + HOH + \text{Light Energy} \rightarrow CH_2 \text{(as part of cells)} + 1.5O_2 + \text{Heat Released} \quad (2\text{-}15)$$

Here, oxygen is released as the primary end product of the energy reaction the same as with algae. Since the blue-green photobacteria are a transition between bacteria and algae, it is not surprising that the blue-green photobacteria can metabolize hydrogen sulfide, the same as the other photobacteria. Over the years, some environmental microbiologists thought blue-green photobacteria produced hydrogen sulfide as part of their metabolism. Actually, the blue-green photobacteria grew because the environment had a high hydrogen sulfide level. It is very important that environmental microbiologists have a good understanding of biochemistry to be able to distinguish between cause and effect of complex environmental reactions. Microbial reactions in the natural environment are far more complex than laboratory reactions and require careful evaluation for proper understanding.

A major problem with evaluating energy reactions in biological systems is the coupling of the energy reactions with synthesis reactions. The energy reactions do not exist as independent reactions. Efforts to use normal chemical equations to express the energy reactions result in gross approximations. The thermodynamic data on biologically important organics are not as extensive as for hydrocarbons that have been used in energy systems. Because of problems using heat energy, most energy measurements for biological systems have used Gibb's free energy measurements. It is readily recognized that bacteria are not heat machines. Yet, bacteria must follow chemical thermodynamics from both a heat energy and a free energy point of view. Fundamental concepts show that heat energy and free energy are related, as shown in Equation 2-16.

$$\Delta H = \Delta G + T\Delta S \qquad (2\text{-}16)$$

where: ΔH = change in heat energy
ΔG = change in free energy
T = absolute temperature
ΔS = change in entropy

Care should be taken when using thermodynamic data in evaluating the energy metabolism of bacterial systems. As knowledge increases, the apparent problems with the thermodynamic data will disappear.

SYNTHESIS REACTIONS

The second part of metabolism is the synthesis of new cell protoplasm. The primary purpose of the energy reaction is to provide the energy for cell synthesis. The bacteria must synthesize hundreds of different chemical compounds to make a cell. It is not surprising to find that the bacteria use the same basic chemical structures repeatedly, greatly simplifying the synthesis of new cells. It is too difficult to analyze every compound produced in each cell; but it is possible to use total measurements of the organic fraction of bacteria for the synthesis reactions. The primary method for evaluation of synthesis has been to develop an empirical equation of the cell mass from the carbon, hydrogen, oxygen and nitrogen analyses of bacteria. The techniques for evaluating the empirical formula of bacteria protoplasm have already been presented. It has been found that the bacteria produce the same cell protoplasm regardless of the chemical nature of the substrate, as long as all nutrients are available for normal cell production. The nutrients and the environment determine the overall synthesis reaction. Under optimum conditions Prochazka, Payne and Mayberry reported in 1973 that 15 different bacteria had a heat of combustion of 22.6 kJ/g cell weight on an ash free basis. At the same time Abbott and Clamen reported that *Pseudomonas fluorescens* had a heat of combustion of 22.5 kJ/g cell mass on an ash free basis. Other investigators have confirmed about the same energy content for the cell mass. The energy in the cell mass must come from the substrate being metabolized. Since the heat of combustion determination for bacteria cell mass oxidizes nitrogen to nitrogen gas, the measured heat of combustion values are high when compared to the energy required by the bacteria to form a unit mass of protoplasm. Most bacteria use ammonia nitrogen as their source of cell nitrogen. Only a few specialized, nitrogen-fixing bacteria are able to use nitrogen gas as their source of cell nitrogen. Based on the use of ammonia nitrogen for cell protoplasm the bacteria protoplasm would contain 19.6 kJ/g cell mass on an ash free basis. The total energy requirement for the synthesis of bacteria protoplasm will be the energy for synthesis of the cell components plus the energy content of the cell mass produced.

COUPLED REACTIONS

The bacteria couple the energy-synthesis reactions into a single set of reactions rather than in separate reactions. Substrate metabolism results in a continuous processing of nutrients to create energy for the manufacture of cell mass and the cell mass produced from the use of that synthesis energy. Since the cell mass contains the same amount of energy per unit weight, the energy requirement to produce that unit of cell mass is also constant. Payne reported that 62% of the substrate energy was conserved as cell mass with 38% being used to produce the energy for synthesis. Published data on cell mass yields, based on glucose metabolism, have shown variations in the conversion of glucose and ammonia to cell mass energy ranging from 50% to 70%. A major part of the problem with glucose metabolism is the production of organic storage products in addition to the production of cell mass. Another part of the variation in energy-synthesis relationships can be traced to analytical problems in harvesting and weighing small quantities of bacteria. Still, another part of the problem is related to maintenance energy expended during the growth period. Maintenance energy, also termed endogenous respiration, is the energy used to maintain cell integrity and depends on the active bacteria mass, time and temperature. The longer the growth period for the bacteria, the greater is the effect of endogenous respiration. The net effect of endogenous respiration is to reduce the proteins contained in the new cell mass. Close coupling of the energy-synthesis-endogenous respiration reactions allows the energy from the endogenous respiration reaction to pass without notice until studies are carried out over long periods under substrate limiting conditions. Since most bacteriological studies are carried out in excess substrate over short growth periods, endogenous respiration is masked. Temperature also affects endogenous respiration. Endogenous respiration increases as the temperature of metabolism increases. For this reason endogenous respiration has a greater impact on bacteria grown at 37° C than on bacteria grown at 20° C. Additional research is needed on evaluation of the impact of endogenous respiration by bacteria.

Based on 62% energy conversion to cell mass, normal metabolism requires a total of 31.6 kJ/g cell mass produced. The energy requirements for cell synthesis are the same for aerobic metabolism as well as anaerobic metabolism. The cell mass contains the same amount of energy and the synthesis reactions require the same amount of energy to produce a unit of cell mass. The difference between aerobic metabolism and anaerobic metabolism lies in the amount of energy produced under the two environments. Aerobic metabolism produces more energy than anaerobic metabolism from the same amount of substrate metabolized. The net effect is that more cell mass is produced aerobically than anaerobically. Aerobic metabolism of glucose should produce 0.49 g VSS cell mass/g glucose metabolized. Anaerobic metabolism of glucose to lactic acid should produce only 0.04 g VSS cell mass/g glucose metabolized. The problem with anaerobic metabolism is increased by the highly varied end products that different bacteria produce. The small quantity of cell mass produced per unit substrate metabolized makes accurate evaluation difficult. Energy-synthesis-endogenous respiration relationships will be a fruitful area for future research.

****

TYPICAL CALCULATIONS:

1. Find the heat of combustion and molecular weight of glucose and lactic acid from a chemical handbook.

Glucose = 2,813 kJ/mol. Lactic acid = 1,363 kJ/mol.

Glucose = 180 GMW Lactic acid = 90 GMW

2. Metabolism based on 62% glucose energy to cell mass, aerobic metabolism.

Cell mass energy = (0.62)(2,813)/(180) = 9.7 kJ/g glucose metabolized

Synthesis energy = (0.38)(2,813)/(180) = 5.9 kJ/g glucose metabolized

3. Bacteria cell mass synthesis in aerobic glucose metabolism.

Bacteria cell mass energy = 19.6 kJ/g VSS

Bacteria cell mass synthesis = (9.7)/(19.6) = 0.49 g VSS/g glucose metabolized

4. Synthesis energy in anaerobic metabolism

Energy released by glucose metabolism to lactic acid =

(2,813 - (2)(1,363))/(180) = 0.48 kJ/g glucose metabolized

5. Bacteria cell mass synthesis in anaerobic glucose metabolism to lactic acid

Energy for synthesis in aerobic metabolism = (0.49)/(5.9) = 0.083 g VSS/kJ
Same energy required for synthesis of a unit mass of cell mass anaerobically as aerobically

Bacteria cell mass synthesis = (0.48)(0.083) = 0.04 g VSS/g glucose metabolized

****

NITROGEN FIXATION

An interesting aspect of bacterial metabolism is related to the fixation of nitrogen. It is not to be confused with nitrification or denitrification. In environments that are deficient in ammonium nitrogen or oxidized nitrogen salts, a few bacteria species can fix atmospheric nitrogen into cellular protoplasm. The free-living *Azotobacter* metabolize carbohydrates aerobically to obtain sufficient energy to fix nitrogen gas into ammonia nitrogen for their cell proteins. Nitrogen fixing bacteria expend part of their synthesis energy for nitrogen fixation and cannot produce as much cell mass per unit substrate metabolized as normal bacteria. When the *Azotobacter* undergo endogenous respiration, ammonia nitrogen is released back into the environment, giving the impression that nitrogen fixing bacteria fix excess nitrogen and excrete it into the outside environment. In the presence of ammonia nitrogen the *Azotobacter* stop fixing atmospheric nitrogen, using the available ammonia nitrogen for cell synthesis. Bacteria are quite efficient in their reactions, not doing any more work than necessary for the desired chemical reactions.

Rhizobium are symbiotic nitrogen fixing bacteria that produce nodules on the roots of legumes. The *Rhizobium* fix atmospheric nitrogen as they grow on the nutrients from the legumes. Nitrogen is fixed into cell mass and eventually released by endogenous respiration for the plants to use for their cell growth. Nitrogen fixing bacteria are important in agriculture, especially in developing countries where chemical fertilizer is limited. The blue-green photobacteria, *Cyanobacteria,* are important nitrogen fixing bacteria. They use light energy for the conversion of nitrogen gas to ammonia for cell synthesis. The *Cyanobacteria* grow on water and soil surfaces where light energy is readily available and other forms of nitrogen are not readily available. On the anaerobic side, *Clostridium* can fix atmospheric nitrogen with the expenditure of energy during cell synthesis. The simplified nitrogen fixing reaction can be expressed as shown in Equation 2-17.

$$N_2 + 6H + \text{Energy} \rightarrow 2NH_3 \qquad (2\text{-}17)$$

The ability of bacteria to fix atmospheric nitrogen illustrates how bacteria have adapted to their environment over the years. The economic importance of nitrogen fixing bacteria has helped focus considerable attention on the biochemistry of these bacteria. Recent studies have shown that many other bacteria can fix atmospheric nitrogen but the groups shown above are the most widespread in the environment and the easiest to grow and evaluate.

ACTINOMYCETES

Actinomycetes are a group of specialized bacteria that are closely related to fungi. The actinomycetes are filamentous microorganisms with true branching like fungi; but lacking a distinct nuclei. The filaments are less than 1.0 μm in width, normally 0.7 to 0.8 μm with cell wall chemistry similar to bacteria. The actinomycetes are single cell microorganisms that reproduce by binary fission. They appear to form spores, but not in the same sense as either bacteria or fungi. The actinomycetes spores appear more as cell fragmentation without any hard protective coatings. The economic importance of the actinomycetes has resulted in their detailed study and separation from the other bacteria.

Actinomycetes are found in soil where they participate in the stabilization of dead plants and animal residues. They are washed from soil into streams, rivers, lakes and reservoirs where they may or may not find suitable nutrients for growth. In the Southwestern part of the United States excessive growths of actinomycetes in reservoirs, used for water supplies, have resulted in taste and odor problems. Actinomycetes produce an "earthy" taste in water. Some of the actinomycetes are pathogenic to humans, animals and plants. The pathogenic actinomycetes are found in the oral cavity of humans and animals.

Although much is known about the actinomycetes, much remains to be learned. The classification of actinomycetes is undergoing rapid change as additional studies are published. The older approach will be taken to the actinomycetes since it provides a larger umbrella to this interesting group of organisms. Waksman demonstrated one of the best approaches to actinomycetes, having studied them in great detail for many years. The optimum pH for actinomycetes is between pH 7.0 and 8.0. Yet, they can grow under acidic conditions at pH levels as low as 4.0. Although the actinomycetes are classified as bacteria, a study of *Streptomyces griseus* found the cell mass contained about 9% nitrogen, lower than most bacteria. Since the mass of actinomycetes is not as important as their end products, very little quantitative data are available on the actinomycetes.

Actinomycetes metabolize most naturally occurring organics. Since they require a little less nitrogen than most bacteria for cell growth, the actinomycetes tend to metabolize the more resistant forms of organic matter. After the normal bacteria and fungi metabolize the readily biodegradable components of dead plant material, the actinomycetes are able to continue to metabolize the residual organics. The actinomycetes can grow on carbohydrates, proteins, lipids and aromatics. Complex lignocellulose compounds and humus can be slowly metabolized by this group of microorganisms. The actinomycetes are facultative organisms with the ability to metabolize either aerobically or anaerobically. Aerobic metabolism gives greater energy yields and more cell mass production than anaerobic metabolism.

Nocardia have been considered as part of the actinomycetes; but may be moved to a different group as more data are collected to show the lack of similarity of the *Nocardia* with the other actinomycetes. *Nocardia* are interesting environmental organisms since they are able to metabolize sterols and steroids found in animal wastes. They are also able to decompose rubber and other complex hydrocarbons found in many industrial wastes. In recent years, *Nocardia* have been implicated in the production of heavy scum on the surface of activated sludge aeration tanks, treating domestic sewage and various industrial wastes. One thing for certain, the heavy scum on the aeration tanks is a good source of *Nocardia* for study when needed. Fortunately, the *Nocardia* growing in the scum layer on aeration tanks have all been non-pathogenic. Unlike the other actinomycetes, *Nocardia* are aerobic organisms, requiring free oxygen to metabolize the lipids. A study on lipid formation by Raymond and Davis indicated that *Nocardia* cells grown on glucose had 28% lipids and produced 56% lipids when grown on octadecane. The hydrophobic characteristics of the *Nocardia* cells assist this microorganism in accumulating in greasy scum on the surface of aeration tanks. As more research develops on *Nocardia,* environmental microbiologists will be able to provide a better understanding of the cause-and-effect relationships in activated sludge.

The major biochemical interest in the actinomycetes has been in their production of various antibiotics. Streptomycin has been produced by various species of *Streptomyces*. Many other antibiotics have been produced over the years from different species of actinomycetes. While the actinomycetes normally grow at ambient temperatures in the soil, pathogenic actinomycetes grow best at body temperature, 37° C. Some actinomycetes grow at thermophilic temperatures to about 70° C. The rates of growth of actinomycetes follow the same general temperature relationships as bacteria with a change in metabolism by a factor of two for each 10° C temperature change. Because actinomycetes play a major role in the stabilization of agricultural residues in the soil, they are easily found in rivers and streams draining from agricultural areas. With increased interest in composting of yard wastes across the United States, there will be a greater spread of actinomycetes in the environment than in the past.

THINGS TO REMEMBER

1. Bacteria are single cell microorganisms that are vital to the recycling of materials in nature.

2. The carbon and hydrogen contents of bacteria cells remain relatively constant with time, while the oxygen increases and the nitrogen decreases as endogenous respiration proceeds.

3. Bacteria metabolize soluble chemical compounds.

4. Insoluble chemicals must be made soluble before bacteria can metabolize them.

5. Bacteria oxidize chemical compounds to obtain energy to synthesize new cell mass.

6. Substrate energy and available nutrients determine the chemical characteristics of the new bacteria cells.

7. Aerobic metabolism produces more energy for the bacteria than anaerobic metabolism.

8. Dissolved oxygen yields more energy for the bacteria than nitrates, sulfates or carbon dioxide.

9. Anaerobic metabolism produces organic acids as end products that must be neutralized with alkalinity prior to forming methane.

10. Nitrogen fixation results in atmospheric nitrogen being converted to protein nitrogen in the bacteria.

11. Bacteria identification is shifting from a combination of physical characteristics and biochemical reactions to 16S rRNA analyses and a new set of nomenclature.

REFERENCES

Abbott, B. J. and Clamen, A. (1973) The Relationship of Substrate, Growth Rate and Maintenance Coefficient to Single Cell Protein Production, *Biotech. & Bioengr.*, **15**, 117.

Alm, E. W., Oerther, D. B., Larsen, N., Stahl, D. A., and Raskin, L. (1996) The Oligonucleotide Probe Database, *Appl. & Environ. Microbiol.*, **62**, 3557.

Battley, E. H. (1995) An Apparent Anomaly in the Calculation of Ash-Free Dry Weights for the Determination of Cellular Yields, *Appl. Environ. Microbiol.,* **61**, 1655.

Balch, W. E., Fox, G. E., Magrum, L. J., Woese, C. R., and Wolfe, R. S. (1979) Methanogens: Reevaluation of a Unique Biological Group, *Microbiol. Reviews*, **43**, 260.

Bottger, Erik C. (1996) Approaches for Identification of Microorganisms, *ASM News*, **62**, 247.

Clark, J. M. (1979) *The Use of Automatic Titrant Addition to Monitor Metabolic Reactions in Activated Sludges and Selected Studies of Acetate Mediated Denitrification*, PhD Thesis, University of Kansas.

Collard, P. (1979) *The Development of Microbiology*, Cambridge University Press.

Doetsch, R. N. (1960) *Microbiology*, Rutgers University Press, New Brunswick, New Jersey.

Goodfellow, M., Bownwell, G. H. and Serrano, J. A., *The Biology of Nocardiae*, Academic Press (1976).

Gupta, R. S. (2002) Phylogeny of *Bacteria*: Are We Now Close to Understanding It ?, *ASM News*, **68**, 284.

Harris, R. F. and Adams, S. S. (1979) Determination of the Carbon-Bound Electron Composition of Microbial Cells and Metabolites by Dichromate Oxidation, *Appl. Environ. Microbiol.*, **37,** 237.

Ingraham, J. L., Maaloe, O. and Neidhardt, F. C. (1983) *Growth of the Bacterial*

Cell, Sinauer Associates, Inc., Sunderland, Massachusetts.

Koch, A. L. (1990) Growth and Form of the Bacterial Cell Wall, *American Scientist*, **78**, 327.

Ludwig, W. and Schleifer, K-H (1999) Phylogeny of *Bacteria* Beyond the 16S rRNA Standard, *ASM News*, **65**, 752.

Mayberry, W. R., Prochazka, G. J. and Payne, W. J. (1967) Growth Yields of Bacteria on Selected Organic Compounds, *Appl. Microbiol.*, **15**, 1332.

Mayberry, W. R., Prochazka, G. J. and Payne, W. A. (1968) Factors Derived from Studies of Aerobic Growth in Minimal Media, *J. Bacteriol.*, **96**, 1424.

Morell, V. (1997) Microbiology's Scarred Revolutionary, *Science*, **276**, 699.

Pace, N. R. (1999) Microbial Ecology and Diversity, *ASM News*, **65**, 328.

Payne, W. A. (1970) Energy Yields and Growth of Heterotrophs, *Annual Reviews of Microbiol.*, **24**, 17.

Polz, M. F. and Cavanaugh, C. M. (1997) A Simple Method for Quantification of Uncultured Microorganisms in the Environment Based on In Vitro Transcription of 16S rRNA, *Appl. & Environ. Microbiol.*, **63**, 1028.

Prochazka, G. J., Payne, W. J. and Mayberry, W. R. (1973) Calorific Contents of Microorganisms, *Biotech. & Bioengr.*, **15**, 1007.

Raymond, R. L. and Davis, J. B. (1960), "n-Alkane Utilization and Lipid Formation by a *Nocardia*", *Appl. Microbiol.*, **8**, 329.

Schaal, K. P. and Pulverer, G. (Editors) (1981) *Actinomycetes - Proc. of the Fourth International Symposium on Actinomycetes Biology, Cologne, Sept. 3 - 7, 1979*, Gustav Fischer Verlag, Stuttgart, Germany.

Slack, J. M. and Gerencser, M. A. (1975) *Actinomycetes, Filamentous Bacteria*, Burgess Publishing Co., Minneapolis, Minnesota.

Venter, J. C., Smith, H. O., and Fraser, C. M. (1999) Microbial Genomics: *in the Beginning*, *ASM News*, **65**, 322.

Waksman, S. (1967) *The Actinomycetes*, Ronald Press, New York.

Chapter 3

BACTERIA GROWTH

Under the proper environmental conditions bacteria follow definite patterns of growth that are highly reproducible. The small size of bacteria and their simplicity of metabolism allow bacteria to reproduce at a rapid rate, compared to higher organisms. Bacterial growth is best examined in completely soluble substrates under optimum environmental conditions. Growth can be measured by turbidity, by direct counts in liquid media, or by plate counts on solid media. Turbidity is useful over a moderate range of bacteria growth. Turbidity can be determined as optical density at 600 nm in a single beam spectrophotometer. Turbidimetric measurements work best for growth studies in dilute nutrient media. As bacteria growth increases significantly, the errors in turbidimetric measurements increase at a rapid rate. The primary error comes from several cells located along the same light path, giving the impression of only a single cell in the turbidimetric measurement. Direct bacteria counts require dilution to reduce the number of bacteria to a reasonable level for accurate counting. Staining is normally used to permit easy observation of the cells. Fluorescent dyes have been used to permit direct counting of specific bacteria in mixed culture systems. Plate counts on solid media have proven to yield the best results and have been widely accepted for measuring bacteria growth. Experience has shown that serial dilutions to produce between 30 and 300 colonies per plate provide the best results. There is no single universal media that permits growth of all bacteria. Standardized protein and carbohydrate media have been used for isolation and growth of most common bacteria. Specialized bacteria require both specific nutrients and the proper environment. Membrane filters have been used for concentrating bacteria from dilute solutions for either direct counting or growth on specific media. Since the bacteria have definite masses, the total mass of bacteria has also been used to measure growth. There is no one recommended method for evaluating bacteria growth. You use the method best suited to your particular study.

Initially, bacterial growth patterns were made on batch-fed systems. A small population of bacteria was introduced into a nutrient solution and counts were made at regular intervals until the bacteria stopped growing and began to die off. Since a single pure bacterial culture was used, the numbers of bacteria produced good results. Mixtures of bacteria created problems since the different bacterial species did not grow at the same rate or use the same amount of substrate per cell. It was not always possible to distinguish the different bacteria, one from the other. Continuous flow systems were developed to examine equilibrium populations under uniform environmental conditions and produced sufficient microbial mass to allow mass units as the measure of growth. Use of the mass of bacteria allowed the study of mixed bacteria populations and pure bacterial cultures on a common basis. The mass of bacteria were separated from the liquid by vacuum filtration through pre-weighed, glass fiber filters, having maximum pore sizes of 0.2μ, dried in a 103°C oven, and reweighed in an analytical balance. The change in microbial mass was determined from the weight difference. Combustion in a muffle furnace at 550°C results in loss of the organic solids, leaving the microbial ash as the weight difference. Analytical technique is very important in obtaining valid mass data.

BATCH-FED GROWTH PATTERNS

Initial bacteria growth patterns were observed in concentrated nutrient solutions under sterile conditions to prevent the growth of extraneous bacteria. A small sample of a pure bacteria culture was inoculated into the sterile liquid and allowed to grow over time. At regular time intervals samples were removed aseptically and plated in solid media for growth and counting. The solid media consisted of a concentrated nutrient solution with agar as the solidifying agent. Agar is a purified polysaccharide from marine algae that is not metabolized by most bacteria. Agar has the unique characteristic of remaining solid until the temperature is raised above 100°C and then not solidifying until the temperature drops below 40°C. When solid media is sterilized, the agar melts and mixes with the concentrated nutrients. When the liquid agar solution cools to about 40°C, a one ml bacteria sample, having between 30 and 300 bacteria, is added to a sterile petri dish with about 10 ml of the liquid agar solution and rapidly mixed. As the temperature drops below 40°C, the agar solidifies. The bacteria are incubated at the desired temperature for growth for a period of 24 to 48 hours. Individual bacteria produce a colony that can be seen with a low-power magnifying glass and counted. Each distinct colony represents the growth from a single bacterium. The bacteria counts can be graphically plotted against time to yield the bacteria growth curve. A typical bacteria growth curve is shown in Figure 3-1. The bacteria growth curve is a complex curve with several distinct phases. The first phase of growth has been designated as the

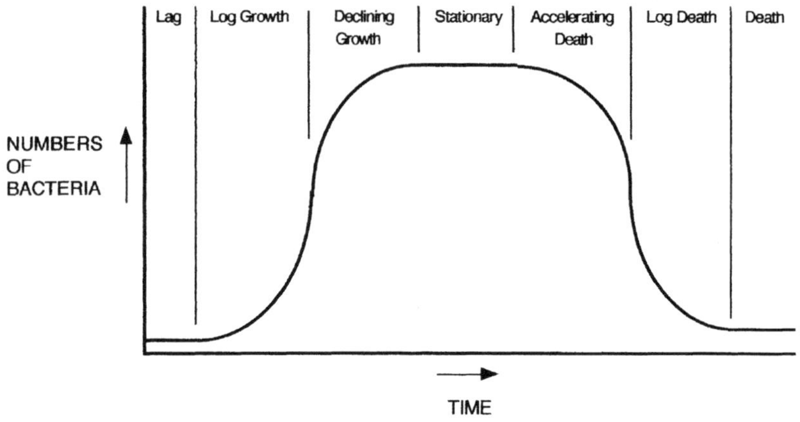

Figure 3-1 TYPICAL BACTERIA GROWTH CURVE

Lag Phase. During the lag phase the bacteria do not increase in numbers, but are adapting to metabolism of the new substrate. Once the bacteria have adapted to the substrate, they begin *Log Growth.* During log growth the bacteria are metabolizing at their maximum rate, doubling at a fixed interval designated as the *Generation Time.* Bacteria continue in the log growth phase until metabolism becomes limiting. The number of bacteria per unit volume often limits continued growth at the log rate. Metabolic end products, accumulating in the liquid around the bacteria, can slow the rate of metabolism by applying backpressure shift from log growth. As metabolism slows, growth shifts to *Declining Growth.* Eventually, the numbers of bacteria reach a maximum and enter the *Stationary Phase* where the bacteria numbers remain constant for a long period of time. As the bacteria begin to die, they enter a period of *Accelerating Death.* The rate of dying soon reaches the *Log Death* rate. Finally, the rate of dying slows as the bacteria reach the final *Death Phase.*

The graph shown in Figure 3-1 is a generalized graph of the numbers of bacteria using both numbers and time on linear scales. Plotting the growth data on semi-log graph paper with the numbers of bacteria shown on the semi-log scale and time on the linear scale, the log growth and the log death phases will plot as straight lines. Figure 3-2 shows the semi-log plot of the bacteria growth pattern. The declining growth phases and accelerating death phases tend to be compressed with the log scale plot while illustrating the full ranges of the log growth and log death phases.

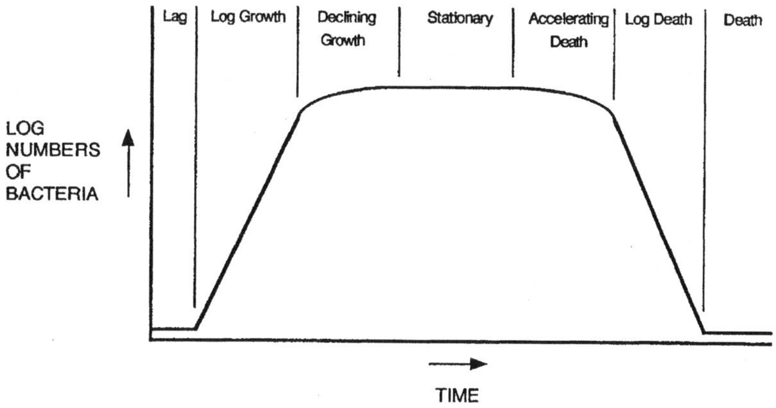

Figure 3-2 SEMI-LOG PLOT OF BACTERIA GROWTH CURVE

In concentrated nutrient substrates the initial growth will be aerobic until the dissolved oxygen (DO) has all been used. Metabolism shifts from aerobic to anaerobic with accumulation of end products that ultimately limits metabolism. Aerobic conditions can be maintained by growing the bacteria in low nutrient concentrations or in thin liquid layers in Erlenmeyer flasks rather than in test tubes or bottles, which have a small surface area to volume ratio. Using thin layers of liquid media in Erlenmeyer flasks on a shaking apparatus can insure aerobic conditions during the growth cycle. Completely anaerobic conditions require purging the media with nitrogen to strip the oxygen, as the first step, and then growth in an anaerobic jar or in an anaerobic chamber. Oxygen can also be removed chemically to produce an atmosphere of nitrogen.

Jacques Monod was one of the first bacteriologists to quantitatively examine the growth of bacteria in dilute organic solutions. His research was published in France in 1942. Because of World War II, it was 1949 before Monod could publish his research in English. Monod's original study is considered a classic in the microbiological literature. Monod used turbidity as the measure of bacterial growth and converted turbidity data to weight of bacteria. By growing bacteria in different concentrations of a simple organic substrate, Monod found that the maximum quantity of bacterial growth was directly proportional to the initial organic concentration. Since the plot of cell mass against organic substrate concentration passed through zero, it was concluded that the bacteria did not require any maintenance energy. Another part of his study showed that the total mass of bacteria produced in a sample that was agitated was the same as a sample that was not agitated. The agitated sample reached its maximum concentration faster than the unagitated sample. The same growth was also taken to suggest that the bacteria did not require maintenance energy during growth. Monod's failure to observe maintenance growth, i.e. endogenous respiration, effectively put a damper on this important concept of bacterial growth for many years. The most significant aspect of Monod's research was establishing the fact that the rate of bacteria growth was a function of the substrate concentration up to a specific concentration where the rate of growth became a constant with increased substrate concentrations. An important part of Monod's work was the introduction of quantitative relationships to predict the rate of microbial growth. He developed a series of equations that could be used to predict the amount of bacteria mass, produced from the metabolism of specific amounts of organic nutrients. Monod's equations started with the previously observed relationship for log growth where the rate of growth was a function of the microbial mass and substrate nutrients were always in excess. The basic log growth relationship for bacteria, based on numbers of bacteria, has been expressed as follows.

$$dN/dt = \mu N \tag{3-1}$$

where: N = number of bacteria
t = time, hrs.
μ = specific growth rate, 1/hr.

Solving Equation 3-1 for the number of bacteria results in Equation 3-2.

$$N = N_o (e^{\mu t}) \tag{3-2}$$

where: N = number of bacteria at time t.
N_o = number of bacteria at initial time, t = 0.

The numbers of bacteria increase very rapidly during log growth.

****

TYPICAL CALCULATIONS:

E. coli at 37° C has a specific growth rate of about 3/hr

If we started with one (1) *E. coli*, in one hour we would expect

$$N = (1)(e^{(3)(1)}) = 20 \text{ bacteria.}$$

In 10 hrs, the number of bacteria would be

$$N = (1)\,(e^{(3)(10)}) = 1.07 \times 10^{13} \text{ bacteria.}$$

If *E. coli* continued to grow at a log rate, it would quickly reach an overwhelming number.

****

It had long been observed that microbial growth slowed when the nutrient substrate became limiting. The shift from log growth to declining growth intrigued Monod. Since the bacteria were growing as fast as they could, Monod recognized that the specific growth rate, μ, shifted from being a constant to being a variable that was related to the remaining substrate concentration. He found that the value of μ could be expressed in terms of substrate concentration, Equation 3-3. Monod determined that μ_{max} for *E. coli* occurred above 25 mg/L glucose with the value of K_s being 4 mg/L glucose. The maximum specific growth rate, μ_{max}, was a constant that could be measured in an excess of substrate during log growth. The saturation constant, K_s, was measured experimentally as

$$\mu = \mu_{max}\,(S/\,(K_s + S)) \tag{3-3}$$

where: μ_{max} = maximum specific growth rate, 1/hr.
S = substrate concentration, mg/l.
K_s = saturation constant, substrate concentration when $\mu = 0.5\mu_{max}$, mg/L.

the substrate and the specific growth rate decreased. Substitution of these data in the initial equation showed μ would be $0.86\mu_{max}$ for 25 mg/L glucose. At 50 mg/L glucose, μ would be $0.93\mu_{max}$. Yet, Monod's data showed a constant growth rate at 25 mg/L glucose and higher. The differences in the measured data and the calculated results reflect the limitations of the empirical equation developed by Monod. The Monod equation should be recognized as an approximation of the data and not as a precise equation to predict the total range of data. The data gave the best results for the equation when the substrate concentration was close to K_s. As the data approached both ends of the equation, the errors increased. It is important to understand the limitations of published equations if they are to be properly applied. When equations become common and are published over and over, the limitations of the equations tend to be overlooked. A second equation, Equation 3-4, developed by Monod showed that the amount of bacteria growth was related to the substrate metabolized.

$$dx/dt = Y(dS/dt) \tag{3-4}$$

where: dx/dt = rate of bacteria mass growth, mg/L/hr
Y = yield factor, mg bacteria mass/mg substrate metabolized
dS/dt = rate of substrate metabolism, mg/L/hr

Monod's data for *E. coli* and glucose showed a yield factor, Y, of 0.233 mg *E. coli*/mg glucose metabolized at glucose concentrations between 25 mg/L and 200 mg/L. His data on *Bacillus subtlis* gave a yield factor, Y, of 0.218 mg *B. subtilis*/mg sucrose metabolized. These two different bacteria had similar yield factors on two related sugars. Later studies showed that these values were low, indicating that the substrate used by Monod may have been oxygen or nutrient deficient.

CONTINUOUS FEED GROWTH PATTERNS

Monod's next major contribution occurred in 1950, when he published his study on continuously fed bacteria growth systems. Monod developed a completely mixed bioreactor that could be maintained under aerobic conditions. By feeding a low concentration of substrate continuously, he found that the growth of bacteria was related to the fluid displacement time as long as it was greater than the generation time of the bacteria. Monod believed that when the fluid displacement was less than the generation time of the bacteria, the bacteria would be completely washed out of the system and there would be no bacterial growth. The effluent nutrient concentration would be the same as the influent nutrient concentration. Monod developed the following mathematical relationship between the rate of bacterial growth in the bioreactor and the displacement rate of bacteria from the bioreactor, Equation 3-5.

$$dx/dt = (\mu - D)x \qquad (3\text{-}5)$$

where: x = microbial concentration in the bioreactor, mg/L.
μ = specific growth rate, 1/hr.
D = displacement rate, Q/V or 1/t, 1/hr.
t = time, hrs.
Q = substrate flow rate, L/hr
V = bioreactor volume, liters (L)

The growth of bacteria will increase until the system comes to equilibrium. At equilibrium the rate of change in bacteria mass in the bioreactor is zero, $dx/dt = 0$. From the above equation, this means that either x or $(\mu - D)$ must be zero. Since the microbial mass concentration, x, is not zero, $(\mu - D)$ must be zero. This means that $\mu = D$ at equilibrium. The rate of bacteria growth in a completely mixed, continuously fed bioreactor is a function of the fluid retention time; i.e., $\mu = 1/t$, until the retention period is so short that the bacteria can no longer divide before being washed from the system.

Examination of the completely mixed bioreactor shows that the substrate will be removed by metabolism and by displacement. Metabolism results in part of the substrate being oxidized for energy and a corresponding part being converted to cell mass. Displacement results in loss of unmetabolized substrate in the effluent. Monod developed the following equation, Equation 3-6, to measure the rate of change in the substrate concentration in the bioreactor.

$$dS/dt = D(S_o - S) - (1/Y)(dx/dt) \qquad (3\text{-}6)$$

where: dS/dt = rate of substrate metabolism, mg/L/hr
dx/dt = rate of bacteria growth, mg/L/hr
S = substrate concentration in bioreactor, mg/L
S_o = influent substrate concentration, mg/L
Y = yield factor, mg/L x produced/mg/L S metabolized
t = time, hrs.

This equation can be solved for several variables. The concentration of bacteria in the bioreactor at equilibrium can be determined from Equation 3-7.

$$x = Y(S_o - S) \qquad (3\text{-}7)$$

In effect, the bacteria concentration depends on the substrate removal and the efficiency of converting substrate into cell mass. When the residual substrate is quite small compared with the influent substrate, the cell mass concentration can be written as a function of the influent substrate, Equation 3-8.

$$x = Y(S_o) \qquad (3\text{-}8)$$

Determination of the residual substrate was made using Monod's original relationship for μ at low substrate concentrations in Equation 3-9.

$$S = K_s/(\mu_{max} \cdot t - 1) \qquad (3\text{-}9)$$

Equation 3-9 shows that the residual substrate in the treated effluent is related directly to the dilution rate D, the reciprocal of the fluid displacement time, 1/t. For a given set of bacteria and substrate, both K_s and μ_{max} are constants at a specific temperature. At long hydraulic retention times there will be very little unmetabolized substrate. As the product of μ_{max} and t approaches one, the concentration of unmetabolized substrate will increase rapidly until the substrate concentration reaches the influent substrate concentration and growth stops.

Both of Monod's studies have had a profound impact on the quantitative aspects of bacteriological growth at low organic substrate concentrations. The basic nature of these studies is such that many investigators are still using them today. It is important to recognize both the value and the limitations of Monod's equations if they are to be properly applied. Monod developed his equations from laboratory data and fundamental concepts. He demonstrated that the growth rate of bacteria becomes a function of the substrate concentration when substrate is limiting. He demonstrated that growth of bacteria was directly proportional to the substrate metabolism. His research also showed that the bacteria in continuous flow reactors always grew at log rates. On the negative side, there is the limitation of the specific growth rate factor, μ , at high and low substrate concentrations. Monod's failure to observe the effect of maintenance energy invalidates his mass calculations except at short detention periods. Monod's data on cell yield per unit substrate indicated that his substrate was limited in trace nutrients. Trace nutrients are essential for the maximum cell yield per unit substrate metabolized. Available oxygen may also have been limiting. The extended value of Monod's research was the stimulation of other investigators to carry out additional research on food limiting conditions. This is very important for environmental microbiologists working in the area of biological wastewater treatment. All of the important biological wastewater treatment systems operate under food limiting conditions.

****

TYPICAL CALCULATIONS:

An aerobic bioreactor, fed 1,000 mg/L glucose and containing *E. coli*, operated with a 6 hr retention period.

1. Determine the specific growth rate of the *E. coli* in the bioreactor.

$$\mu = D = 1/t = 1/6 = 0.17/\text{hr}$$

2. Determine the maximum mass of *E. coli*.

Information from Chapter 2 indicated Y = 0.49 mg VSS/mg glucose

$$x = Y(S_o) = 0.49(1000) = 490 \text{ mg/l VSS}$$

3. Determine effluent glucose concentration.

$K_s = 4.0$ mg/L glucose and $\mu_{max} = 3.0$/hr

$S = K_s/(\mu_{max}\, t - 1) = 4.0/((3)(6) - 1) = 4.0/(18 - 1) = 0.24$ mg/L glucose

Monod's equations indicate that most of the glucose would have been metabolized with 6 hrs aeration.

4. Determine the effluent glucose concentration with one hr retention period.

$S = 4.0/((3)(1) - 1) = 4.0/2 = 2.0$ mg/l glucose

Monod's equation indicates 99.8% glucose metabolism. With the same microbial mass the key would lie in the ability of the system to transfer sufficient oxygen for aerobic metabolism.

5. Determine the oxygen demand rate at both 6 hrs aeration and one hr aeration.

Glucose has a COD of 1.07 mg COD/mg glucose
In Chapter 2 metabolism of glucose used 0.38 for energy

6 Hrs Aeration: Oxygen Used = 1.07(1000 – 0.24) (0.38)/6 = 68 mg/L/hr for energy

1 Hr Aeration: Oxygen Used = 1.07(1000 – 2) (0.38)/1 = 406 mg/L/hr for energy

The rate of oxygen demand increases rapidly as the detention time in the bioreactor is shortened. If the system is unable to transfer the oxygen, oxygen transfer will be the limiting factor controlling metabolism. Monod's equations were developed for aerobic systems with the substrate limiting.

****

Several studies followed publication of Monod's research on continuous flow systems. One of the more detailed studies was made by Herbert, Elsworth and Telling in 1956. Their study attempted to determine the validity of Monod's theory and to show how easily continuously fed studies could be made. They used a 20-liter bioreactor fed 2,500 mg/L glycerol as the organic substrate and *Aerobacter cloacae* as the bacteria. Batch fed growth studies on this substrate and organism produced a μ_{max} of 0.85/hr, a K_s value of 12.3 mg/L and a Y value of 0.53 mg bacteria mass/mg glycerol metabolized. Data were collected on a continuous flow basis with hydraulic retention times varying from 4.35 hrs down to 0.89 hrs at an operating temperature of 37°C. They were strongly interested in gathering data near the generation time of the bacteria, 0.82 hrs. Glycerol is a 3 carbon organic compound that is quite soluble with a COD of 1.22 mg/mg glycerol. At the longer HRT values the bacteria metabolized the glycerol with the maximum production of new cell mass. As the HRT was reduced, the total amount of glycerol fed to the bioreactor increased. The growth of bacteria increased linearly at the rate of 0.50 g bacteria mass/g glycerol fed up to a flow rate of 0.8 the theoretical volumetric displacement. As the addition of glycerol continued to be increased the production of bacteria mass began to decrease and the concentration of glycerol in the effluent began to increase. Variations in the data indicate that there were problems measuring the bacteria mass produced with incomplete metabolism. In spite of the analytical problems, it became apparent that the bacteria were unable to completely metabolize the glycerol in the bioreactor. The bacteria production per unit of glycerol removed decreased. At very high organic loading rates the substrate metabolism was incomplete. The importance of this study was twofold. It confirmed Monod's idea that stable growth conditions could be maintained with continuous feeding as long as the fluid displacement time was longer than the generation time of the bacteria. It also showed that Monod's equations were not valid at the high fluid displacement rates, only at low fluid displacement rates. One of the major findings of this research was the value of good laboratory experimental data in evaluating theoretical concepts. Although Herbert, Elsworth, and Telling demonstrated the shortcomings of Monod's equations, their research had no measurable negative impact on the acceptance of Monod's equations by microbiologists.

A study was made by Robert J. Ooten at the University of Kansas in 1968 in an effort to understand how mixed microbial populations operated in a simple, aeration only bioreactor similar to the one used by Herbert, Elsworth, and Telling. The substrate was a 1,000 mg/L glucose and mineral salt solution. The bioreactor was operated at decreasing HRT values from 24 hrs down to 1.3 hrs. The microbial population increased as the HRT was reduced to 4 hrs. As the HRT was reduced below 3 hrs, microbial metabolism dropped off rapidly. Maximum total bacteria growth occurred at 3 hrs HRT, well above the time observed by Herbert *et al.* The primary difference between the two studies was temperature. The Herbert *et al* study was made at 37°C in contrast to 20°C for the Ooten study. The slower rate of metabolism at 20°C made the data easier to collect as the bioreactor approached failure. Ooten observed the same general relationships as Herbert *et al* with one exception. By using a longer time span for his study, Ooten's data clearly demonstrated the endogenous respiration effect that was missed by Monod and by Herbert *et al.* The microbial mass concentration increased as the aeration time was shortened to the critical aeration time, 2.8 hrs. Reducing the aeration time below 2.8 hrs resulted in decreased metabolism and ultimate failure of the system to maintain microbial growth.

During the 1950s environmental microbiologists began detailed examination of the mixed microbiology and biochemistry of the various biological wastewater treatment processes. The initial laboratory studies dealt with batch fed systems before giving way to continuous flow systems. The continuous flow, aerobic bioreactors used in the biological wastewater treatment studies were similar to the continuously fed bioreactors that Monod and Herbert, Elsworth, and Telling had used in their studies. The major differences in bioreactor operations were in the use of sedimentation tanks to separate flocculated bacteria for return back to the bioreactor, the complex nature of the substrates, and the aeration systems. The use of sedimentation tanks following the aeration tanks and returning the settled microbial solids resulted in more microbial solids than were needed for substrate metabolism. The additional active microbes kept the residual organic substrate in the aeration tank at the lowest possible concentration for the aeration time. The substrates were normally domestic wastewaters or complex mixtures of different industrial organic compounds, both soluble and suspended. The aeration systems were primarily diffused aeration with sufficient air added for good mixing, as well as, for oxygen transfer. In an effort to understand the fundamental relationships in aerobic metabolism, the environmental microbiologists examined numerous pure organic compounds under uniform operating conditions using mixed microbial cultures rather than pure cultures of specific bacteria. The use of mixed bacteria cultures allowed the optimum bacteria to grow, producing the best treatment results possible. The pure culture studies and the mixed culture studies both contributed to a better understanding of bacteria metabolism in aerobic treatment systems.

VARIABLE LOADING RATES

Wastewater loading rates to biotreatment facilities are seldom uniform. Wastewater flows tend to vary over definite time periods in both municipal wastewater treatment plants and in industrial wastewater treatment plants. The variable wastewater flows produce variable organic loads on the microorganisms responsible for metabolizing the biodegradable materials in the wastewaters. It is important to understand how the bacteria respond to the changes in organic loading rates if the wastewater treatment plant is to be properly designed and operated for the maximum possible effluent quality.

In 1972 Standing, Fredrickson, and Tsuchiya published the results of their study on the effect of changing the rate of substrate flow in a continuously fed reactor. This study was similar to the one by Ooten, except they examined the metabolism of 200 mg/L glucose by a pure culture of *E. coli.* They found that complete substrate utilization produced a culture having an optical density, OD, of 0.240 with 85 mg/L suspended solids and a total cell count of 7×10^8 /ml. The fluid retention time in the bioreactor was allowed to come to steady state conditions at 24 hours. On Day 7 they changed the rate of feed to give a hydraulic retention time (HRT) of 6 hours at the same glucose concentration. The increased flow rate had two physical effects on the bioreactor. It increased the rate of displacement of bacteria from the bioreactor and the rate of glucose to the bioreactor. The bacteria population quickly dropped and the glucose concentration increased before the bacteria responded and acclimated to the new retention period at a higher OD. The *E. coli* population had initially established equilibrium at 85 mg/L TSS with the addition of 8.3 mg/L glucose/hr. The specific growth rate factor, μ, was 0.042/hr. Changing the flow rate to 6 hours increased the rate of glucose addition to 33.3 mg/l/hr, a four-fold increase. The specific growth rate was required to increase to 0.17/hr. The data showed that the hydraulic washout affected the bacteria more than the increased glucose addition. The bacteria population fell to 50 mg/l with the glucose concentration increasing to 20 mg/l in the bioreactor. The change in retention time initially exceeded the bacteria's ability to adjust to the increased nutrients. The bacteria changed their growth rate and began to increase in numbers after the initial drop. The unmetabolized glucose quickly decreased and the system established a new equilibrium level. It was apparent that there was a little more microbial concentration at the shorter detention period, 90 mg/l in contrast to 85 mg/l. The impact of endogenous respiration between 24 hrs HRT and 6 hrs HRT with *E. coli* produced the difference in microbial solids.

These results showed that sudden changes in the substrate flow rates resulted in a period of transition before the bacteria population in the bioreactor stabilized at the new level. It took about 43 hrs for the system to return to equilibrium, seven times the new theoretical displacement time. The change in retention time did not change the concentration of influent glucose, but changed the total quantity of glucose added each day. The effluent glucose concentration, 12 hours after the HRT change, showed that 90% of the glucose was metabolized, but there was only 56% of the initial cell concentration. It appeared that the glucose was partially metabolized to intermediate organic compounds rather than to normal cell mass. The effluent analyses were not designed to measure intermediate organic compounds. The data showed that the glucose was quickly changed; but new cells were not produced. With time the bacteria were able to metabolize the intermediate organic compounds to new cell mass. These data indicated that the rate of bacteria synthesis limited the metabolism of glucose initially. As the bacteria population increased, the metabolism of glucose increased. The system shifted from a substrate limited operations to a bacteria limited operations for a short period of time and then shifted back to a substrate limited operations.

Variable flow rates have not been of much interest to general bacteriologists; but they are important to environmental microbiologists and environmental engineers. Municipal wastewater treatment plants have continuously varying wastewater flows, producing a cycle every 24 hours. Small wastewater treatment plants show the widest flow variations, varying from zero flow early in the morning to 3 or 4 times the average daily flow in the afternoon. Large wastewater treatment plants have less flow variations than small plants. The large collection systems tend to level out the wastewater flows. The flow variations in large wastewater treatment plant may range from a minimum of 90 percent of the daily average flow to a maximum of 110 percent of the daily average flow. Some industrial wastewater treatment plants are faced with both variable flow rates and variable nutrient concentrations. Equilibrium conditions may not exist for extended periods of time in full size wastewater treatment plants as they do in controlled laboratory systems. The study of varying operating conditions in continuous flow reactors has been quite productive for environmental microbiologists.

Variations in organic concentration occur in wastewaters the same as variations in flow. Robert W. Agnew completed a study of organic load variations in a simple aerobic bioreactor at the University of Kansas in 1968. One part of his study dealt with a mixed microbial population and a soluble substrate that was 40% glucose, 50% glutamic acid, and 10% acetic acid with mineral salts to provide excess N and P for bacteria metabolism. The synthetic substrate was fed to a completely mixed bioreactor at 3 days HRT and 25°C. In one experiment the substrate concentration COD of the substrate was shifted from 240 mg/L COD to 1,460 mg/L COD. The 240 mg/L COD operation showed an oxygen uptake rate of 28 mg/L/hr with 120 mg/L microbial solids. Three hrs after the substrate change the oxygen uptake rate had increased to 55 mg/L, indicating rapid metabolism of the organic substrate. The microbial solids had increased to 170 mg/L. The soluble COD jumped from 24 mg/L to 630 mg/L, indicating incomplete metabolism of the added substrate. At the end of 6 hrs after the substrate change the oxygen uptake rate was 142 mg/L/hr; and the microbial solids were up to 410 mg/L. The soluble COD was down to 530 mg/L. At 9 hrs after the substrate change the oxygen uptake rate was at 200 mg/L/hr with the microbial solids at 670 mg/L. The soluble COD was down to 67 mg/L, indicating most of the substrate was being metabolized. At the end of 12 hrs after the substrate had been changed, the oxygen uptake rate was steady at 180 mg/L/hr. The soluble COD was at 32 mg/L with 720 mg/L microbial solids. These data showed that a sharp increase in BCOD resulted in the bacteria responding as quickly as they could. Since more substrate was added than could be metabolized, the excess substrate accumulated in the bioreactor and in the treated effluent. It took three displacement times to build up sufficient numbers of bacteria to metabolize the higher substrate COD plus the accumulated substrate.

Initially, metabolism of 240 mg/L COD resulted in 83 mg/L oxygen uptake, 120 mg/L microbial solids, and 24 mg/L soluble COD in the effluent. The metabolism of the substrate resulted in 38% oxidation and 62% cell synthesis. After 4 displacement periods since increasing the substrate COD to 1,460 mg/L, the oxygen uptake was 540 mg/L with 720 mg/L microbial solids and 32 mg/L soluble COD in the effluent. The substrate metabolism resulted in 38% oxidation and 62% cell synthesis. The one factor missing in the metabolic relationships was the oxygen utilized for endogenous respiration. Even though the aeration time was only 3 hrs, a portion of the oxygen utilized by the bacteria was expended for endogenous respiration, an item given little consideration in making material balances.

ENDOGENOUS RESPIRATION

Endogenous respiration is an important factor that was not readily recognized by most bacteriologists until after 1950. No one was quite certain what endogenous respiration really was or even if it existed. Initially, endogenous respiration was believed to occur only after growth ceased and was considered as the cell maintenance energy; *i.e.*, the energy used by the bacteria to remain alive when there was no external source of nutrients. Growth of large masses of bacteria in batch fed reactors showed that the bacterial mass increased until the substrate was essentially all metabolized and then slowly decreased. The rate of cell mass decrease was much slower than the rate of cell mass increase, showing that the rate of endogenous respiration was slower than the rate of synthesis. Oxygen respiration measurements of bacteria, fed an organic substrate and unfed, confirmed that endogenous respiration did exist. Unfed bacteria exerted a decreasing rate of oxygen demand with time. The fed bacteria exerted a rapid rate of oxygen demand until the substrate was metabolized and then gave a slow rate of oxygen demand, similar to the unfed bacteria. It appeared that synthesis occurred first, followed by endogenous respiration.

The development of C^{14} tracers allowed studies to be made on endogenous respiration. One of the first studies was by Moses and Syrett in 1955. They used radioactive carbon to make several different organisms radioactive. Once the microbes were radioactive from normal metabolism, they changed to non-radioactive substrates and measured the evolution of radioactive carbon during the metabolism of the non-radioactive substrates. The evolution of $C^{14}O_2$ showed that endogenous respiration was continuous during normal metabolism of external substrates and was not suppressed. The radiocarbon studies changed the understanding of endogenous respiration and how it affected metabolism.

A study by Warren, Fells, and Campbell in 1960 showed that endogenous respiration by *Pseudomonas aeruginosa* resulted entirely from the metabolism of proteins with release of ammonia nitrogen into solution. Adding glucose, as a source of nutrients, allowed the ammonia nitrogen to be reincorporated into cell protoplasm. Clifton and Sobek examined endogenous respiration of *Bacillus cereus* harvested from agar plates and also found ammonia nitrogen release. They noted from another study on *Bacillus cereus* by Urba that protein metabolism progressed at the rate of 0.07/hr based on the protein remaining in washed cell suspensions. Clifton and Sobek, in a previous paper, showed that the use of uniformly labeled C^{14} glucose gave 50% conversion to cell mass with 20% to 40% suppression of endogenous respiration. Gronlund and Campbell examined nine different bacteria and found that all but one released ammonia nitrogen over time, confirming that proteins were being used as substrates for endogenous respiration. It appeared from oxygen uptake rate data that the bacteria underwent endogenous respiration at about 0.01/hr at 30°C. Although radiotracers were important in understanding endogenous respiration, oxygen uptake data confirmed that endogenous respiration occurred during substrate metabolism the same as during substrate absence. It is important to recognize that endogenous metabolism is part of the normal synthesis metabolism with proteins being formed and degraded continuously. In the presence of external substrates endogenous respiration is masked by the synthesis reactions. When the external substrates have been completely metabolized, endogenous respiration is the primary reaction keeping the bacteria alive and functioning until a new external substrate can be found. Since many wastewater treatment plants operate under substrate limiting conditions with a large mass of bacteria under aeration, endogenous respiration is more important in environmental microbiology than in conventional microbiology.

L. J. Pirt was one of the first microbiologists to modify Monod's equations of bacterial metabolism to include endogenous respiration. In 1990 Pirt's research on endogenous respiration was recognized in a special symposium on microbial growth dynamics by the Society for General Microbiology in England. While it took a while for the microbiologists to fully accept the quantitative relationships in endogenous respiration, sanitary engineers and environmental microbiologists concerned with biological wastewater treatment systems had long recognized and accepted endogenous respiration. The differences between the two points of view were bacteria concentrations and mixed microbial populations. The microbiologists worked with small bacteria populations. Normal errors in data collection and evaluation made it difficult to measure endogenous respiration very accurately. It was Pirt who provided the data to confirm the validity of endogenous respiration for microbiologists. In 1971 Carter, Bull, Pirt, and Rowley found that the endogenous respiration rate of the fungi, *Aspergillus nidulans*, was about 0.024/hr at 30° C. Sam Hoover and Nandor Porges showed endogenous respiration in 1953 during the treatment of synthetic dairy wastewater in the laboratory with activated sludge. They observed that the large bacteria populations in activated sludge quickly metabolized the skim milk and then underwent rapid endogenous respiration with a significant decrease in bacterial mass. My own research on activated sludge in the 1950s showed that endogenous respiration could be measured with both laboratory studies and field studies in full-size WWTP. By the end of the 1950 decade endogenous respiration was well established with environmental microbiologists.

One of the basic issues with environmental endogenous respiration was the oxygen demand created by protozoa and higher animals in mixed bacteria systems. There was no way to separate the endogenous respiration of the bacteria from the normal synthesis metabolism of the protozoa, rotifers, and higher animals. It simply suffices to recognize that the endogenous respiration of pure culture bacteria is less than the total endogenous respiration of mixed microbial populations in wastewater treatment systems.

OXYGEN METABOLISM

Aerobic microbial metabolism utilizes dissolved oxygen and produces more energy for synthesis than anaerobic microbial metabolism. The uptake of dissolved oxygen by bacteria is a direct measure of the energy used by the bacteria in metabolism of the substrate plus endogenous respiration. The total quantity of oxygen utilized is indicative of the amount of substrate metabolized, while the changing rate of oxygen uptake can be used as a measure of the state of growth of the bacteria. Correlation of the changes in substrate, oxygen uptake and bacteria mass can provide a total picture of metabolism. The initial problem was the development of a method for measuring oxygen uptake rate by the bacteria. In 1931 Butterfield, Purdy, and Theriault made one of the first studies on the growth of bacteria and oxygen utilization. They used three different pure bacteria cultures in sterilized DO bottles. The growth of bacteria was measured with standard pour plate counts. The corresponding DO concentration was measured chemically. They found that the rapid uptake of oxygen correlated directly with the bacteria growth. Mixtures of bacteria were also examined with similar results. Unfortunately, the techniques used for measuring the changes in DO and in the growth of the bacteria were not as accurate as desired. Overall, the results were good; but better analytical techniques were needed for quantitative results.

The Warburg respirometer was developed by Otto Warburg in 1926 to examine respiration of various microbes and enzymes. Like many new pieces of equipment that are complex and expensive, it took time before the Warburg respirometer was used extensively. The development of commercial units and the publishing of *Manometric Techniques* by Umbreit, Burris, and Stauffer in 1945 stimulated interest in using the Warburg respirometer for measuring oxygen utilization on a continuous basis in batch cultures. I first used the Warburg apparatus at MIT in 1950 to study the metabolic activity of floc-forming bacteria isolated from activated sludge. The application of the Warburg respirometer as a tool in applied environmental microbiology was different from its application by conventional microbiologists. The microbiologists used pure bacteria cultures in concentrated nutrient substrates over a short time period, two to four hours. Environmental microbiologists used both pure and mixed microbial cultures in dilute organic substrates over a long period, 24 to 48 hours. Each group of microbiologists developed their own techniques to fit their specific needs for information.

The Warburg apparatus consists of a series of flasks attached to U-shaped manometers. The flasks are submerged in a temperature controlled water bath with the manometers mounted around the outside of the water bath on a shaking mechanism. Figure 3-3 shows a schematic diagram of the Warburg apparatus that was widely used in the 1950s and 1960s. The water bath could maintain a temperature to within 0.1° C of the set temperature. The rate of shaking could be controlled over a wide range to match the desired oxygen transfer. Since the metabolic studies were normally designed to be aerobic, it was important that oxygen transfer not be the limiting factor. This meant that the oxygen demand over the study period had to be less than the total oxygen available in the flask. Warburg flasks came in 15 ml and 125 ml capacity. The gas volume in 15 ml flasks with a 2 ml sample and 0.1 ml KOH solution contained about 3.9 mg oxygen in air under STP. This permitted measuring a total oxygen demand of over 1,900 mg/L. The 125 ml flasks normally contained 20 ml total sample and 1 ml KOH solution, giving a gas volume with 31.2 mg oxygen at STP for a maximum oxygen demand of 1,560 mg/L. If mixing was not rapid enough, the rate of initial metabolism could exceed the rate of oxygen transfer, causing a shift from aerobic metabolism to partial aerobic metabolism. As the rate of oxygen demand

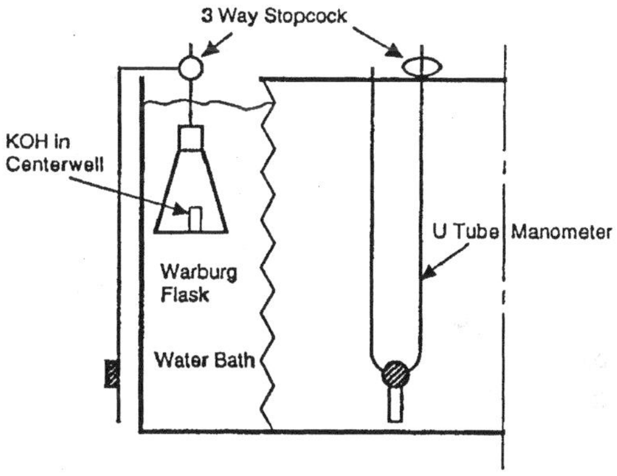

Figure 3-3 SCHEMATIC DIAGRAM OF THE WARBURG APPARATUS, SHOWING SUBMERGED FLASK AND U-TUBE MANOMETER

decreased, the system would return to a completely aerobic environment. With good shaking Warburg flasks could transfer over 300 mg/L O_2/hr at room temperature. Overall, the Warburg respirometer could sustain maximum oxygen utilization for up to 5 hours without becoming oxygen deficient.

The Warburg respirometer measures the total pressure of gas change in the flasks at a constant volume. Under aerobic conditions oxygen is removed and carbon dioxide is produced. The use of a strong KOH solution in the center well of each flask permits the carbon dioxide to be absorbed and removed from the gas phase, leaving oxygen as the only gas undergoing significant change in the flask gas volume. The gas pressure in the Warburg flask decreases as oxygen is removed for bacteria metabolism. The attached manometer reflects the change in gas pressure in the flask. The manometer fluid is Brodie's solution with a specific density of 1.033. By always setting the closed leg of the manometer to a constant value, 150 mm, the open leg of the manometer reflects the change in pressure in the closed flask. A flask constant is used to convert the change in mm on the manometer to mg/l oxygen utilized in the flask. Because the system is subject to changes in atmospheric pressure, two flasks containing only clean water are used as thermobarometers. The change in pressure in the thermobarometers (TB) is used to correct for variations in atmospheric pressure and temperature variations. Normally, one TB flask is placed first on the Warburg apparatus and the other is placed at the end of the series of test flasks. A laboratory Warburg unit can hold 18 flask and manometer combinations, allowing 16 different conditions to be tested at one time. When using pure bacteria cultures, the flasks are sterilized by heat. Often, the substrate is sterilized and added aseptically to the flasks just before the bacteria are added.

Once the substrate and bacteria are added to a flask, the flask is attached to the corresponding manometer and placed into the temperature controlled water bath. A three-way stopcock, located at the top of the closed leg of the manometer, is left in the open position. The other flasks are prepared and loaded, one by one. The flasks are shaken for several minutes to allow them to come to temperature equilibrium in the water bath. Failure to allow the flasks to come to temperature equilibrium can produce large errors in the initial readings. Once the flasks have reached temperature equilibrium, the manometers are adjusted to 150 mm on both legs by twisting the screw clamps at the bottom of each manometer; and the three-way stopcock on the closed leg is closed. The time of closure is noted when the metabolic run began. At regular time intervals, the shaking is stopped and the closed leg of the manometer is adjusted to 150 mm. The scale indicating the fluid height on the open leg of the manometer is recorded along with the time. Each unit is read in sequence to provide a uniform time lag in reading and recording the data. If the rate of oxygen uptake is so great that it is not possible to obtain a suitable reading at the next time interval, the three-way stopcock is opened after taking the reading and both legs are readjusted to 150 mm before closing the stopcock. This procedure allows the pressure in the system to readjust to atmospheric pressure without significant addition of air to the system.

By the 1950s bacteriologists were beginning to recognize the quantitative aspects of metabolism and found the Warburg apparatus was a good tool to determine the oxygen utilization and the carbon dioxide production. In 1954 Wilner and Clifton examined the oxidative assimilation of *Bacillus cereus* with the Warburg apparatus. They used washed cells in large quantities to create a rapid oxygen demand. Varying concentrations of glucose were added to separate flasks, all containing the same quantities of bacteria. The glucose concentrations varied from 0 mol. in the control flask to a maximum of 0.02 mol. Data were collected over 170 minutes. Their data showed that metabolism of the glucose by the bacteria created a greater oxygen demand than theoretically existed in the substrate itself. Wilner and Clifton found that subtracting the endogenous respiration measured by the same concentration of bacteria in water without any glucose yielded oxygen uptake data in direct proportion to the original glucose concentrations. Their study was very important in showing how to use data generated from Warburg measurements. The average substrate metabolism showed about 56% oxidation of the glucose.

In 1955 Anthony F. Gaudy used the Warburg respirometer in his SM thesis research at M.I.T. to examine the effect of mixing several pure bacteria cultures together in the metabolism of glucose and other carbohydrates. The bacteria were selected for their metabolic characteristics and their ability to be counted on surface spot plates. The bacteria included: *Alcaligenes faecalis, Pseudomonas fluorescens, Serratia marcescens* and *Flavobacterium esteroaromaticum. Alcaligenes faecalis* are small, rod shaped bacteria that grow with a white color on the surface of tryptone-glucose-agar (TGA) plates and do not metabolize carbohydrates. *Pseudomonas fluorescens* are rod shaped bacteria that grow as translucent colonies on the surface of TGA plates and metabolize carbohydrates. *Serratia marcescens* are rod shaped bacteria that produce a red pigment on the surface of TGA plates and metabolize carbohydrates. *Flavobacterium esteroaromaticum* are rod shaped bacteria that produce a yellow color on the surface of TGA plates and do not metabolize carbohydrates.

Small numbers of the different bacteria were placed in sterile Warburg flasks containing 500 mg/l glucose as the organic substrate. The oxygen uptake was measured over 51 hours. At the end of the 51 hours, the bacterial populations were measured by direct counting using the surface spot plate technique to permit recognition of the different bacteria. Three runs were made with glucose and the data averaged for evaluation. The data confirmed that *Alcaligenes faecalis* and

Flavobacterium esteroaromaticum did not metabolize glucose, while *Pseudomonas fluorescens* and *Serratia marcescens* metabolized glucose. Yet, both *Alcaligenes* and *Flavobacterium* grew to a limited extent in the presence of either *Pseudomonas* or *Serratia.* Mixtures of all four bacteria showed that *Pseudomonas* and *Serratia* grew on the glucose while the *Alcaligenes* and *Flavobacterium* grew on intermediates released from the bacteria that metabolized glucose. The total oxygen uptake after 51 hours growth was slightly greater in the mixtures of bacteria than in individual cultures of the bacteria. The total oxygen uptake for *Pseudomonas* was 167 mg/l, 31% of the theoretical oxygen demand of the glucose fed. The change in bacteria population was 12.5×10^8/ml, giving a cell yield of 7.5×10^{12} cells/g oxygen used. The total oxygen uptake for *Serratia* was 153 mg/l, 29% of the theoretical oxygen demand of the glucose fed. The change in bacteria population was 9.9×10^8/ml, giving a cell yield of 6.5×10^{12} cells/g oxygen used. Overall, there was no significant difference in cell yield for these two bacteria. The rates of oxygen uptake showed that *Serratia* grew at a faster rate than the *Pseudomonas*, allowing *Serratia* to predominate when the two bacteria were grown in mixtures. The mixture of the four bacteria had an oxygen uptake of 170 mg/l, 32% of the theoretical oxygen demand of the glucose fed. The *Alcaligenes* increased 1.02×10^8 cells/ml; the *Flavobacterium* increased 2.28×10^8 cells/ml; the *Pseudomonas* increased only 0.96×10^8 cells/ml; while the *Serratia* increased 5.80×10^8 cells/ml. The total population was 10.06×10^8 cells/ml, giving a cell yield of 5.9×10^{12} cells/g oxygen used. These results showed that the metabolism of a simple organic substrate in a mixed bacterial culture was not a simple reaction. The bacteria that could metabolize the substrate grew in proportion to their ability to obtain and process the substrate. Secondary growth also occurred as the primary bacteria released organic intermediates that could be used by the bacteria that could not metabolize the glucose directly. The survival of each species of bacteria depended upon their ability to obtain nutrients from the environment.

The oxygen uptake data represented the total energy used by all of the metabolic reactions. The rate of oxygen uptake showed the phase of growth in the substrate. The oxygen uptake data for the metabolism of the glucose by *Serratia* alone during Run 1 is shown in Figure 3-4. Overall, the shape of the oxygen uptake curve resembles a normal growth curve with an initial lag phase, a log growth phase, and a declining growth phase. The run did not last long enough to show the endogenous respiration phase. This is typical for bacteria grown on one organic compound and then transferred to a second organic compound. The substrate carryover from the initial growth media allowed the bacteria to continue to metabolize until the carryover organic compounds were consumed, requiring a maximum of 6 mg/l oxygen uptake. This was followed by adaptation to the new substrate and log growth before slowing the rate of metabolism in proportion to the amount of substrate remaining unmetabolized. Finally, the bacteria approached endogenous

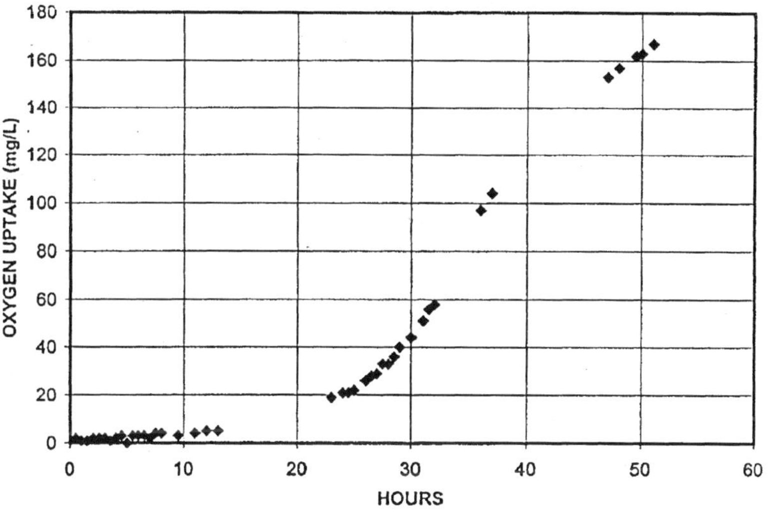

Figure 3-4 OXYGEN UPTAKE IN METABOLISM OF GLUCOSE BY *SERRATIA MARCESCENS*

respiration. Figure 3-4 also illustrates the problems in periodic data collection. Long term Warburg runs create blank spaces in the data, even though the data generation is continuous between readings. This set of data illustrates why microbiologists liked larger initial microbial populations and short time runs

Actual data do not always produce smooth lines since errors can occur in the recording the time of readings, variations in the temperature cycling of the water bath, and in manometer readings. Since data evaluation requires the use of a thermobarometer correction, data reading errors can exist for two consecutive sets of data readings. As long as the flasks have not been opened to adjust for the changing atmospheric pressure, an error in one reading is usually compensated for in the next reading. The shape of the Warburg oxygen uptake curve can provide considerable information on the bacteria metabolism of the substrate. A long lag period in oxygen uptake means that the bacteria are adapting their metabolism to the new organic substrate. The rapidly increasing oxygen uptake after the lag period indicates the log growth period of the bacteria. The log growth period lasts as long as the organic substrate is in excess compared to the bacteria population. Eventually, a point is reached where the oxygen uptake rate shifts from an increasing rate to a decreasing rate, signaling a shift from log growth to declining growth. The declining growth phase indicates that the concentration of organic substrate is limited in relation to the concentration of bacteria. The oxygen uptake rate in the declining growth phase drops sharply at a log rate until the endogenous phase is reached. Endogenous respiration produces a declining oxygen uptake rate over time at a lower log rate than the declining growth phase. A simple plot of the total oxygen uptake against time will show the overall metabolic activity of the bacteria being examined. The log growth phase will produce a concave shape that shifts to convex in the declining growth phase. The endogenous respiration phase occurs as the oxygen uptake line tends to form a slow curvature.

A concern that should be noted lies with periodic opening of the three-way stopcock on the manometers to adjust for pressure changes. The sensitivity of the Warburg manometers to pressure changes requires that the stopcock be opened and the manometer fluid adjusted to equilibrium conditions when it is not possible to obtain readings on the manometer. Opening the stopcock allows the pressure in the flasks to equal the atmospheric pressure and adjustment back to the initial equilibrium reading allows normal data collection to continue. While the pressure in the flasks is allowed to increase, the amount of air entering the flasks is not sufficient to change the overall oxygen availability for metabolism. Pressure adjustments are made very quickly and should have no impact on the data collected. With mixed microbial systems nitrifying bacteria can generate part of the oxygen uptake, making it difficult to observe the carbonaceous metabolism without running nitrogen analyses and correcting the data for nitrification.

It is important to recognize that the Warburg metabolism studies simulate batch-fed biological reactors. The high substrate concentrations may result in the production of different metabolic end products than would be seen if the bacteria were in a continuously fed system where the substrate around the bacteria can be maintained at a low concentration. High concentrations of organic substrate may be toxic or partially toxic. The bacteria may shift the excess organics into internally stored nutrients or may generate large quantities of external polysaccharide slime. The metabolic data obtained from Warburg studies are best used when evaluating batch-fed systems operating under similar environmental conditions. The oxygen uptake measurements show the overall energy transfer in the biological system. Unfortunately, the high rate of oxygen transfer in the Warburg apparatus can produce results that cannot be obtained in large-scale systems. In spite of these limitations, Warburg respiration data have proven very useful in developing a better understanding of bacterial metabolism. Care must be used with Warburg data to insure that the data are properly evaluated and not misinterpreted.

EFFECT OF TEMPERATURE

The growth of bacteria is affected by temperature the same as other chemical reactions. As the temperature increases, the rate of metabolism increases. Bacteria grow faster and die faster. Lowering the temperature causes metabolism to slow. The temperature effect is important in environmental microbiology since temperature varies with the seasons of the year and with location between the equator and the poles. Some bacteria can survive at low temperatures, around 0°C. The bacteria that grow best at low temperatures have been designated as *psychrophilic*. The problem at low temperatures is the increased viscosity of water, making it difficult for the bacteria to obtain nutrients. The ability of bacteria to grow at temperatures below 0°C depends upon the substrate remaining fluid, rather than freezing solid. Once the liquid forms ice, the bacteria cannot obtain nutrients and cease growing. Bacteria can survive at low temperatures unless ice forms inside the cell, causing rupture and death.

Most of the natural bacteria grow at temperatures between 5°C and 40°C. Bacteria that grow between 5°C and 40°C have been designated as *mesophilic*. A close mathematical approximation of bacteria growth rates between 5°C and 35°C is a doubling of the growth rate for each 10°C temperature increase. As the temperature rises above 35°C, the proteins in enzymes begin to breakdown by a process termed *denaturation*, adversely affecting bacteria metabolism. The mesophilic bacteria die off rapidly as the temperature rises above 40°C. A few mesophilic bacteria can survive temperatures as high as 45°C, but die off as the temperature increases above 45°C.

The bacteria that survive at temperatures above 45°C have been designated as *thermophilic*. The most common groups of thermophilic bacteria have a survival range up to about 70°C. Some thermophilic bacteria have been found that can survive at temperatures close to boiling in hot springs. The thermophilic bacteria also grow at mesophilic temperatures. Unfortunately, the thermophilic bacteria are not as efficient as the mesophilic bacteria in their metabolism and cannot compete effectively with the mesophilic bacteria at mesophilic temperatures. Since the mesophilic bacteria cannot keep the thermophilic bacteria from obtaining some substrate, normal bacteria populations at mesophilic temperatures in the natural environment will be predominantly mesophilic with a small fraction of thermophilic bacteria. The rate of metabolism by thermophilic bacteria also increases with temperature, doubling with each 10°C temperature increase to 65° - 70°C when the protein enzymes begin to breakdown. At the maximum temperatures thermophilic bacteria metabolize at very high rates. Both growth and death occur at rapid rates, making it difficult to maintain thermophilic bacteria in an active state of metabolism.

EFFECT OF pH

The growth rate of bacteria is affected by the pH of the environment. Technically, pH is the log of the reciprocal of the molar hydrogen ion concentration as shown in Equation 3-9. The pH scale provides an easy to use number for hydrogen ion

$$pH = \log (1/H^+) \tag{3-9}$$

concentration. The pH scale operates from 0 to 14 with 0 to 7 being acidic and 7 to 14 being basic for water. The pH scale was derived from the ionization of water as shown in Equation 3-10. The hydrogen ion concentration decreases as the pH rises

$$HOH \leftrightarrow H^+ + OH^- \tag{3-10}$$

from 0 to 14; while the hydroxyl ion concentration decreases as the pH drops from 14 to 0. In water pH 7.0 is considered neutral since the concentration of hydrogen ions equals the concentrations of hydroxyl ions. When strong acids, such as hydrochloric acid, sulfuric acid or nitric acid, are added to water, they ionize completely, as shown in Equation 3-11. When a strong base, such as sodium

$$HCl \rightarrow H^+ + Cl^- \tag{3-11}$$

hydroxide is added to water, it ionizes completely with the formation of hydroxide ions, as shown in Equation 3-12. On the other hand, weak acids such as organic

$$NaOH \rightarrow Na^+ + OH^- \tag{3-12}$$

acids, do not ionize completely. Acetic acid, $C_2H_4O_2$, ionizes to form hydrogen ions and acetate ions, as shown in Equation 3-13. Weak organic acids and weak

$$CH_3COOH \rightarrow H^+ + CH_3COO^- \tag{3-13}$$

organic bases have *ionization constants* to help in the calculation of the extent of ionization. The ionization constants can be found in various chemical handbooks, such as *Lange's Handbook of Chemistry* or the *Handbook of Chemistry and Physics*. The ionization constant, K, is the product of the molar concentration of the ionized products divided by the unionized concentration remaining in solution, as shown in Equation 3-14. The handbook shows that the ionization constant, K,

$$K = [H^+][CH_3COO^-]/[CH_3COOH] \tag{3-14}$$

for acetic acid is 1.75×10^{-5} at 25° C. At a pH of 7.0 the hydrogen ion concentration is 1.0×10^{-7}. From Equation 3-13 the acetate ions are 175 times the acetic acid concentration. For practical purposes the acetic acid is almost completely ionized. At a pH of 5.0 the acetate ions are 1.75 times the acetic acid concentration. Equal concentrations of acetate ions and acetic acid occur at pH 4.8. At pH 3.0 the molar ratio of acetate ions to acetic acid drops to 0.0175, about 2%. With bacterial metabolism around neutral pH values, the bacteria metabolize the ionized acetate rather than the free acetic acid.

While acetic acid ionization is important in many environmental systems, the most important system is the carbonic acid system. Carbonic acid is formed by dissolving carbon dioxide, CO_2, in water, as shown in Equation 3-15. Carbonic

$$CO_2 + H_2O \rightarrow H_2CO_3 \tag{3-15}$$

acid has two ionization constants, $K_1 = 4.30 \times 10^{-7}$ and $K_2 = 5.61 \times 10^{-11}$. The initial ionization causes the carbonic acid to break into hydrogen ions and bicarbonate ions, as shown in Equation 3-16. The second ionization results in the bicarbonate

$$H_2CO_3 \rightarrow H^+ + HCO_3^- \tag{3-16}$$

ion breaking into hydrogen ions and carbonate ions, as shown in Equation 3-17.

$$HCO_3^- \rightarrow H^+ + CO_3^= \tag{3-17}$$

At pH 7.0 the bicarbonate ion concentration is 4.3 times the carbonic acid concentration on a molar basis with the carbonate ion concentration 0.00056 times the bicarbonate ion concentration. Essentially, there are very few carbonate ions at pH 7.0 or less. At pH 4.0 the bicarbonate ion concentration is 0.0043 times the carbonic acid concentration, less than 1.0%. At pH 9.0 the bicarbonate ion concentration is 430 times the carbonic acid concentration with the carbonate ions 0.056 times the bicarbonate ion concentration. As the pH rises above pH 9.0, the carbonate ions become significant and the carbonic acid becomes insignificant.

The addition of carbon dioxide to water creates carbonic acid and depresses the pH into the acidic range. If sodium hydroxide is added to the carbonic acid solution, sodium bicarbonate is produced, as shown in Equation 3-18.

$$NaOH + H_2CO_3 \rightarrow NaHCO_3 + HOH \tag{3-18}$$

Further addition of sodium hydroxide would push the bicarbonates to carbonates, as shown in Equation 3-19. The pH is pushed higher as carbonates are formed.

$$NaOH + NaHCO_3 \rightarrow Na_2CO_3 + HOH \tag{3-19}$$

The addition of carbon dioxide to the carbonates would depress the pH and push the equilibrium back toward bicarbonates. The bicarbonates are very important in maintaining proper pH levels in biological systems. Essentially, bicarbonates are *buffers,* designed to keep the pH close to the 7 – 8 range.

There are other weak bases and weak acids than the carbonate system. Weak bases are primarily ammonia related compounds, such as the amines. The amino group can absorb hydrogen ions, pushing the pH higher. The addition of hydrogen ions causes the amine to become positively charged as shown in Equation 3-20 for methyl amine.

$$CH_3NH_2 + H^+ \rightarrow CH_3NH_3^+ \tag{3-20}$$

Amino acids are important components of proteins and contain an amino group, -NH_2, and a carboxyl group, -COOH. Both chemical groups are capable of ionizing, depending upon the environmental pH. At low pH values, high H^+ concentrations, the amino group will be positively charged and the carboxyl group will be unionized. At high pH values the amino group is not ionized and the carboxyl group is ionized. The neutral point for amino acids is not at pH 7.0, the neutral point for water. The neutral point for the amino acids is not a single value. Each amino acid has its own separate neutral point. The neutral points for the amino acids vary between pH 5 and pH 6. Below pH 5.0 the amino acids tend to be positively charged. Above pH 6.0 the amino acids tend to be negatively charged. This charge relationship is very important in microbial systems. Since the outer surface of bacteria is a lipo-protein-polysaccharide complex, the electrical surface charge on the bacteria will be determined by the environmental pH. At low pH values the bacteria will be positively charged. At high pH values the bacteria will be negatively charged. At pH 7 to pH 8 the bacteria will be negatively charged. The negative charge helps to attract the positively charged metallic elements used in enzymes the bacteria produce for metabolism.

Phosphates are essential nutrients for all bacteria. The phosphate system starts as ionized phosphoric acid as shown in Equation 3-21. Phosphoric acid has three hydrogen atoms that ionize at different pH levels. The three ionization constants for

$$H_3PO_4 \rightarrow H^+ + H_2PO_4^- \rightarrow H^+ + HPO_4^= \rightarrow H^+ + PO_4^{-3} \tag{3-21}$$

phosphoric acid are $K_1 = 7.5 \times 10^{-3}$, $K_2 = 6.2 \times 10^{-8}$, and $K_3 = 4.8 \times 10^{-13}$. At pH 7.0 the dihydrogen phosphate ions are 7,500 times the molar phosphoric acid concentration and 1.6 times the molar monohydrogen phosphate. In biological systems phosphates exist as a mixture of dihydrogen phosphate ions and monohydrogen phosphate ions. Adding sodium hydroxide, to raise the pH, will cause the dihydrogen phosphate ions to shift to monohydrogen phosphate ions. The ionization relationships indicate that complete ionization does not occur except at very high pH levels. Phosphates have been used to hold pH levels at levels suitable for bacteria growth. Unfortunately, high phosphate media produce environments that do not normally exist in the real world and can result in biochemical reactions that are not representative of normal environmental reactions. Bicarbonates are the natural buffers in environmental systems; but they are not as effective as phosphates in maintaining relatively constant pH values during biochemical reactions. Care should be taken when evaluating data collected in high phosphate buffer systems.

Metabolism of carbohydrates results in the formation of intermediate organic acids that must be neutralized if the pH is to be maintained at a proper level for continued metabolism. Bicarbonate salts, such as sodium bicarbonate, ammonium bicarbonate, calcium bicarbonate and magnesium bicarbonate, react with the organic acids, as shown in Equation 3-22, to form acid salts with the release of carbon dioxide. The pH will be depressed slightly as carbonic acid is formed. When the organic acid salts are metabolized, sodium bicarbonate is reformed to

$$NaHCO_3 + CH_3COOH \rightarrow CH_3COONa + CO_2 + HOH \qquad (3\text{-}22)$$

maintain the pH at the proper level. In some batch-fed experiments, pH data collected at the start of the experiment and at the end of the experiment can be misleading. The initial pH value at the start of the experiment is normal, ~ pH 8. As the glucose is metabolized with the formation of organic acids, the pH drops and then recovers as the organic acids are metabolized. The final pH is back to normal, ~ pH 8, giving the impression that nothing happened as far as pH was concerned. The pH could have dropped below pH 6, slowing bacteria metabolism or allowing different bacteria to predominate. Completely mixed, continuously fed bacteria experiments will not give misleading pH results as the bioreactor will either have an adequate buffer to maintain the pH or the pH will drop. Alkalinity must be continuously added in glucose fed bioreactors or the pH will drop and adversely affect bacteria metabolism. NaOH should not be added for alkalinity as it will cause the pH to shift too much and will adversely affect the bacteria. Understanding the basic alkalinity-acidity relationships in bacteria metabolism is very important for environmental microbiologists.

Metabolism of lipids results in the production of organic acids, primarily acetic acid. The metabolic end products of lipids require neutralization the same as carbohydrates or the pH will drop to low levels. Protein metabolism results in the release of ammonia as a metabolic end product. The excess ammonia reacts with carbon dioxide and water to produce ammonium bicarbonate, increasing the available supply of bicarbonates in the environment. Ammonium bicarbonate is one of the primary buffer in natural environmental systems. Unfortunately, nitrifying bacteria can metabolize the ammonium bicarbonate under aerobic conditions and produce nitric acid as the ultimate end product. The nitrifying bacteria not only destroy the ammonium bicarbonate, but they also produce a strong acid that can depress the pH unless additional bicarbonate salts are available for neutralization. Two mols of alkalinity are required for every mol of ammonia nitrogen oxidized. Denitrifying bacteria can utilize nitrate salts as electron acceptors for metabolism under the anaerobic environmental conditions. Anaerobic metabolism of nitrate salts results in the production of bicarbonate salts and the release of the nitrogen as nitrogen gas. One mol of alkalinity is produced for every mol of nitrate salt reduced to nitrogen gas. Knowledge of the impact of metabolism on pH and alkalinity in natural systems is required for understanding changes in environmental systems.

Studies have shown that the optimum pH for bacteria growth is between pH 7 and pH 8. Many bacteria can spread the pH range to between 6.5 and 8.5. As the pH drops below 6.0, metabolism begins to decrease at a rapid rate. Few bacteria can grow at pH levels below 5.0. As the pH rises above 8.5, phosphates can be precipitated along with essential trace metals, slowing metabolism. As the pH rises above 9.0, the rate of bacteria growth slows. Excess hydrogen ions at low pH levels and excess hydroxyl ions at high pH levels result in denaturation of the protein enzymes responsible for metabolism. Studies on methane producing bacteria have shown that they stop producing methane when the pH drops below 6.5. The methane bacteria are among the most sensitive groups of bacteria to low pH values.

MIXING AND TURBULENCE

Rapid growth of bacteria depends upon their ability to obtain all elements required to produce cell protoplasm. As bacteria increase in numbers, crowding and accumulation of end products adversely affect further growth. Mixing to create turbulence in the growth environment disperses the bacteria, dilutes the end products throughout the liquid, and brings nutrients to the bacteria. Mixing can be accomplished by using mechanical shakers, mechanical mixers or gas mixers.
Mechanical shakers are widely used for the growth of aerobic bacteria in shallow liquid layers in Erlenmeyer flasks or in special flat culture bottles. Back and forth

agitation provides the mixing necessary to transfer oxygen to the bacteria for maximum growth rates. Displacement of the air in the gas space above the culture by nitrogen gas can be used to create an environment for the growth of anaerobic bacteria. At least six theoretical displacements of the gas volume by nitrogen should insure displacement of the oxygen. Use of sodium sulfide to create a strongly reduced media will result in removal of the small amounts of oxygen left in the gas space and will produce a good environment for growing anaerobic bacteria.

Mechanical mixers have also been used in bacteria cultures. Variable speed motors allow the speed of mixing to be adjusted to the desired level for the specific bioreactor. Different impeller designs produce different mixing patterns. The mixing can be down to the bottom of the bioreactor with the flow being forced across the bottom to the vessel walls, up the outer wall to the surface, back to the center and then down to the mixer. Reversal of the impeller produces an upward flow around the impeller shaft to the surface, outward to the bioreactor wall, down the wall to the bottom of the tank and back across the bottom of the bioreactor to the impeller. Other impeller designs can produce different mixing patterns. If gas transfer is desired from the gas space to the liquid, an impeller can be located at the media surface to pump some of the liquid into the gas space for more rapid dispersion of the gas.

Gas diffusers can be used for both mixing and gas addition. Air or pure oxygen can be used to transfer oxygen to the bioreactor. If mixing is desired, as well as oxygen transfer, air is the better gas to use since the nitrogen gas in air will assist in mixing before being discharged from the bioreactor. Gas diffusers are normally placed around the outer wall of the bioreactor to produce a gas lift pumping action up the wall to the surface. The fluid flow will move back to the center and down to the bottom before returning to the gas diffusers. Gas diffusers can be combined with mechanical mixers to increase the rate of gas transfer.

Since metabolism in aerobic bacteria can be complete, mixing is important for oxygen transfer to insure maximum metabolism. With continuously fed systems the maximum rate of substrate addition is often controlled by the rate of oxygen transfer to insure an excess of oxygen at all times. In most bacteria systems 0.1 mg/l DO in the liquid will be adequate. If the bacteria clump or become attached to the bioreactor walls, the DO in solution will have to be raised to keep oxygen from becoming the limiting factor affecting metabolism in the bacteria floc. Since anaerobic bacteria have limited levels of metabolism, complete metabolism in anaerobic systems requires several different groups of bacteria working together. Mixing is essential for maximum growth of anaerobic bacteria to insure that each group of bacteria obtains its nutrients as rapidly as possible.

DEATH

Endogenous respiration continues until the bacteria are unable to control the enzyme, *lysozyme,* any longer. Lysozyme is an enzyme that hydrolyzes the protein enzymes in the cell wall and allows the remaining cell proteins to be released to the environment. Only the collapsed cell wall material and the slime material remain to accumulate in the environment. The remaining bacteria can metabolize the hydrolyzed proteins and stay alive a little longer. Under quiescent conditions the cell wall material settles out and accumulates on the bottom of the container. The inability of the bacteria to metabolize the cell wall material and the slime material allows the accumulation of these organic residues in the aqueous environment. Approximately, 10 percent of the energy in the original organic compounds metabolized by the bacteria will remain as dead cell residues. The dead cell material can be collected, hydrolyzed by strong acids to their original components, and metabolized by bacteria after neutralization. The bacterial enzymes that produce the cell walls are broken down before the cells die, preventing them from reversing the cell wall synthesis reactions. If the dead cell material is placed in biologically active soil, the cell wall material and the slime material will be slowly decomposed by fungi. Thus, the dead cell mass is not completely stable. It is only stable in the aqueous environment where the bacteria are grown.

THINGS TO REMEMBER

1. Bacteria growth is limited by a single controlling factor.

2. In an excess of nutrients, bacteria growth is limited by their ability to process nutrients.

3. In an excess of bacteria, growth is limited by the availability of the controlling nutrient.

4. Oxygen transfer often limits bacteria growth in strong, organic solutions.

5. In a continuously fed, completely mixed bioreactor, the rate of bacteria growth is controlled by either the hydraulic displacement time or oxygen transfer.

6. Variable flow rates to a bioreactor create periods of unstable operations.

7. Steady flow operations must be maintained for at least 6 hydraulic

displacement periods to produce stable conditions in a bioreactor.

8. Endogenous respiration occurs continuously in bacteria.

9. Endogenous respiration reduces the measured bacteria mass in a bioreactor in proportion to the bacteria retention time and the temperature in the bioreactor.

10. Oxygen utilization is an excellent measure of metabolism under aerobic conditions.

11. Endogenous respiration can be measured in a large bacteria population in the absence of external substrates by either oxygen uptake or mass change.

12. In the presence of external substrates the rate of oxygen uptake is a measure of energy used for bacteria synthesis plus the energy used for endogenous respiration.

13. Temperature changes the bacteria metabolism rate by a factor of 2 for each 10°C change over specific temperature ranges.

14. Mesophilic bacteria grow best between 5°C and 40°C.

15. Thermophilic bacteria grow best between 45°C and 70°C.

16. Bacteria normally grow at pH ranges from 6.5 to 8.5.

17. Alkalinity is essential for normal bacteria metabolism.

18. Without adequate alkalinity organic acids will accumulate, dropping the pH and slowing metabolism.

19. When bacteria die, cell wall materials accumulate as dead cell residue that cannot be metabolized further in the aqueous environment.

REFERENCES

Agnew, R. W. (1968) *The Oxygen Uptake of a Biological Culture Subjected to TransientOrganic Loadings*, PhD Thesis, University of Kansas, Lawrence, KS.

Carter, B. L. A., Bull, A. T., Pirt, S. J., and Rowley, B. I. (1971) Relationship Between Energy Substrate Utilization and Specific Growth Rate in *Aspergillus nidulans*, *J. Bacteriol.*, **108**, 309.

Clifton, C. E. and Sobek, J. M. (1961) Oxidative Assimilation of Glucose and Endogenous Respiration of *Bacillus cereus*, *J. Bacteriol.*, **81**, 284.

Clifton, C. E. and Sobek, J. M. (1961) Endogenous Respiration of *Bacillus cereus*, *J. Bacteriol.*, **82**, 252.

Gronlund, A. F. and Campbell, J. J. R. (1961) Nitrogenous Compounds as Substrates for Endogenous Respiration in Microorganisms, *J. Bacteriol.*, **81**, 721.

Gaudy, A. F. (1955) *Mode of Bacterial Predomination in Aerobic Waste Disposal Systems*, SM Thesis, Mass. Inst. of Tech., Cambridge, Massachusetts.

Herbert, D., Elsworth, H. and Telling, R. C. (1956) The Continuous Culture of Bacteria; a Theoretical and Experimental Study, *J. Gen. Microbiol.*, **14**, 601.

McKinney, R. E. and Ooten, R. J. (1969) Concepts of Complete Mixing Activated Sludge, *Trans. 19th Annual San. Engr. Conf.*, p. 32, Univ. of Kansas, School of Engineering.

Monod, J. (1942) *Recherches sur la Croissance des Cultures Bacteriennes*, Hermann & Cie, Paris, France.

Monod, J. (1949) The Growth of Bacterial Cultures, *Ann. Rev. Microbiol.*, **3**, 371.

Monod, J. (1950) La Technique de Culture Continue, Theorie et Applications, *Annales de l'Instit. Pasteur*, **79**, 390.

Moses, V. and Syrett, P. J. (1955) The Endogenous Respiration of Microorganisms, *J. Bacteriol.*,**70**, 201.

Poole, R.K., Bazin, M.J. and Keevil, C.W. (editors) (1990) *Microbial Growth Dynamics*, **28**, Soc. of Gen. Microbiol., IRL Press, Oxford University, England.

Standing, C. N., Fredrickson, A. G. and Tsuchiya, H. M. (1972) Batch- and Continuous-Culture Transients for Two Substrate Systems, *App. Microbiol.*, **23**, 354.

Umbreit, W. W., Burris, R. H. and Stauffer, J.F. (1964) *Manometric Techniques*, 4th Ed., Burgess Publishing Co., Minneapolis.

Warren, R. A. J., Ells, A. F. and Campbell, J. J. R. (1960) Endogenous Respiration of *Pseudomonas aeruginosa*, *J. Bacteriol.*, **79**, 875.

Wilner, B. and Clifton, C. E. (1954) Oxidative Assimilation by *Bacillus subtilis*, *J. Bacteriol.*, **67**, 571.

Chapter 4

FUNGI AND YEASTS

While the bacteria are the largest group of microorganisms, they are not the only microorganisms of interest to environmental microbiologists. The fungi and yeasts are important groups of non-photosynthetic microorganisms that exist widely in nature and are active in the stabilization of organic residues in both the soil environment and the aqueous environment. Unfortunately, less quantitative data are available about the fungi and yeasts than about bacteria. It may well be related to the fact that these microorganisms have played a lesser role in diseases than bacteria. It could also be that bacteria were easier to isolate and to study in pure culture. The limited quantitative biochemical information does not make the fungi and yeasts less important, it only makes them more difficult to study and evaluate. The role of yeasts in alcoholic fermentation and in nutritional supplements has produced considerable information about some of the yeasts. This information can be of value to environmental microbiologists and provides a place to start when looking at yeasts in industrial wastewater treatment systems.

FUNGI

Fungi are non-photosynthetic, multicellular microorganisms that metabolize organic matter in a similar manner as bacteria. Fungi are primarily strict aerobes, requiring dissolved oxygen and soluble organic compounds for metabolism. They are predominately filamentous and reproduce by producing large numbers of spores. Unlike bacteria, fungi have a nucleus that is self-contained within each cell. They are larger than bacteria and can produce true cell branching in their filaments. Fungi have more complex phases in their life cycle than bacteria. The identification of fungi has been based entirely upon their physical characteristics and on the different phases of their life cycle. Current emphasis on genetic structure may result in significant changes in the identification and classification of fungi. Over the years mycologists have placed greater emphasis on identification of fungi than on their biochemistry. There have been more than 100,000 species of fungi identified. Currently, fungi are classified in the *Eucarya* domain. While fungi are very important in applied environmental microbiology, it is not essential to know the names of the fungi in order to recognize their value. Fortunately, most fungi are non-pathogenic and play an important role in the degradation of dead plant tissue and other organic residues. Anyone involved in organic waste processing needs to have a general knowledge of fungi and their metabolic characteristics.

DESCRIPTION

Fungi look similar to filamentous bacteria. The most apparent differences are in the width of the filaments and the presence of a defined nucleus. Bacterial filaments are less than 1.0 μ wide. Fungi filaments are more than 1.0μ wide. Figure 4-1 illustrates a comparison between bacteria filaments and fungi filaments. Bacteria filaments may be a chain of cells or a chain of cells within a sheath. The fungi filaments may be divided into separate cells or may be long continuous filaments with the nuclei spaced at intervals along the filaments. The fungi filament is termed a *hypha*. Several filaments are called *hyphae*. A large number of hyphae are known as a *mycelium*. The cross membrane in the filaments is a *septum*; and several cross members are *septa*. Fungi reproduce by means of *spores*. Some bacteria also produce spores. The bacteria spores are a mechanism

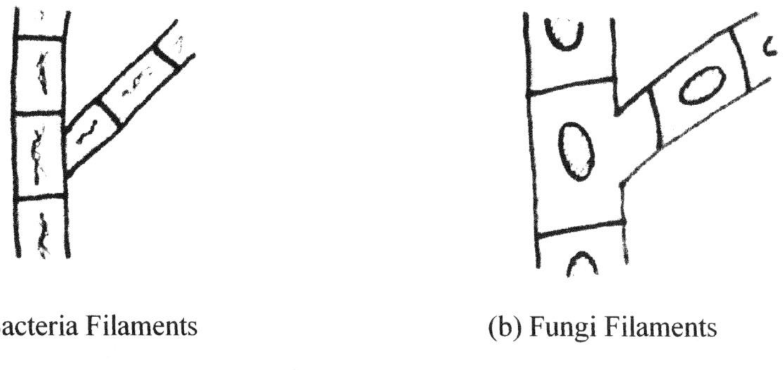

(a) Bacteria Filaments (b) Fungi Filaments

Figure 4-1 DIAGRAMS OF BACTERIA AND FUNGI FILAMENTS

for survival of individual cells in adverse environments. The fungi produce many spores, each of which can reproduce the fungi by generating a new mass of cells. Most of the fungi spores are asexual; but some fungi produce sexual spores. The fungi spores are contained on specific cell structures. Some fungi produce spores on the end of special cell structures, as shown in Figure 4-2. The fungi spores contained in a large sac at the end of a hypha are known as *sporangiospores.* The sac that holds the spores is the *sporangium* and the hypha holding the spores is the *sporangiophore.* Other fungi produce spores on stem-like structures that are called *conidiophores.* The spores on the end of the conidiophores are

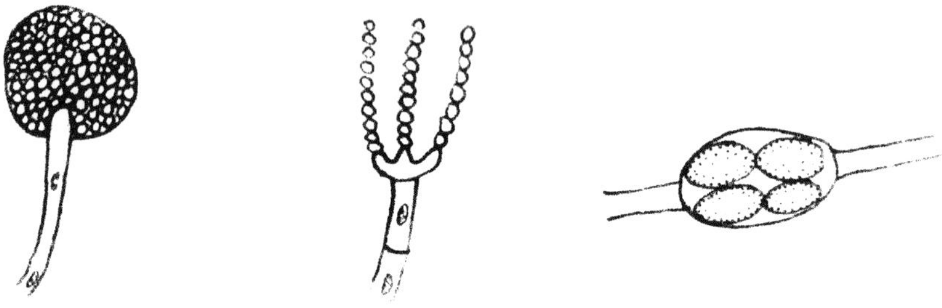

(a) *Sporangiospores* (b) *Conidiospores* (c) *Ascospores*

Figure 4-2 DIAGRAMS OF FUNGI SPORE STRUCTURES

conidiospores. One of the classes of fungi is the *Ascomycotina.* The *Ascomycotina* are grouped together because their sexual spores are encased in a sac called an *ascus.* The spores are called *ascospores.* Unfortunately, there are other forms of spores by specific fungi. The complexity of the life cycles of the different fungi is such that most environmental microbiologists will not be concerned with the details of specific fungi and will depend upon mycologists for precise identification. The fungi that grow entirely in liquid media will release their spores when the spores mature. If sufficient nutrients are available in the liquid medium, the spores will start to germinate and produce new hyphae. Fungi grown in the terrestrial environment depend on the wind currents to disperse their spores. When the wind currents stop moving, the fungi spores will settle onto the land and plant surfaces. Most of the fungi spores fall on barren ground and never germinate. The fungi spores that land on soil suitable for growth determine which fungi will survive and predominate in the environment.

Fungi grown in commercial liquid media will quickly use all the dissolved oxygen and stop further metabolism, except at the liquid-air interface. As oxygen becomes limiting within the liquid, metabolism will be incomplete. There simply will not be sufficient dissolved oxygen to oxidize all of the organic compounds that the fungi have started to metabolize. Metabolic intermediates will accumulate in the liquid, creating the impression of anaerobic metabolism. In recent years a few fungi species have been designated as being anaerobic. The anaerobic fungi have been reported from studies on animal rumen cultures. The concentrated nutrient environment in the rumen provides sufficient nutrients for both anaerobic bacteria and anaerobic fungi. More research is needed to determine the extent of anaerobic fungi and their value in environmental microbiology. To date, anaerobic fungi have not been shown to have any impact in anaerobic waste treatment systems. As additional high rate anaerobic treatment plants are constructed, anaerobic fungi may be observed in these treatment units. The ability of fungi spores to survive in anaerobic environments has permitted fungi to be isolated from anaerobic fluids, creating the impression that the fungi are actively metabolizing in the anaerobic systems. Careful study will show if fungi are strict aerobes or strict anaerobes. It may well be that some fungi are facultative like many bacteria.

Fungi grow best at liquid-air interfaces to maximize oxygen transfer from the air and form a dense mycelium that can be removed from the liquid surface as a single layer. The surface growth has two components, the *vegetative mycelium* that is composed of the hyphae extending into the liquid medium and the *aerial mycelium* that is composed of the hyphae extending into the air above the liquid. The vegetative mycelium absorbs the nutrients from the liquid media for the growth of both the vegetative mycelium and the aerial mycelium. The aerial mycelium contains the fungi spores that are easily released into the air. Fungi also grow on the surface of solid media with the vegetative mycelium seeking nutrients from the moist media.

CHEMICAL COMPOSITION

Based on information by Cochrane, fungi contain 85% to 90% water. The dry matter is about 95% organic compounds with 5% inorganic compounds. Growth of fungi in a high salt environment will have a greater inorganic fraction than fungi grown in normal salt media, the same as bacteria. The organic fraction of fungi contains between 40% and 50% carbon and between 2% and 7% nitrogen. Protein analyses show that the fungi cell mass contains only 20% to 25% proteins. Proteins are a major difference between fungi and bacteria. Fungi produce protoplasm with less protein than bacteria and require less nitrogen per unit cell mass synthesized. Fungi also have less phosphorus than bacteria, containing from 1.0% to 1.5% P in the fungi cell mass. Fungi do not produce significant amounts of lipids, usually less than 5.0%. The fungal protoplasm is largely polysaccharide. The fungi cell wall structure is a lipo-protein-polysaccharide complex. Lipids make up less than 8% of the fungi cell walls and proteins are less than 10%. The majority of the cell wall composition is chitin, a polysaccharide composed of N-acetylglucosamine. Since fungi must hydrolyze complex organic solids the same as bacteria, the lipids and proteins in the cell wall are very important for metabolism. J. W. Foster gave a chemical analysis of *Aspergillus niger* as 47.9% carbon, 5.24% nitrogen, 6.7% hydrogen and 1.58% ash. By difference oxygen was 38.6%. The chemical analysis of *Aspergillus niger* appears to be typical for fungi as a group, yielding an empirical analysis of $C_{10.7}H_{18}O_{6.5}N$ for the organic solids when N = 1.0 or $CH_{1.7}O_{0.61}N_{0.093}$ when C = 1.0. Although the empirical formula for fungi protoplasm is quite different from bacteria protoplasm, the metabolic energy requirements for fungi are essentially the same as for bacteria, 31.6 kJ/g VSS.

METABOLISM AND GROWTH

Aerobic metabolism permits the fungi to obtain the maximum energy from the substrate for synthesis of new cell protoplasm. Surface enzymes allow the fungi to hydrolyze complex organic compounds to simple soluble organics prior to entering the cell, the same as bacteria. Hydrophobic organic compounds enter through the lipids in the cell wall structure. Inside the cell, enzymes oxidize the organic compounds to organic acids and then to carbon dioxide and water. Since the fungi have less protein than bacteria, metabolism of protein substrates by fungi results in more ammonia nitrogen being released to the environment than during bacteria metabolism of the same quantity of proteins in the substrate. Without sufficient dissolved oxygen fungi metabolism results in the release of organic acid intermediates into the environment and a decrease in pH. The low protein content of fungi allows them to be more tolerant of low pH environments than bacteria. Fungi have the ability to grow at pH levels as low as 4.0 to 4.5 with an optimum pH between pH 5.0 and 7.0. From a temperature point of view, fungi grow between 5° C and 40° C with an optimum temperature around 35° C. There are a few thermophilic fungi that grow at temperatures up to 60° C. The low oxygen solubility at high temperatures limits the growth of fungi at thermophilic temperatures.

One of the more interesting aspects of fungi metabolism is the ability of some fungi to metabolize lignin. The white rot fungi, *Phanerochaete chrysosporium*, have been studied in detail because of its ability to metabolize substituted aromatic compounds that accumulate from industrial wastes and its ability to metabolize lignin. Lignin is a complex plant polymer that protects plant cellulose from attack by bacteria. Terrestrial fungi have the ability to metabolize lignin and cellulose, recycling all the dead plant tissue back into the environment. Unfortunately, there are no aquatic fungi capable of metabolizing lignin. Efforts to develop aquatic fungi capable of metabolizing lignin in the aqueous environment have all been unsuccessful. Terrestrial fungi also have the ability to degrade bacteria cell wall polysaccharides in the soil environment. The dead cell mass of fungi and the non-biodegradable plant tissue forms a complex organic mixture that has been designated as *humus*. Humus comprises an important part of soil. It helps soil hold moisture and nutrient salts and allows plant roots to move around dense soil particles.

Since fungi do not produce dispersed cells, growth cannot be measured by numbers of cells. Growth is measured by dry weight mass, the same as bacteria. Righelato found that the μ_{max} for fungi was 0.3/hr, giving a doubling time of about 2 hours. Using glucose as a substrate and *Penicillium chrysogenum* as the fungi, it was found that 0.45 g cell mass was produced per g glucose metabolized. In terms of energy used, 1.55 g cell mass was produced per g oxygen utilized with 0.29 g oxygen being used for the synthesis of the 0.45 g measured cell mass. Approximately, 27 percent of the glucose was oxidized and 73 percent was converted to cell mass. The net result would have been a COD of 1.74 g/g VSS cell mass. The data indicated the cell mass had a higher lipid fraction than normal. It appeared that oxygen was limiting and metabolism was not complete in this study. Based on the normal carbohydrate content of cell mass, fungi should have COD/VSS ratios closer to 1.36/1 than the 1.74/1 from Righelato's data. Carlile and Watkinson indicated that *Candida utilis* produced 0.51 g cell mass/g glucose metabolized. It appears that the quantity of fungi cell mass and the quantity of bacteria cell mass from the metabolism of glucose is essentially the same. With the energy content of two groups of microorganisms the same, it is natural to expect that the energy required for cell protoplasm production to be the same. This is an area where more basic research could be productive.

Competition With Bacteria

In the natural environment fungi compete with bacteria for nutrients to survive. Bacteria normally have the advantage over fungi in the natural environment. Bacteria simply have the ability to obtain more nutrients and can process their nutrients at a faster rate than fungi. Since both groups of microorganisms metabolize soluble nutrients, both groups survive according to their ability to obtain and process nutrients. The greater surface area/mass ratio permits bacteria to obtain nutrients at a faster rate than fungi under normal metabolic conditions. The presence of higher animal forms in the environment favors the fungi since the higher animals can eat bacteria easier than they can consume fungi. The filamentous fungi are difficult for the microscopic animals to metabolize. The environment has a number of factors that allows the fungi to be competitive with bacteria. By understanding the various factors affecting the growth of the bacteria and fungi, the environmental microbiologist can recognize how to adjust the environment of different treatment systems to favor one group of microorganisms or the other.

Moisture

Moisture content is very important for the growth of fungi. Unlike bacteria, fungi can grow in environments with limited amounts of water. Since fungi must have soluble nutrients, the same as bacteria, there must be sufficient water for the nutrients to be dissolved and transported inside the cells. Yet, the cells do not have to be completely immersed in water. Fungi can grow on cotton fibers with a moisture content of 10 percent or on wood particles with a moisture content as low as 26 percent. The ability of fungi to grow at low moisture levels is the reason why fungi grow so readily in damp basements and in the soil. Bacteria cannot grow at the very low moisture levels that fungi can grow, giving the fungi an advantage over the bacteria in the competition for food. Water is one of the end products of metabolism. As the fungi grow on organic surfaces, they tend to release water from their cell mass, raising the moisture level of their environment. Dry soils favor fungi over bacteria, while wet soils favor bacteria. The low solubility of oxygen in water allows bacteria metabolism in water saturated soils to create an anaerobic environment and stop further metabolism by the fungi, except at the air-water interface. Moisture levels in the environment play a major role in the rate of metabolism. Metabolism slows at low moisture levels and increases with higher moisture levels.

Cell N & P

The lower nitrogen and phosphorus content of fungi protoplasm than bacteria protoplasm gives the fungi an advantage over bacteria when metabolizing organic compounds in low nitrogen and low phosphorus environments. Fungi protoplasm contains about half the nitrogen and phosphorus concentrations as bacteria. This allows the fungi to produce normal protoplasm in high carbon environments and forces the bacteria to produce less than normal protoplasm. Fortunately, the bacteria can survive with low nitrogen and phosphorus protoplasm. In low nitrogen and phosphorus environments bacteria tend to produce large quantities of extracellular polysaccharide slime. The ability of the microorganisms to adapt to adverse environments is essential for their survival.

pH

The ability to grow at low pH levels also favors the growth of fungi over bacteria. Organic acids have less effect on fungi because of their lower protein content. One way to isolate fungi without significant bacteria is to use a nutrient media having a pH between 4.5 and 5.5. The fungi can use the nutrients quite easily, while the bacteria cannot. Fungi can be isolated on solid media at pH 7 without bacteria by adding antibiotics to the media that stop bacteria growth.

Oxygen

Both bacteria and fungi grow under aerobic conditions. As the dissolved oxygen is used up, the fungi cannot continue normal metabolism; but the facultative bacteria shift from aerobic metabolism to anaerobic metabolism. Under anaerobic conditions the bacteria predominate. While the fungi cannot normally grow under anaerobic conditions, fungi spores can survive in the anaerobic environment. When the environment shifts from anaerobic to aerobic, the fungi spores are able to germinate and grow. Controlling the oxygen level can be an important tool for environmental microbiologists to minimize the growth of fungi. As previously indicated, a few species of fungi are able to grow anaerobically. Much more research is needed to determine the environmental conditions that favor the growth of anaerobic fungi.

Antibiotics

A few fungi have developed the ability to produce antibiotic substances that prevent the bacteria from growing near these fungi. The antibiotics give the fungi an advantage over certain bacteria, allowing the fungi to grow in normal environments without bacteria competition. Discovery of antibiotics was a major medical advance in controlling some bacteria infections in animals. The useful antibiotics prevented the growth of pathogenic bacteria in people and domestic animals, stopping the spread of diseases caused by the affected pathogens. The success of antibiotics resulted in large scale manufacturing plants, producing antibiotics for wide distribution. Antibiotic production has become a major industry around the world. The small yield of antibiotics requires growth of large quantities of fungi and the subsequent extraction of the antibiotics from the liquid media in which the fungi have been grown. Disposal of the excess fungi and the residual feed solutions from antibiotic manufacturing is a major environmental problem. These waste materials cannot be discharged into adjacent rivers without creating significant water pollution. Economics permit separation of the fungi for protein recovery. The liquid nutrients are too dilute for recovery and too strong for direct discharge to the environment without pretreatment. Biological treatment of the liquid residues is required before final discharge to the environment. The residual antibiotics in the wastewater should not be sufficient to adversely affect the bacteria in the wastewater treatment system.

A second environmental problem from antibiotics has arisen from excessive use of antibiotics to control infections in people and in domestic animals. Destruction of pathogenic bacteria by antibiotics has allowed competing pathogenic bacteria, not affected by the antibiotics, to grow to higher numbers and pose new health threats in the environment. The presence of large quantities of specific antibiotics in the environment has also stimulated the growth of non-pathogenic bacteria that metabolize the antibiotics, reducing their effectiveness. Continuing development of new antibiotics and more economical methods for their manufacture will pose challenges for future environmental microbiologists.

YEASTS

Yeasts are part of the fungi family that has proven of such value that they have their own detailed study. Yeasts belong to the *Ascomycotina.* They are similar to bacteria in that they are single cells, but their other characteristics favor fungi. Yeasts are non-photosynthetic microorganisms that have a separate nucleus and a complex life cycle. They are larger than bacteria and appear to be spherical to egg shaped. They are non-motile and reproduce asexually by budding. Sexual reproduction results in the formation of ascospores. Most yeasts are non-pathogenic; but a few yeast can grow parasitically in the right environment. The primary role for yeast is in alcoholic fermentation and in bread manufacturing, although they have been used for enzyme and vitamin production. The *Saccharomycetes cerevisiae* have been studied the most from a biochemical point of view. The ability of yeasts to metabolize natural sugars has resulted in their wide distribution throughout the environment.

A drawing of yeast cells undergoing budding is shown in Figure 4-3. The new cell expands as it develops its chemical structure. As soon as the new cell has sufficient chemical composition, it breaks free, creating two separate cells. Although yeast cells are non-motile, they tend to remain dispersed until the

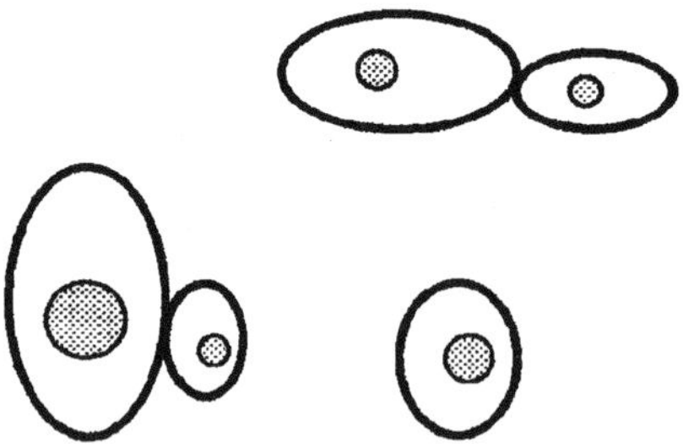

Figure 4-3 SCHEMATIC DRAWING OF BUDDING YEAST CELLS

substrate has been metabolized. The yeast cells then flocculate into large masses of cells that settle out under quiescent conditions. The flocculating characteristics of yeasts are very valuable in the fermentation industry. When the sugars have been converted to alcohol and new cell mass, the cells flocculate and settle out, leaving a clear liquid above the yeast. Calleja indicated that the material around the yeast cells that produces the flocculation is primarily carbohydrate material. It appears that yeasts and bacteria have similar mechanisms for flocculation. The accumulation of polysaccharide material around older cells provides surfaces with few ionizable groups, giving a large surface area with a low surface charge. The cells tend to form aggregates with strong Van der Waal surface attraction forces holding them together. Young, rapidly growing yeast cells have little accumulated polysaccharides and remain dispersed with the older cells forming floc and settling out.

CHEMICAL COMPOSITION

Yeasts have a chemical composition between fungi and bacteria. The yeasts have more protein than fungi, but less than bacteria. An analysis of baker's yeast by Harrison indicated 91% volatile organic compounds and 9% non-volatile inorganic compounds. The organic fraction contained 49% to 50% carbon, 6.6% hydrogen, 34% oxygen and 9.9% nitrogen. The empirical formula for yeast is $C_{5.8}H_{9.3}O_3N$, based on N = 1.0. The empirical formula for yeasts indicates a closer relationship to bacteria than to fungi. The high protein content of the yeasts has attracted attention as a potential source of food. The ability of yeasts to produce many of the organic growth accessory substances for microorganisms and higher animals has made yeast extract a normal part of growth media and animal feeds. These factors have stimulated considerable quantitative data on yeast metabolism.

METABOLISM AND GROWTH

The primary substrate for yeasts is fermentable sugar. Yeasts can metabolize most natural organic compounds under the proper environmental conditions. The *Saccharomycetes* metabolize sugar to ethyl alcohol, carbon dioxide and cell mass under oxygen limiting conditions. Because most of the energy from the sugar is transferred to ethyl alcohol, growth of yeast cells is minimized. If adequate oxygen is supplied to the yeast cells, metabolism will be complete to carbon dioxide, water and cell mass. The release of additional energy results in greater cell yield. A study reported by Fiechter et al indicated that *Saccharomycetes cerevisiae* produced 0.10 to 0.15 g cell mass/g glucose metabolized to alcohol when grown in batch cultures. The synthesis increased to 0.35 g cell mass/g glucose when additional oxygen was supplied. *Trichosporon cutaneum* does not produce alcohol from glucose, but requires sufficient oxygen for metabolism to carbon dioxide and water. *Trichosporon* has been shown to produce 0.55 g cell mass/g glucose metabolized. Continuous cultures of *Trichosporon* produced 0.57 g cell mass/g glucose at a 0.4/hr dilution rate. *Saccharomycetes* in continuous culture produced

0.16 g cell mass/ g glucose when alcohol was the end product and 0.50 g cell mass/g glucose when adequate oxygen was supplied. Several investigators have confirmed that yeast produce about 0.5 g cell mass/g sugar metabolized. In a study measuring heat released during metabolism, Harrison reported that 3.7 Kcal (15.5 kJ) heat was released/g cell mass produced during the metabolism of sucrose. Changing the substrate to a saturated alkane produced 7 Kcal (29.3 kJ) heat release/g cell mass produced. The heat release data are important since this heat must be removed in any large-scale yeast production systems. Prochazka et al measured the heat of combustion of eight different species of yeasts, averaging 5.32 Kcal (22.3 kJ)/g VSS cell mass. As previously indicated, correcting the heat of combustion for the differences in end products during normal metabolism resulted in 4.8 Kcal (20.1 kJ)/g VSS cell mass from the substrate. With the production of 0.5 g cell mass/g glucose, 2.4 Kcal (10.1 kJ) from the glucose went to cell mass and 1.3 Kcal (5.4 kJ) was oxidized for energy. Endogenous respiration by yeast cells accounted for the additional energy released. Righelato indicated an endogenous respiration rate of 0.019 to 0.021 g O_2 used/g cell mass/hr. The major problem in measuring growth and metabolism lies in the changing relationships with time. Substrate metabolism is balanced between cell mass production, energy production required for the cell mass synthesis and endogenous respiration. Under a given environmental condition, the microbes attempt to optimize their metabolism with the maximum possible production of new cell mass. Once formed, the cell mass is continuously degraded by endogenous respiration. The cell mass is decreased by the amount of energy expended. Measured cell mass and oxygen utilization is the sum of the synthesis reaction and the endogenous respiration reaction for the culture over the time period studied. It is not surprising that the literature reports different values for cell mass production and energy utilization since measurements are not taken at the same time with the same microbial mass under the same environmental conditions. Yet, the reported data are all valid for the conditions under which measurements were made.

Temperature is an important environmental parameter for yeast cells, the same as for other microorganisms. Increasing the temperature results in increased rates of metabolism until the optimum temperature is reached. At temperatures above 30° C the rate of metabolism begins to slow. When heat begins to denature the proteins in the enzymes, metabolism decreases at a rapid rate. Yeasts that are pathogenic to humans grow very well at 37° C, body temperature. Some yeast cells also grow at thermophilic conditions. The yeast cells react the same as bacteria and fungi to changes in temperature.

Yeast cells grow in low pH environments, the same as fungi. A study of *Candida utilis* by Paredes-Lopez et al indicated optimum growth at a pH between 3.5 and 4.5. Yeast cells also grow well at neutral pH; but cannot compete very well against bacteria because of their size and their specific substrate preferences. Yeasts can be found in soil environments as well as in aqueous environments.

THINGS TO REMEMBER

1. Fungi are primarily strict aerobes; but some anaerobic fungi have recently been isolated.

2. Fungi grow best under acid conditions and in carbonaceous substrates with little nitrogen or phosphorus.

3. Fungi reproduce by spores that are easily spread by wind currents or air movement.

4. Fungi produce wider filaments than bacteria.

5. Fungi require the same energy for cell synthesis as bacteria.

6. Terrestrial fungi can metabolize lignin and other complex aromatic compounds.

7. Terrestrial fungi can also metabolize the bacteria cell walls that remain after the bacteria die.

8. Fungi can grow at lower moisture levels than bacteria.

9. Bacteria outgrow fungi in normal environments.

10. Yeasts are non-filamentous fungi that grow in a dispersed state.

11. Yeasts flocculate in the same way as bacteria, when metabolism is complete.

12. Yeast cells contain micro-nutrients that are used to stimulate microbial and animal growth.

13. Yeasts are grown in concentrated nutrient solutions with the production of considerable heat.

14. Yeasts are a good source of proteins.

REFERENCES

Calleja, G. B. (1987) Cell Aggregation, *The Yeasts, Vol. 2, Yeasts and the Environment*, Rose, A. H. and Harrison, J. S. (Editors), 2nd Ed., 165, Academic Press, New York.

Carter, B. L. A., Bull, A. T., Pirt, S. J. and Rowley, B. I. (1971) Relationship Between Energy Substrate Utilization and Specific Growth Rate in *Aspergillus nidulans*, *Jour. Bacteriol.*, **108**, 309.

Cochrane, V. W. (1958) *Physiology of Fungi,* John Wiley & Sons, New York.

Davenport, R. R. (1980) An Introduction to Yeasts and Yeast-like Organisms, *Biology and Activities of Yeasts*, Skinner, F. A., Passmore, S. M. and Davenport, R. R. (Editors), Academic Press, New York.

Deacon, J. W. (1948) Introduction to Modern Mycology, *Basic Microbiology*, *Vol. 7*, 2nd Ed., Blackwell Scientific Publications, Oxford, England.

Fiechter, A., Kappeli, O and Meussdoerffer, F. (1987) Batch and Continuous Culture, *The Yeasts, Vol. 2, Yeasts and the Environment*, Rose, A. H. and Harrison, J. S. (Editors), 2nd Ed., pp 99, Academic Press, New York.

Foster, J. W. (1949) *Chemical Activities of Fungi*, Academic Press, New York.

Funder, S. (1954) *Practical Mycology*, Broggers Boktr. Forlag, Oslo, Norway.

Garraway, M. O. and Evans, R. C. (1984) *Fungal Nutrition and Physiology*, John Wiley and Sons, New York.

Harrison, J. S. (1987) Food and Fodder Yeasts, *The Yeasts, Vol. 2, Yeasts and the Environment*, Rose, A. H. and Harrison, J. S. (Editors), 2nd Ed., 399, Academic Press, New York.

Hudson, H. J. (1986) *Fungal Biology*, Edward Arnold Publishers, London, England.

Lilly, V. G. (1965) Chemical Constituents of the Fungal Cell 1. Elemental Constituents and Their Roles, *The Fungi, an Advanced Treatise, Vol. 1, The Fungal Cell*, Ainsworth, G. C. and Sussman, A. S. (Editors), 163, Academic Press, New York.

Litchfield, J. H. (1992) Single-Cell Proteins, *Encyclopedia of Microbiology, Vol. 4*, Academic Press, pp 11.

Paredes-Lopex, O., Camargo-Rubio, E. and Ornelas-Vale, A. (1976) Influence of Specific Growth Rate on Biomass Yield, Productivity, and Composition of *Candida utilis* in Batch and Continuous Culture, *Appl. Environ. Microbiol.,* **31**, 487.

Righelato, R. C. (1975) Growth Kinetics of Mycelial Fungi, *The Filamentous Fungi, Vol. 1,* Smith, J. E. and Berry, D. R. (Editors), 79, Edward Arnold Publishers, London, England.

Chapter 5

ALGAE

Algae are photosynthetic microorganisms containing chlorophyll. They can be single cell or multicell, motile or non-motile. Algae use light as their source of energy for cell synthesis and inorganic ions as their source of chemicals for cell protoplasm. They are found where there is a reasonable amount of light and nutrient elements. Algae have both a positive and a negative impact on the environment. One of the positive aspects of algae is the production of oxygen in proportion to the growth of new cells. On the negative side, algae are responsible for tastes and odors in many surface water supplies. Algae also take stable inorganic ions and convert them into organic matter that ultimately must be stabilized. The new organic matter can be used as a source of nutrients for both plants and animals. Unfortunately, stabilization of dead algae uses most of the oxygen produced during growth.

DESCRIPTION

Like fungi, algae are identified by their physical characteristics. The primary characteristic of algae is pigmentation. Colors range from green to yellow-green to brown to red. Algae are larger than bacteria and can be easily seen and identified by microscopic examination at 100 X magnification. Like fungi, the algae have a clearly defined nucleus and are classified as *Eucarya*. The filamentous algae resemble fungi with photosynthesis providing the major difference between these two groups of microorganisms. Most of the filamentous algae are larger than fungi. Some of the motile algae resemble protozoa. The single cell, motile algae look like green, flagellated protozoa. Current classification systems used to group microorganisms are arbitrary systems that do not delineate all the differences in the organisms. Transition organisms fall between two groups instead of being clearly defined by any one group. The lack of a defined nucleus caused a change in the classification of the blue-green microorganisms from algae to bacteria. The Eighth Edition of *Bergey's Manual of Determinative Bacteriology* classified the blue-green organisms as *Cyanobacteria.* The blue-green organisms have a mixture of photosynthetic pigments that produce the various shades of color. They grow as single cells, like bacteria, in colonial masses, or as filaments. Some blue-green algae have the ability to fix atmospheric nitrogen, the same as the free-living, nitrogen-fixing bacteria. While the colonial masses of blue-green organisms lack motility, the filamentous organisms can be motile. Motility in the filamentous blue-green organisms is much slower than bacteria. Since the blue-green organisms were classified as algae for so many years, it took time before everyone adjusted to the change in classification. Genetic research on algae may stimulate future changes in the classification of algae. It is important to keep an open mind on microbial classifications and recognize that changes will occur as new research develops a better understanding of the various differences that are used to base microbial classification.

The green algae are largely grouped as *Chlorophycophyta.* The *Euglenophycophyta* are a special group of motile, green algae that have different characteristics than the *Chlorophycophyta.* Differences in the organisms are of significance only to phycologists. Both groups of algae use chlorophyll as their photosynthetic pigment and metabolize inorganic nutrients in the same general pattern. The *Chlorophycophyta* grow as individual cells, motile and non-motile, and as filaments. *Chlorella* have been the most widely studied green algae. *Chlorella* are small, spherical, non-motile cells, about 5 to 10 μm in diameter. *Chlorella* can be found as dispersed cells or as clumps of cells, depending on the growth environment. The ease of isolation and growth of *Chlorella* in pure culture has assisted in their study in the laboratory under artificial light conditions. *Chlamydomonas* is the smallest motile green algae, commonly found in the natural environment. It is a round cell with two flagella that provide motility. The flagella cause the *Chlamydomonas* to move in an undulating path, making it easy to identify under the microscope. The *Euglena* are also motile green algae that exist as individual cells. *Euglena* has a distinctive shape, tending to be long and narrow and quite flexible. The *Euglena* cells have a single flagella. They also have a red eyespot that is readily apparent under the microscope. The eyespot allows the motile *Euglena* to respond to light. *Phacus* is another member of the *Euglenophycophyta.* With its large flat flexible body

Phacus is easily recognizable under the microscope. Some of the typical small, green algae are shown in the sketches in Figure 5-1.

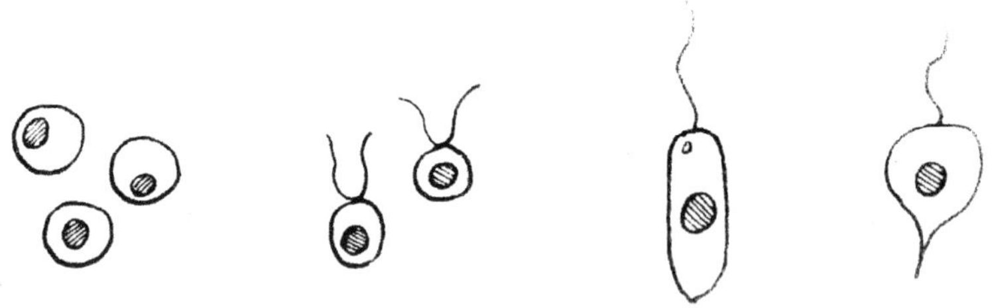

(a) *Chlorella* (b) *Chlamydomonas* (c) *Euglena* (d) *Phacus*

Figure 5-1 SKETCHES OF SMALL, GREEN ALGAE

Motile green algae also grow as colonies. *Pandorina* grows as dense masses of cells with their flagella creating a rolling motion for the colony. *Volvox* is a motile green algae, growing as large spherical colonies, but with the individual cells spaced apart. The large *Volvox* colony also moves with a rolling motion. Non-motile colonies include *Pediastrum* and *Scenedesmus. Pediastrum* grows in a circular fashion as a flat plate. The new cells form circular rows around the outer periphery of the colony. In a turbulent environment *Pediastrum* fragments into separate cells, giving rise to many new colonies. *Scenedesmus* grows in bundles of four cells with the two end cells having sharp points. All four cells may have sharp points, depending upon the species. As the cells age, they fragment into separate cells that can grow into new colonies. Figure 5-2 shows typical, non-filamentous, colonial, green algae. Green algae also grow as filaments. There are many different groups of filamentous green algae. *Spirogyra* is one of the most interesting groups of green algae, producing a

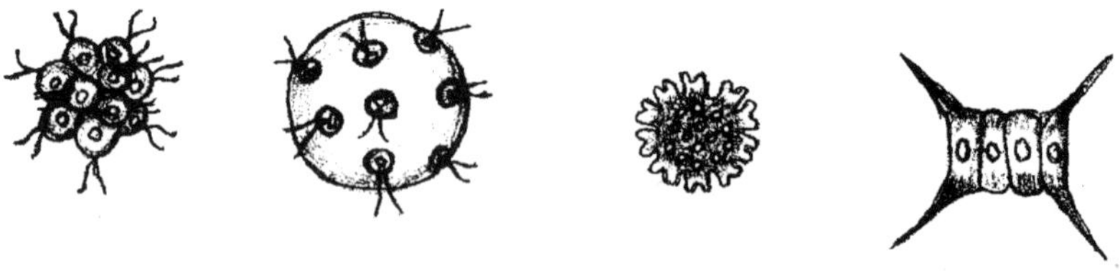

(a) *Pandorina* (b) *Volvox* (c) *Pediastrum* (d) *Scenedesmus*

Figure 5-2 SKETCHES OF NON-FILAMENTOUS, COLONIAL, GREEN ALGAE pattern of spiral chloroplasts around the cells. There are several different species of *Spirogyra*, but the spirals of chloroplasts make them easy to identify. Some of the filamentous algae form natural branches, the same as fungi, and some form only simple filaments. *Stigeoclonium* is a branching algae that is widely distributed in nature. Figure 5-3 illustrates two common, blue-green, filamentous algae, *Oscillatoria* and *Anabaena*, together with the two green, filamentous algae, *Spirogyra* and *Stigeoclonium*.

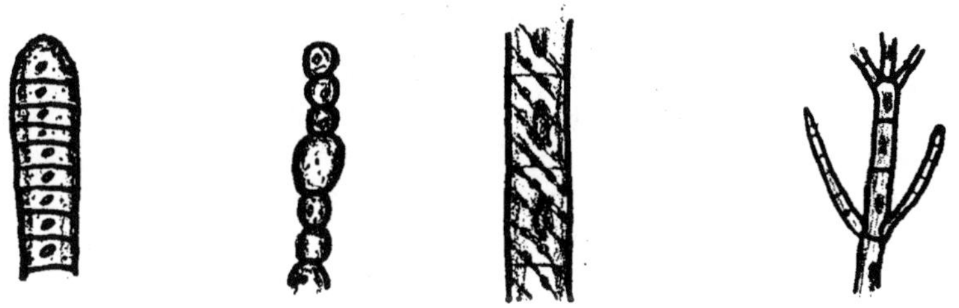

(a) *Oscillatoria* (b) *Anabaena* (c) *Spirogyra* (d) *Stigeoclonium*

Figure 5-3 SKETCHES OF FILAMENTOUS ALGAE

The *Chrysophycophyta* are yellow-green to brown algae. They have a mixture of photosynthetic pigments that create the yellow-brown colors. They grow as filaments as well as individual cells. *Ochromonas* is a motile, single cell algae with two flagella. Diatoms are the most common *Chrysophycophyta*. The most important characteristic of diatoms is their ability to create a silica shell around the cell. Diatoms tend to be small, elongated cells. Many of the diatoms are motile, moving smoothly along the longitudinal axis. A sketch of *Ochromonas* and the diatom, *Navicula*, are given in Figure 5-4.

(a) *Ochromonas* (b) *Navicula*

Figure 5-4 SKETCHES OF TWO *CHRYSOPHYCOPHYTA*

The *Pyrrophycophyta* tend to be yellow-brown. They include the dinoflagellates, small cells with two flagella and rigid bodies. The *Rhodophycophyta* are red marine algae and the *Phaeophycophyta* are the brown marine algae. These latter two groups of algae form the seaweeds that grow along the coasts in saline water.

METABOLISM

Algae use light as their source of energy for synthesizing cell protoplasm. Sunlight furnishes the light in the natural environment for algae. Since light energy is absorbed by water, growth of the algae occurs near the water surface. The photosynthetic pigments in the algae convert light energy into chemical energy by electron transfer. The rate of energy production in algae is a function of the surface area of photosynthetic pigments and the light intensity. Research has shown that algae have a light saturation limit around 600 ft-candles. Above the light saturation limit, the algae cannot use the additional light. Below the light saturation limit, the light level limits algae metabolism in direct proportion to the available light. Because of the importance of light intensity, algae have the ability to adjust to varying light levels in their aquatic environments. In excess light, the motile algae move deeper into the water, reducing the light in contact with the photosynthetic pigments. Non-motile algae adjust by releasing some of their internal gases and sinking to the desired light level. With limited light, the algae move closer to the light by motility or by accumulating additional gas and floating to the surface. Light is also absorbed by suspended particles in water. Turbid waters absorb light in proportion to the turbidity, limiting the availability of light for the algae. As algae grow, they become part of the turbidity and limit further penetration of light. Unrestricted algae growth is limited to clear water with few nutrients to stimulate excessive growth.

CELL NUTRIENTS

In addition to light, algae need a source of carbon for cell protoplasm. Carbon dioxide is the primary carbon source for algae. The air atmosphere contains about 0.03 percent carbon dioxide, giving very little pressure for transferring carbon dioxide into natural waters. Alkalinity forms the primary source of carbon dioxide in natural waters. Algae grow much better in waters containing high concentrations of bicarbonate alkalinity than in waters with low bicarbonate alkalinity. Water is the source of hydrogen for algae. Removal of hydrogen from water leaves the oxygen to form dissolved oxygen in the water. Interestingly, the reduction of carbon dioxide produces water as a metabolic end product. The net result makes it look like the oxygen is released from the carbon dioxide when it is actually released from water used as the source of hydrogen.

Nitrogen is important to form proteins for algae protoplasm. Ammonia nitrogen is the primary source of nitrogen for algae with nitrates as the secondary source. Nitrates must be reduced to ammonia for incorporation into protoplasm. Part of the light energy must be expended in the nitrate reduction, limiting the amount of potential synthesis. The blue-green bacteria, still considered by some phycologists with the algae, have the ability to use atmospheric nitrogen as their nitrogen source, when nitrates or ammonia nitrogen are not readily available. Photosynthesis furnishes the energy necessary for the blue-green bacteria to fix gaseous nitrogen dissolved in the water. The solubility of nitrogen in water at one atmosphere air pressure and 20°C is close to 15 mg/l. While this concentration of soluble nitrogen is not large, it is sufficient for considerable nitrogen fixation under the proper environment.

Phosphorus is a critical element for the growth of algae. Although algae do not need a large quantity of phosphorus, it is important in energy transfer for the algae, the same as for other microorganisms. Phosphates are the primary source of phosphorus for the algae. Since phosphates are limited in the natural environment, phosphorus availability is often the limiting factor in the growth of algae. Eutrophication of lakes and reservoirs has been caused by the discharge of excess phosphates from domestic wastewater and fertilizer runoff with the subsequent algae growth. Algae also need sulfates and trace metals. The sulfates in natural waters are adequate to supply the algae demands. The trace metal needs are quite small compared with the other elements; but they are essential if normal algae growth is to occur. Iron is needed for electron transfer. Magnesium is required for chlorophyll. Other important trace metals include calcium, potassium, zinc, copper, manganese and molybdenum. The lack of sufficient trace metals will limit the magnitude of algae growth.

CHEMICAL COMPOSITION

Examination of algae cell mass by a number of investigators indicates algae have 45% to 50% carbon in their organic fraction. Hydrogen ranges from 6.8% to 9.0%. The nitrogen content of algae shows wide variation from 2% to 11%. The other major element, oxygen, averages 32% to 37%. Since *Chlorella* are the easiest algae to grow in mass culture, most of the chemical analyses have been done on *Chlorella*. In 1953 H. W. Milner reported 49.5% carbon, 6.8% hydrogen and 9.3% nitrogen on an ash free basis for *Chlorella*. He also reported that *Chlamydomonas* contained 47.5% carbon, 6.8% hydrogen and 5.8% nitrogen. The algae were similar to bacteria with the same carbon-hydrogen concentrations, but less nitrogen. The empirical formula for the organic fraction of *Chlorella* was $C_6H_{10}O_3N$ and $C_{10}H_{17}O_6N$ for *Chlamydomonas*. It appeared that the *Chlamydomonas* had more polysaccharide materials than the *Chlorella*. B. Richardson *et al* published data on the chemical composition of *Chlorella*, when grown at three different levels of nitrogen on a continuous flow basis. The carbon concentration was 50.5%, when grown in an excess of nitrogen. The hydrogen concentration was 7.4%; the oxygen concentration was 31.4%; and the nitrogen concentration was 10.7% on an ash free basis. The empirical formula from these data was $C_6H_{10}O_3N$, the same as previously reported. Later, Goldman reported data that gave a rounded formula of $C_6H_{10}O_2N$. Rounding the empirical formula to the nearest units is probably reasonable, considering the analytical variations. A general report on mass culturing of algae indicated that algae contained 52.1% carbon, 7.9% hydrogen, 33.3% oxygen and 6.7% nitrogen for an empirical formula of $C_9H_{16}O_4N$. The primary differences in the formula were related to the nitrogen concentration in the cells. With increased endogenous respiration, the nitrogen fraction of cell mass decreased the same as in bacteria. Thus, the age of the culture is a major factor in chemical analysis of algae protoplasm. The true formula of algae protoplasm requires analysis of young cells. The same is true of bacteria and other microorganisms. The empirical formula for algae cell mass VSS appears to be close to $C_6H_{11}O_3N$ until more accurate data are made available. The percentages of the various elements are 49.7% carbon, 7.6% hydrogen, 33.1% oxygen and 9.7% nitrogen.

When algae are grown in a deficiency of nitrogen, they produce more lipids and fewer proteins in their protoplasm. Lipid production is simply a way to use the energy trapped by the chlorophyll. While bacteria store excess energy as polysaccharides, algae store energy as lipids. A number of investigators have measured the energy content of the organic fraction of algae and found that it ranged from 4.6 Kcal/g to 5.8 Kcal/g when measured in a calorimeter. The chemical composition of algae is very close to that of bacteria when algae are grown in an excess of nutrients.

GROWTH

The growth of algae follows the same general pattern as the growth of bacteria and fungi. In an excess of light and nutrients, growth is restricted only by the ability of the algae to process the nutrients. Unrestricted growth results in a log increase of cell mass. Eventually, either nutrients or light becomes limiting. Growth slows and reaches a maximum. Studies on *Scenedesmus* by Fogg indicated a μ_{max} of 2.83/d at 25° C. Gons found a μ_{max} of 1.14/d. Goldman and Graham reported a μ_{max} of 1.59/d at 20° C. These data show that the algae grow slower than bacteria. Different species and different environments make it difficult to obtain similar results.

Algae undergo endogenous respiration the same as bacteria. In the presence of light the nutrients released by endogenous respiration are immediately metabolized back to normal protoplasm. Although it might appear that endogenous respiration is not occurring, the accumulation of inert cell wall components indicates that endogenous respiration is occurring. It is easy to demonstrate endogenous respiration by placing the algae in the dark. The rate of endogenous respiration is directly proportional to the active mass of algae. Proteins are the primary materials undergoing endogenous respiration, leaving cell wall polysaccharides and lipids as the non-biodegradable residue, accumulating as dead cell mass suspended solids. Some of the polysaccharides will wash off the cell surfaces and appear as inert, soluble organics in the liquid. The important aspect about algae growth is the production of organic cell mass and the release of oxygen from stable inorganic compounds in proportion to the light energy absorbed. Unfortunately, the complete stabilization of the algae cell organic matter will require the same amount of oxygen as the algae produced during growth. Only the inert dead cell residue allows the demand for oxygen to be less than the oxygen production. Equation 5-1 illustrates the simplified reaction produced by the algae in the synthesis of new cells, using $C_6H_{11}O_3N$ as the formula for the organic fraction of algae.

$$6\ CO_2 + 4\ HOH + NH_3 + \text{Light Energy} \rightarrow C_6H_{11}O_3N + 6.5\ O_2 \qquad (5\text{-}1)$$

This equation indicates the production of 2.9 mg oxygen for each mg carbon converted to algae mass. A laboratory study by Ammann and Fraser-Smith, using *Chlorella* over 11 months continuous operation, indicated 3.0 mg oxygen production per mg carbon utilized from carbon dioxide. Another study by Hannan and Patouillet with *Chlorella* resulted in 2.8 mg oxygen for each mg carbon used. It appears that Equation 5-1 is in reasonable agreement with the experimental data for algae growth under optimum conditions. A slight variation in the water used during algae metabolism could lower the hydrogen in the cell mass from H_{11}, 7.6% hydrogen, to H_{10}, 6.9% hydrogen. The oxygen production per unit carbon would decrease to 2.8 mg oxygen/mg carbon metabolized. If there was a little less oxygen in the cell mass, there could be an increase in oxygen production to 3.0 mg oxygen/mg carbon metabolized. These variations illustrate the sensitivity of metabolic data. Small differences in the data can create significant differences in data interpretation.

Endogenous respiration by the algae results in the oxidation of cell mass with the use of oxygen and the release of carbon dioxide, water and ammonia. In the presence of light, the end products of endogenous respiration are metabolized back to new algae cell mass as indicated in Equation 5-1. In effect, the system appears to remain static when it is far from being static. In the dark the algae are unable to synthesize the end products of endogenous respiration to new cell mass. Algae endogenous respiration in the dark is the same as bacteria endogenous respiration. Equation 5-2 illustrates algae endogenous respiration in a dark environment.

$$C_6H_{11}O_3N + 4.9\ O_2 \rightarrow 4.8\ CO_2 + 3.2\ HOH + 0.8\ NH_3 + \text{Dead Cell Mass} \quad (5\text{-}2)$$

The dead cell mass is about 20% of the original organic fraction of algae mass. The dead cell mass is not metabolized in the aqueous environment and slowly settles as inert mass. When light returns to the environment, the algae are able to reuse the carbon dioxide, water, and ammonia as shown in Equation 5-3.

$$4.8\ CO_2 + 3.2\ HOH + 0.8\ NH_3 + \text{Light Energy} \rightarrow 0.8\ C_6H_{11}O_3N + 5.2\ O_2 \quad (5\text{-}3)$$

Since the nutrients are reduced by the fraction of dead cell mass, the regrowth mass is also reduced. The new algae undergoes the same cycle with about 20 percent of the living mass being removed as dead cell mass. It is readily apparent that it will take a number of cycles before all the nutrients are tied up as dead cells. Since endogenous respiration is not readily apparent except under dark environments, very few measurements have been made on the endogenous respiration rate.

In a study at the University of Kansas, E. Nelson's data indicated the rate of endogenous respiration of *Chlorella* was close to 0.9%/hr at 25° C. Nelson used a Warburg apparatus with lights to evaluate the metabolism of *Chlorella.* In one study, concentrated algae were placed in one flask with water in the center well and in a second flask with KOH in the center well. The flask with water in the center well allowed carbon dioxide to remain in the water and in the gas phase. The flask with KOH in the center well allowed the carbon dioxide released by the algae into the gas phase to be absorbed and not be available for metabolism in the light. The Warburg apparatus started on the dark cycle to allow endogenous respiration to occur. After 24 hours of darkness, the lights were turned on for two hours to allow metabolism of the nutrients released during endogenous respiration. Data collected at half-hour intervals showed a straight-line generation of oxygen, Figure 5-5. The pressure in the flask with water in the center well returned to a level just below the initial level, but the pressure in the KOH flask reacted only slightly. The algae responded to the light immediately without a significant lag. They reused the carbon dioxide that they had released during endogenous respiration. The system was made dark for another 24 hours and then the lights were turned on for two hours. This dark-light cycle was repeated for a total of four complete cycles over 105 hours. The flask with KOH indicated a small amount of carbon dioxide remained in the water and was metabolized during the light phase. The flask with water in the center well allowed the system to return to the same pressure in the light, confirming that oxygen was released as the carbon dioxide and water were metabolized. Each 24-hour period for the water flask showed about 81% metabolism as during the previous period, indicating the accumulation of inert material. Together, the two flasks showed an average of 17% reduction in pressure change as inert materials

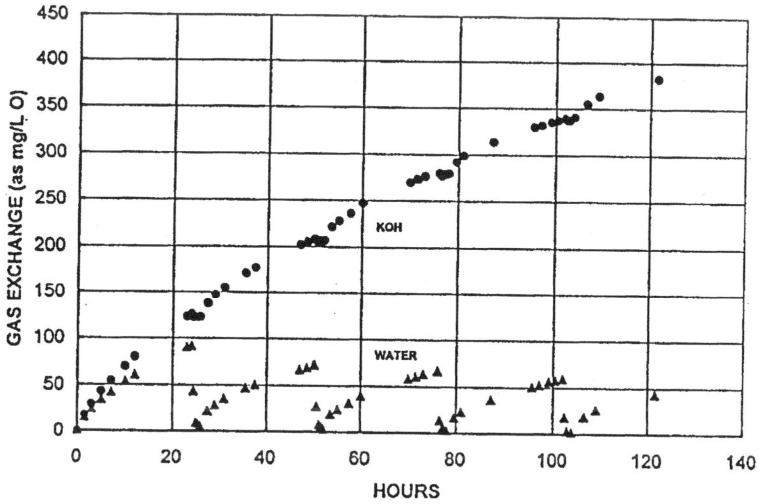

Figure 5-5 METABOLISM BY *CHLORELLA* IN LIGHT-DARK CYCLES

accumulated. The flask with the KOH allowed the total pressure change to reflect oxygen uptake by the algae. The lower pressure change for each successive 24-hour dark period indicated a smaller amount of active algae mass remaining as a result of endogenous respiration. The flask with water in the centerwell was actually a closed system, keeping all nutrients in the system. The flask with the KOH in the centerwell simulated an open system with loss of carbon dioxide to the atmosphere.

It is important to recognize the limitations of laboratory studies when dealing with the real environment. Failure to properly understand the boundary conditions of controlled experiments can result in the wrong answers. Basic quantitative studies on microbial metabolism are essential to provide data on synthesis, endogenous respiration and dead cell mass accumulation with time on different substrates and under different environmental conditions. Further metabolic studies with algae should add more important information on the rates of metabolism, synthesis and endogenous respiration.

In the natural environment the amount of available carbon dioxide for algae metabolism is related to the pH-alkalinity relationships in surface waters. The carbon dioxide in the atmosphere is about 0.03% by volume. At equilibrium pure water will contain only 0.45 mg/L carbon dioxide. Needless to say, pure water will not grow any algae. As previously indicated, algae cells require a variety of chemical elements in their protoplasm. Fortunately, carbon dioxide reacts with water to form carbonic acid. The carbonic acid reacts with a number of minerals to form soluble bicarbonate salts. As the formation of bicarbonate salts increases, the pH in the water also increases and the concentration of carbonic acid decreases. The natural minerals from soil in contact with the surface water determine the ultimate chemical characteristics of the surface water. Some minerals are quite resistant to dissolution, leaving the surface water with few dissolved chemical elements. Even in the presence of sunlight, there will not be sufficient chemical elements to allow the algae to grow to any significant extent. In other areas the minerals dissolve easily, creating highly mineralized surface waters. The mineralized surface waters allow algae to grow readily in sunlight. As the algae remove carbon dioxide for cell growth, the remaining cations form carbonates with the residual bicarbonates. The pH of the water increases, reducing the carbon dioxide in solution. Calcium ions can form insoluble precipitates with carbonate ions and phosphate ions, reducing the mineral content of the surface water. To a large extent algae depend on bacteria to produce sufficient carbon dioxide for maximum growth. Since the algae produce oxygen as an end product of metabolism, the algae and the bacteria form an excellent symbiotic relationship. It is important for environmental engineers and environmental microbiologists to understand the growth characteristics of both bacteria and algae in order to use these microorganisms to their maximum effectiveness in wastewater treatment systems.

ENVIRONMENTAL CONCERNS

Algae have created a number of environmental concerns in rivers and lakes throughout the world. The discharge of untreated wastewaters into rivers and lakes stimulated excessive growths of algae and created some environmental problems. The excessive algae growths adversely affected most of the normal uses of the river and lake waters. People avoided swimming in the green colored waters containing significant quantities of algae. Fishermen did not like the excessive algae growths as they caused fish kills. Even boaters did not like the algae. The algae created water treatment problems by increasing treatment costs. When the algae died in the fall season, some algae would release chemicals that created offensive tastes and odors in drinking water. The addition of activated carbon to the water treatment process would be required to remove the obnoxious organic compounds and reduce customer complaints concerning the water quality.

The negative aspects of algae in the surface waters in the United States helped the public to demand improved wastewater treatment throughout the entire country to reduce water pollution and the excessive growth of algae. It took many years to construct all the wastewater treatment plants to handle the wastewaters produced in the United States. Unfortunately, the wastewater treatment plants failed to solve the algae problems. The organic matter was largely removed by the wastewater treatment plants. The treated effluents were quite clear. Research soon showed that excess nitrogen and phosphorus in the treated effluents stimulated the excessive algae growths in receiving waters. Further research indicated that phosphorus was the critical element affecting the algae growth. Commercial detergents were determined to be a major source of phosphorus in municipal wastewater. Efforts were directed to reduce the phosphates in commercial detergents in critical areas having serious algae pollution. In recent years the federal EPA has developed regulations requiring municipal wastewater treatment plants to reduce the concentration of both nitrogen and phosphorus to the lowest practical levels in their treated effluents. Removing these two nutrient elements will require more time and effort before all municipal wastewater treatment plants meet the new effluent requirements.

Algae also play a positive role in wastewater treatment for many small communities. Wastewater stabilization ponds have been widely used to treat municipal wastewaters from small communities. While bacteria play the primary role in stabilization ponds, algae play an important secondary role. The algae metabolism at the pond surface provides oxygen to keep the bacteria aerobic and ties up some nitrogen and phosphorus in the dead algae cell mass that accumulates along with the dead bacteria cell mass on the bottom of the stabilization pond. Research on the role of algae in stabilization ponds began at the University of Kansas in 1962 and led to the development of *activated algae*, a high rate algae-bacteria treatment system. The activated algae treatment system depended on the balanced growth of both bacteria and algae. In a shallow treatment system the bacteria and algae worked together to produce a high quality effluent. The bacteria aerobically stabilized the biodegradable organic matter in the wastewater with the oxygen produced by the algae. It was found that the bacteria were able to flocculate the non-biodegradable suspended solids and the algae and produce a high quality effluent. While the concept of activated algae appears sound, there are a number of engineering concepts that need to be worked out to make the system practical in the field. This is an area where research could be quite productive. The keys to using algae in wastewater treatment systems lie in understanding the basic biochemistry of the algae, the wastewater characteristics, and the practical engineering design concepts.

THINGS TO REMEMBER

1. Algae are photosynthetic microorganisms using sunlight as their source of energy.

2. Algae generate cell protoplasm from inorganic compounds.

3. Growth of algae causes the pH of the water to rise as carbon dioxide is removed.

4. Oxygen is an important end product of algae metabolism.

5. Algae protoplasm is similar to bacteria protoplasm.

6. Algae undergo endogenous respiration the same as bacteria, using most of the oxygen released during synthesis.

7. The endogenous end products may be reused for cell synthesis in the presence of light.

8. Algae depend on bacteria to generate carbon dioxide for maximum growth.

9. Algae and bacteria form a symbiotic relationship in surface waters.

10. In hard waters algae can produce natural softening by precipitating calcium carbonate.

11. When algae die, they can generate tastes and odors in the water from cell organic compounds released back into the water.

12. Although excessive algae growth has caused pollution problems in surface waters, algae can play a positive role in wastewater stabilization lagoons.

REFERENCES

Ammann, E. C. B. and Fraser-Smith, A. (1968) Gas Exchange of Algae IV. Reliability of *Chlorella pyrenoidosa*, *Appl. Microbiol.*, **16**, 669.

Bold, H. C., Alexopoulos, C. J. and Delevoryas, T. (1980) *Morphology of Plants and Fungi*, 4th Edition, Harper & Row, New York.

Fogg, G. E. (1975) *Algal Cultures and Pytoplankton Ecology*, 2nd Ed., Univ. of Wisconsin Press, Madison, WI.

Goldman, J. C. (1979) Outdoor Algal Mass Cultures - II. Photosynthetic Yield Limitations, *Water Research*, **13**, 119.

Goldman, J. C. and Graham, S. J. (1981) Inorganic Carbon Limitation and Chemical Composition of Two Freshwater Green Microalgae, *Appl. Environ. Microbiol.*, **41**, 60.

Gons, H. J. (1977) On Light Limited Growth of *Scenedesmus protaberans Fritech*, Thesis, Univ. of Amsterdam.

Hannan, P. J. and Patouillet, C. (1963) Gas Exchange with Mass Cultures of Algae II. Reliability of a Photosynthetic Gas Exchanger, *Appl. Microbiol.*, **11**, 450.

Milner, H. W. (1953) The Chemical Composition of Algae, *Algae Culture,* John Burlew, Editor, pp 285 - 302, Carnegie Institution of Washington Publication 600, Washington, D.C.

Nelson, E. (1964) Manometric Observations of Algal Endogenous Metabolism, MS Thesis, Univ. of Kansas.

Prochazka, G. J., Payne, W. J. and Mayberry, W. R. (1973) Calorific Contents of Microorganisms, *Biotech. & Bioengr.,* **15**, 1007.

Richardson, B., Orcutt, D. M., Schwertner, H. A., Martinez, C. L. and Wickline, H. E. (1969) Effects of Nitrogen Limitation on the Growth and Composition of Unicellular Algae in Continuous Culture, *Appl. Microbiol.*, **18**, 245.

Richmond, A. (Editor)(1986) *CRC Handbook of Microalgal Mass Culture*, CRC Press, Boca Raton, FL.

Chapter 6

PROTOZOA AND OTHER ANIMALS

Microscopic animals differ from microscopic plants by their ability to metabolize solid particles directly. Actually, the protozoa and higher animals hydrolyze the solid organics internally rather than externally. The animals have complex metabolic systems that allow them to metabolize the nutrients and release the inert portions of the suspended organic solids back into the environment. Microscopic animals range from single cells organisms to multicell animals that approach the macroscopic size. The microscopic animals are a part of the organic matter that forms a link in the food chain for macroscopic organisms. The larger organisms use smaller organisms as their source of nutrients. Although the microscopic animals do not metabolize waste materials, they play an important role in the organic waste stabilization process. For this reason it is important for environmental microbiologists to learn to recognize the different groups of microscopic animals and the role they play in maintaining the environmental balance in both aqueous systems and soil systems.

PROTOZOA

Protozoa are the simplest microscopic animals, being single cell organisms. In nature, bacteria form the major food supply for protozoa. The bacteria concentrate various nutrients into their protoplasm, making them the perfect food for the protozoa. A portion of the organic matter from the bacteria is oxidized to yield energy for the synthesis of new protoplasm from the remaining organic matter. The energy-synthesis relationships for protozoa are similar to the bacteria energy-synthesis relationships, 38% oxidation and 62% new cell mass. The large protozoa can also eat small algae. Most of the protozoa are aerobic, requiring dissolved oxygen as their electron acceptor. There are a few anaerobic protozoa. The problem with anaerobic protozoa is even more acute than with anaerobic bacteria. Anaerobic organisms must process considerable quantities of organic matter for energy since most of the energy remains in the partially metabolized organic compounds. Anaerobic protozoa will only be found in environments having very high organic concentrations and high concentrations of bacteria. Until more research is carried out on anaerobic metabolism in protozoa, environmental microbiologists will deal with protozoa as if they are strict aerobic microorganisms.

DESCRIPTION

Protozoa are identified entirely from their physical characteristics. Microscopic examination of the protozoa at 100 X allows observation of the major characteristics used to identify the different organisms. Protozoa are much larger than bacteria, ranging in size from about 10 μm to several hundred microns. Since the protozoa have a discernable nucleus, they are classified as *Eucarya.* There are five families of protozoa: (1) *Sarcodina*, (2) *Mastigophora*, (3) *Sporozoa*, (4) *Ciliata* and (5) *Suctoria.* Figure 6-1 shows sketches of the five families of protozoa, illustrating their major physical characteristics.

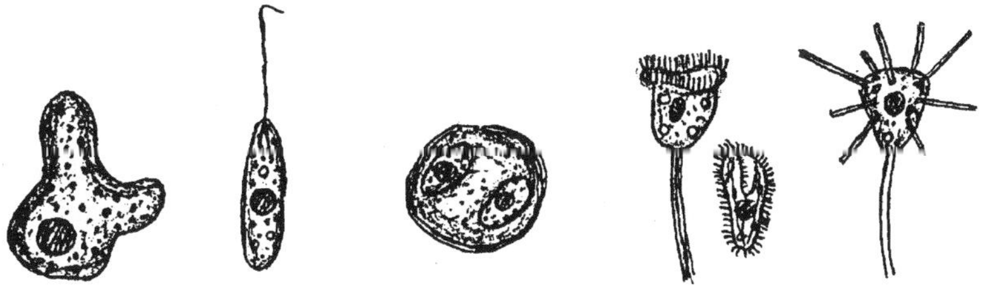

(a) *Sarcodina* (b) *Mastigophora* (c) *Sporozoa* (d) *Ciliata* (e) *Suctoria*

Figure 6-1 SKETCHES OF THE FIVE FAMILIES OF PROTOZOA

The *Sarcodina* are the simplest protozoa. They have flexible bodies and move by pseudopodia created by streaming protoplasm within the cell while attached to a surface. The *Sarcodina* must live on solid surfaces in order to move under control. If they lose contact with a solid surface, the *Sarcodina* have no control over their movements and simply drift with the fluid currents. The nucleus and food vacuoles are easily observed in the *Sarcodina*. They eat by engulfing their food. As the cell wall moves over the solid food, the food goes from outside the cell to inside the cell where it can be solubilized by enzymes. The soluble nutrients are taken inside the cell and used for energy and synthesis. The *Amoeba* is the most common *Sarcodina* and is widely distributed in the environment. Unfortunately, the *Sarcodina* are not efficient food gatherers compared to the other protozoa and are not able to compete efficiently against the other protozoa. They will be found growing on solid surfaces where bacteria and algae are attached. Some of the *Sarcodina* have the ability to create solid shells that can protect them from small predators. A few of the *Sarcodina* are pathogenic. The *Entamoeba histolytica* is one of the most famous pathogens. It was responsible for an epidemic at the 1933 Chicago World's Fair. *Entamoeba* were carried by a cross-connection in a large hotel in Chicago to a number of rooms from the sewage of a contaminated guest. The cross-connection allowed the sanitary sewage to enter the water distribution system within the hotel by mistake. The newly contaminated guests of the hotel carried the protozoa back home when they left Chicago, making it very difficult to trace the magnitude of the epidemic. One of the positive aspects of this epidemic was to focus attention on the elimination of cross-connections between sanitary sewage pipes and water distribution pipes. The *Sarcodina* form cysts when the environment becomes unfavorable. The cysts are quite similar to bacterial spores and protect the nucleus with a hard coating. When the cysts return to a favorable environment, the nucleus stimulates normal protozoa growth. Growth of *Entamoeba* inside animals results in the discharge of large numbers of cysts in feces from the infected animal. In countries where untreated sewage is applied directly to agricultural fields as a fertilizer, the cysts become attached to the crops. If the crops are eaten without adequate treatment, the cysts are ingested and grow again, allowing the cycle to continue unabated. The *Endamoeba* are parasitic pathogens, drawing all their nutrients from their hosts. The parasites sap the strength of people and reduce their ability to work. *Endamoeba* are seldom fatal, except for people who have a damaged immune system. Sewage treatment can remove the parasites from human wastewaters and break the growth cycle of this parasite. Medical treatment of the infected individual can also destroy the pathogen. Individual treatment is a more difficult and expensive way to control the spread of the pathogen in large populations than wastewater treatment.

Mastigophora are the flagellated protozoa. They have from one to four flagella that are used for motility and for gathering food. The *Mastigophora* are divided into two groups, *Phytomastigophora* and *Zoomastigophora*. The *Phytomastigophora* are flagellated protozoa that are the transition phase between bacteria and algae. Like bacteria, the phytoflagellates metabolize soluble nutrients. Because of their large size the phytoflagellates cannot compete against the bacteria and can survive only in concentrated organic environments before the bacteria begin to grow and predominate. Protozoologists consider flagellated algae as a major part of the phytoflagellated protozoa; but environmental microbiologists have kept the photosynthetic phytoflagellates with the algae and the non-photosynthetic phytoflagellates with the protozoa. Growth of bacteria permits the zooflagellates to grow since they eat bacteria for their food. The inefficiency of the zooflagellates in obtaining nutrients keeps their populations low except in very high bacteria populations. *Mastigophora* are easily recognized under the microscope because of their size and their slow undulating motion. *Mastigophora* range in size from 10 μm to over 100 μm. It is possible to see the flagella under the optical microscope.

The *Sporozoa* are parasitic protozoa that have complex life cycles. The formation of spores is the chief characteristic of the *Sporozoa. Plasmodium vivax* is the most common *Sporozoa.* It is the causative agent for malaria, one of the most common diseases around the world. Malaria is transmitted by mosquitos from person to person. The mosquito plays an important part in the growth of this parasitic protozoa, as well as in its movement in the environment. Limiting mosquito populations has provided control over this disease since it cannot be transmitted without the host mosquito.

The *Ciliata* use short cilia for motility and gathering food. They are grouped as free-swimming ciliated protozoa, crawling ciliated protozoa, and stalked ciliated protozoa. The free-swimming ciliates move very quickly and require lots of food for energy. Dispersed bacteria are the primary source of food for the free-swimming ciliated protozoa. They come in many sizes and shapes, ranging from 20 μm to 300 μm in length. *Paramecium* is the typical free-swimming ciliated protozoa that most people recognize; but the smaller *Tetrahymena* is more common, since it requires less food for survival. *Blepharisma* is an easy free-swimming ciliated protozoa to recognize because of its pink color. *Paramecium bursaria* is also interesting with the algae, *Chlorella,* growing inside the protozoa. The protozoa and algae find the relationship suitable for both organisms. *Stylonychia* and *Euplotes* are more complex free-swimming ciliates with cirri on their underneath side. The cirri allow these free-swimming ciliates to crawl over solid surfaces in search of food. The crawling ciliates require less energy than the free-swimming ciliates and survive better as food becomes limiting. The stalked ciliates are ciliated protozoa that have stalks to permit them to attach to surfaces. The cilia located near the mouth are primarily for food gathering, but can be used for motility. The stalked ciliates can be found as single cells or as colonies of cells. Some of the stalked ciliates have stalks that contract and some have rigid stalks. Stalked ciliated protozoa and crawling ciliated protozoa require the least amount of food for survival as far as ciliated protozoa are concerned. Under adverse environmental conditions, the stalked ciliated protozoa form cilia around the bottom of the cell near where the stalk is attached. The stalked ciliated protozoa can detach itself from the stalk and becomes a large, free swimming ciliated protozoa. It can become attached again when environmental conditions are suitable. Figure 6-2 shows sketches of the three different groups of ciliated protozoa.

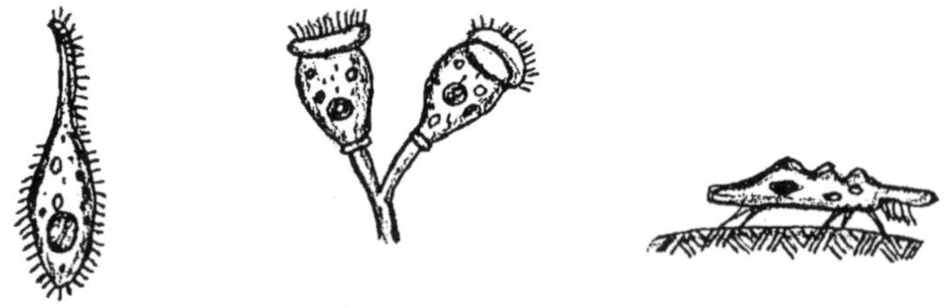

(a) Free-Swimming Ciliate (b) Stalked Ciliate (c) Crawling Ciliate

Figure 6-2 SKETCHES OF THREE MAJOR GROUPS OF CILATED PROTOZOA

Suctoria are interesting protozoa that look like stalked ciliated protozoa. Instead of having open mouths for feeding, the *Suctoria* use hollow tubes to suck their nutrients inside the cell. *Suctoria* are parasites, using free-swimming ciliates as their source of food. The *Suctoria* are more complex protozoa, having two phases in their life cycle. The stalked growth is one phase and a free-swimming ciliate is the other phase. Since the free-swimming ciliated protozoa used for nutrients are quite large, it takes considerable time for the *Suctoria* to capture and to eat a free-swimming ciliate. *Suctoria* will be observed only when there are large numbers of free-swimming ciliated protozoa in the environment. The growth of protozoa is similar to the growth of the other microorganisms. Nutrients and environment determine which protozoa grow and the extent of that growth.

METABOLISM AND GROWTH

Protozoa are primarily aerobic organisms, requiring dissolved oxygen as their electron acceptor. Although protozoa can be grown in concentrated, complex nutrient media, they prefer to use bacteria as their source of nutrients. The bacteria are concentrated nuggets of nutrients. The protozoa metabolize the biodegradable portion of the bacteria for energy and synthesis and excrete the non-biodegradable fraction back into the environment. The protozoa must continuously ingest nutrients or they will have to consume their own cell mass and die. A number of studies have been carried out over the years evaluating the relationships between the growth of bacteria and protozoa.

Butterfield, Purdy, and Theriault carried out one of the earliest studies on bacteria-protozoa metabolism in 1931. They were able to obtain a pure culture of *Colpidium*, a small, free swimming ciliated protozoa, varying in size from 50 to 70 μm in length. They used a dilute glucose-peptone solution as a growth media and quickly found that *Colpidium* could not survive in a 5 mg/L solution of each nutrient, a total of 10 mg/L nutrients, unless the media became contaminated with bacteria. Increasing the glucose-peptone concentration to 5,000 mg/L of each nutrient, 10 g/L total nutrients, provided sufficient nutrients for the *Colpidium* to grow quite well without bacteria in a batch-fed bioreactor. It took the *Colpidium* 23 days to reach its maximum population, 15,600/ml, when grown at 20°C in the concentrated nutrients. The *Colpidium* used 245 mg/L oxygen, about 0.016 μg/cell. Once the *Colpidium* reached their maximum population, they began to slowly die while using dissolved oxygen to remain alive by endogenous respiration. They used 55 mg/l DO over the next 4 days, 3.9 x 10^{-5} μg/hr/cell. It appeared that oxygen transfer limited metabolism in both the growth and the endogenous phases. Using a dilute glucose-peptone solution containing 5 mg/l of each nutrient and *Aerobacter aerogenes* as the bacteria together with *Colpidium*, they found that the bacteria metabolized the organics to new cells with the utilization of oxygen in the batch-fed bioreactor. The *Colpidium* growth lagged the bacteria growth, but quickly began to reduce the bacteria population. The protozoa reached its maximum population in 5 days incubation at 20°C. Both the protozoa and bacteria populations slowly decreased after 5 days. The numbers of *Colpidium* reached 180/ml in 5 days and dropped to 10/ml by Day 10. The bacteria population reached 6.9 x 10^6/ml after one day and was down to 0.7 x 10^6/ml by Day 10. Their study gave some additional data. The growth of *Aerobacter aerogenes* in the dilute glucose-peptone solution used 3.0 mg/l DO in 5 days incubation at 20°C. Adding the protozoa, *Colpidium*, to the *A. aerogenes* gave an oxygen uptake of 4.8 mg/L in 5 days. Using a mixture of several different bacteria in the same substrate gave 4.3 mg/L oxygen uptake. With *Colpidium* the mixture of bacteria used 5.2 mg/L DO. Finally, river water with mixed bacteria and mixed protozoa used 6.4 mg/L DO under the same conditions. These data showed that the protozoa were dependent upon the bacteria to concentrate nutrients in dilute organic environments and that mixtures of microorganisms were more efficient at metabolism than pure cultures. This is not surprising since the most efficient microorganisms grow and provide the greatest stabilization in the shortest time.

In 1972 Tsuchiya, Drake, Jost, and Fredrickson published the results of their study on the interaction of the amoeboid protozoa, *Dictyostelium discoideum*, and *Escherichia coli*. In a continuously fed bioreactor with 500 mg/l glucose as the substrate, *E. coli* metabolized the glucose and produced about 1.5 x 10^9 bacteria/ml. The amoeboid protozoa began to eat the *E. coli* and increased in numbers. The protozoa metabolism produced a major drop in the *E. coli* population and a rise in glucose concentration after several days of operation. As the *E. coli* population decreased, the protozoa population also began to decrease. The increased glucose concentration stimulated the *E. coli* to grow. The glucose concentration soon dropped quite low and became growth limiting for the *E. coli*. The protozoa found a ready supply of food again and began to grow at the expense of the *E. coli*. The oscillations in bacteria and protozoa populations eventually became damped and the system operated with a balance between substrate fed and the growth of both microorganisms. At 25°C the μ_{max} for *D. discoideum* was 0.24/hr and 0.25/hr for *E. coli*. With continuous feeding of nutrients, one would expect that the μ for both organisms growing together would be the same. The half-saturation constant, K_s, was 4×10^8 bacteria/ml for the protozoa and 0.5 mg/l glucose for the bacteria. Quantitative evaluation of the data indicated that it took 1.4×10^3 bacteria to create an amoeba and 3.3×10^{-10} mg glucose to produce a bacterium.

This same group continued their study of competition between protozoa and bacteria by examining two bacteria, *E. coli* and *Azotobacter vinelandii*. *Azotobacter vinelandii* are nitrogen-fixing bacteria that can use glucose as its substrate, the same as *E. coli*. The two bacteria are quite different in size. *E. coli* has a mean cell volume of 0.3 μm^3; while the *Azotobacter* has a mean cell volume of 3.0 μm^3. *Azotobacter* has to metabolize much more glucose to produce a single cell than *E. coli*. Size differential is a major factor affecting competition between microorganisms. The microorganisms with the greatest surface area to mass ratio have a distinct advantage over the other microorganisms. Growth of the two bacteria in a simple, continuous feed system without the protozoa resulted in the *E. coli* displacing the *Azotobacter* in a short period of time. Theoretically, the *E. coli* should not completely displace the *Azotobacter* since both can compete for the soluble substrate. The smaller, faster growing *E. coli* should have been and was the predominant bacteria. In this study, a free-swimming ciliated protozoa, *Tetrahymena pyriformis*, was used, since they were far more efficient at gathering food than the amoeboid protozoa used in the previous study. In the presence of the ciliated protozoa, both bacteria groups survived. *E. coli* could not displace *Azotobacter* when the protozoa ate the bacteria. Together, the *E. coli* had a population around 1×10^9 cells/ml with *Azotobacter* around 3×10^7 cells/ml and the *Tetrahymena* around 6×10^3 cells/ml. The faster growing *E. coli* appeared to provide most of the nutrients for the protozoa, depending upon the accuracy of the bacteria numbers and their corresponding volumes. Predator-prey relationships and competition between microorganisms for nutrients are very important in allowing the various groups of organisms to survive in the real world. An interesting reaction was observed when they attempted to grow the microorganisms at fluid retention periods greater than 15 hours. They noted that the bacteria aggregated, making it impossible to distinguish between the two organisms. Aggregation of bacteria is very important in environmental microbiology and has an important impact on bacteria survival. This study clearly demonstrated that flocculation occurred when food became limiting in the environment. In an excess of nutrients the two bacteria species remained dispersed with the protozoa.

Growth of *Tetrahymena pyriformis* on pure cultures of bacteria was the subject of a study by Holm and Smith. They used the bacteria, *Citrobacter*, which contained 8.6 x 10^{-11} mg carbon/cell. The *Tetrahymena* contained 1.1 x 10^{-6} mg carbon/cell. The free-swimming ciliated protozoa required 3 x 10^{4} bacteria to produce a single new cell. The overall metabolic efficiency was about 42%. The same year, Sudo and Aiba reported on the isolation and growth of the stalked ciliated protozoa, *Vorticella microstoma*. They used *Alcaligenes faecalis* as the source of food for the *Vorticella*. They found that the weight of *Vorticella* averaged 3.85 x 10^{-6} mg/cell and had a μ_{max} of 2.2 days. The cell yield of protozoa was about 47%, based on mass of bacteria metabolized.

The metabolic data from these studies demonstrated that the aerobic growth of the protozoa followed the same general relationships of metabolism as the other microorganisms. The difference was the bacteria supplied both the energy and the components for cell synthesis. The energy content of bacteria is less than 100% based on VSS. The net result is less synthesis of protozoa than would be expected. As large organisms, the protozoa must metabolize large numbers of bacteria to make a new cell. Fenchel reported protozoa used 50% to 60% of their nutrients for cell synthesis. The endogenous respiration rate proceeded at 2% to 5% of the normal growth rate in protozoa. The dispersed bacteria in the environment form the best source of nutrients for the protozoa; but bacteria on the surface of soil particles or on the surface of bacteria floc also can be used for nutrients.

Although the majority of protozoa are aerobic organisms, there are anaerobic protozoa. Like their bacteria counterparts, the anaerobic protozoa must eat tremendous quantities of nutrients in order to obtain sufficient energy for cell synthesis. The low bacteria growth in anaerobic environments means anaerobic protozoa will only be found in high organic concentration environments. Fenchel and Finlay reported that anaerobic protozoa had an overall yield of about 10% of their nutrients. They indicated that there were anaerobic protozoa with methane bacteria growing inside the protozoa. As the protozoa produced organic acids, the methane bacteria converted the organic acids to new cells and methane gas. The methane metabolism removed potentially toxic organics from the protozoa and supplied the protozoa with additional nutrients. There are both flagellated and free-swimming ciliated, anaerobic protozoa with the predominant numbers being ciliated protozoa. The free-swimming ciliated protozoa predominate over the flagellated protozoa in anaerobic environments for the same reason that they predominate in aerobic environments. The free-swimming ciliated protozoa are simply more efficient in capturing bacteria for food than the flagellated protozoa. Because of the limited environments for anaerobic protozoa, protozoa were considered as being strictly aerobic for many years. More research is definitely needed on anaerobic protozoa to establish their relationships in the environment.

POPULATION DYNAMICS

In the natural environment the different groups of protozoa compete for nutrients. All of the major groups of protozoa will be found living together in numbers proportional to their ability to obtain nutrients. The natural environment is more dynamic than static and does not allow a static equilibrium to exist for any extended period of time. The addition of organic nutrients to the aquatic environment stimulates the growth of bacteria best equipped to metabolize the specific organic compounds. The growth of the bacteria will be aerobic as long as there is sufficient dissolved oxygen in the water. If the bacteria remove the dissolved oxygen, metabolism shifts from aerobic to anaerobic. Since most protozoa will not grow under anaerobic conditions, there will be no significant growth of protozoa until the rate of bacteria metabolism slows and the system becomes aerobic again. Figure 6-3 is a schematic diagram of the population dynamics of microbial growth in a batch fed system following the addition of organic nutrients to stimulate the bacteria. Both the time scale and the numbers of

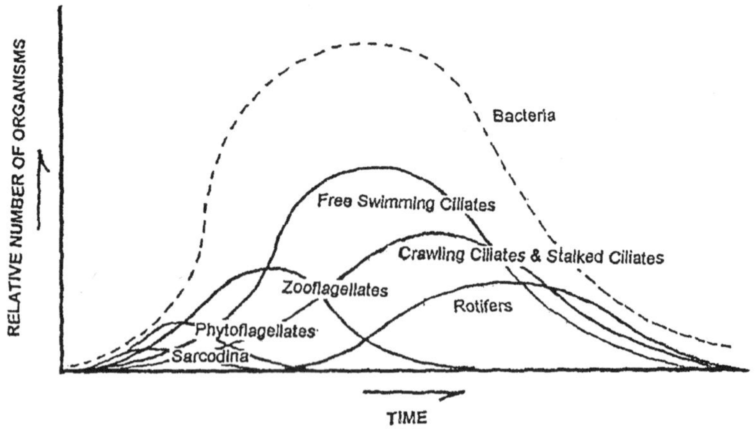

Figure 6-3 SCHEMATIC DIAGRAM OF POPULATION DYNAMICS OF MICROBIAL GROWTH IN A BATCH FED SYSTEM

organisms are distorted to show relative growths of the different groups of organisms. Small, flagellated protozoa will appear first since there are many bacteria to eat. Larger flagellated protozoa appear next as sufficient food is available growth. The amoeboid protozoa appear on solid surfaces where there are attached bacteria for them to use as food. The amoeboid protozoa do not approach the numbers of the flagellated protozoa as the amoeboid protozoa are not as efficient food gatherers as the flagellated protozoa. The small ciliated protozoa appear quickly as the DO rises. The small, free-swimming ciliated protozoa move quickly through the solution, harvesting as many bacteria as possible. The numbers of free-swimming ciliated protozoa increase rapidly with larger species appearing in smaller numbers. The numbers of bacteria decrease as the ciliated protozoa grow. If there are enough small free-swimming ciliated protozoa, a few *Suctoria* will appear. As the bacteria decrease, the free-swimming protozoa give way to the crawling ciliated protozoa that find their food on solid surfaces. The crawling ciliated protozoa are more efficient food gatherers than the amoeboid protozoa at finding the bacteria on the solid surfaces. Stalked ciliated protozoa appear when the bacteria numbers drop lower and lower. The free-swimming protozoa rapidly decrease as they expend too much energy trying to find enough bacteria to remain alive. The stalked ciliated protozoa require many bacteria to grow; but they expend less energy in obtaining those bacteria. As the bacteria population drops to very low levels, the stalked ciliates die off.

REPRODUCTION AND SURVIVAL

Protozoa undergo reproduction by fission, splitting into two cells along the longitudinal axis. Division starts with the nucleus splitting and creating the basis for two separate cells. It takes several hours for the two cells to completely split. Growth continues as long as environmental conditions are favorable. When environmental conditions begin to turn bad for continued growth of the protozoa, they form cysts. Each cyst is produced by coating the nucleus with a hard shell, allowing the nucleus to survive in adverse environments. The rest of the cell tissues become nutrients for additional bacteria growth. When the cyst finds a reasonable environment for growth, the nucleus begins to expand, creating a new protozoa.

Environmental factors such as pH and temperature have the same relative effect on protozoa as on bacteria. Protozoa grow best at pH levels between 6.5 and 8.5. Strongly acidic or strongly alkaline conditions are toxic to the protozoa. As far as temperature is concerned, protozoa can be either mesophilic or thermophilic, the same as bacteria. Most protozoa are mesophilic, having a maximum temperature for growth around 40° C. Fenchel indicated that a few protozoa have been found in hot springs at 50° C. There do not appear to be many thermophilic protozoa. Part of the problem is the lower solubility of oxygen at higher temperatures. Protozoa change their rate of metabolism by a factor of two for each 10° C temperature change, the same as the other organisms. Protozoa have difficulty surviving at temperatures below 5° C because the viscosity of the water increases, making it more difficult for the protozoa to move and obtain food.

ROTIFERS

Rotifers are multicellular, microscopic animals with flexible bodies. They are larger than protozoa and have complex metabolic systems. Like the other microscopic animals, the rotifers prefer bacteria as their source of food, but can eat small algae and protozoa. The rotifers have cilia around their mouths to assist in gathering food. The cilia also provide motility for the rotifers if they do not remain attached to solid particles with their forked tails. The flexible bodies allow the rotifers to bend around and feed on bacteria and algae attached to solid surfaces. A typical rotifer is shown in Figure 6-4. *Philodina* is one of the most common rotifers. It is about 400 μm long, making it easy to see under the microscope at 100X magnification. The cilia give the appearance of two rotating wheels at the head of the rotifer. *Epiphanes* is a large rotifer, reaching 600 μm in length. Some rotifers are as small as 100 μm. Rotifers are all strict aerobes and must have several mg/L dissolved oxygen in order to grow. They can survive for several hours in low DO environments, but not for long periods. In the presence of large bacteria populations and adequate DO the rotifers will quickly eat most of the bacteria, even if the bacteria are flocculated. In a suitable environment the rotifers can quickly metabolize all the bacteria and then starve to death. Excessive growth of rotifers can be controlled by reducing the dissolved oxygen to prevent them from growing so rapidly. The DO can be reduced to around 1.0 mg/L to favor the metabolism of aerobic bacteria and protozoa and slow the growth of rotifers. As large, complex

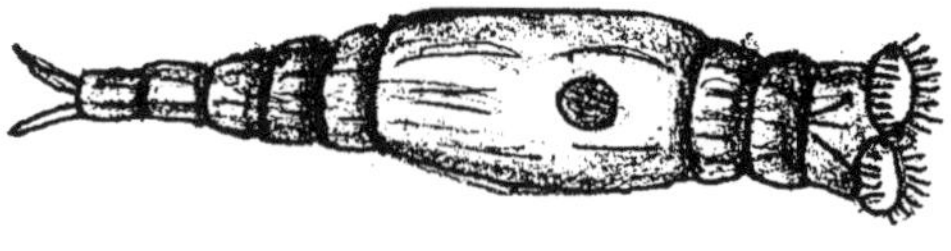

Figure 6-4 SCHEMATIC DIAGRAM OF A TYPICAL ROTIFER

organisms, rotifers require lots of bacteria in their growth. Rotifers can remove the bacteria attached to solid surfaces and can ingest small, flocculated masses of bacteria. They are more sensitive to environmental stresses than either bacteria or protozoa. Temperature affects rotifers the same as temperature affects the other microorganisms. Their metabolism slows as the temperature decreases and increases as the temperature rises. There do not appear to be any thermophilic rotifers. Reproduction in rotifers occurs through egg formation rather than by binary fission. Rotifer eggs can remain dormant for a considerable period of time if environmental conditions are not satisfactory for growth. It has been difficult to study the quantitative growth characteristics of the rotifers since they cannot be grown free of bacteria.

Recently, Walz reported on some studies using the rotifer, *Brachionus angularis*, when grown on algae. By using algae as the food for the rotifers, it was possible to minimize extraneous organic compounds that would affect the growth of the rotifers. The algae were grown entirely on inorganic components in the media, using light as the source of energy for the algae. One and two stage chemostats were used to evaluate the rotifer growth. It was found that the rotifers grew very nicely as long as the liquid displacement time was under 0.31/d. When the rate of dilution reached 0.34/d, the rotifers were washed out of the system and the algae population increased. The rotifers had to consume 20% of their body weight each day, just to remain alive without reproducing. Maximum growth occurred when the rotifers consumed 70% of their body weight each day. About 43% of the algae consumed were converted to cell mass. The rest of the algae cell mass metabolized was oxidized for energy as well as excreted as inert waste products. Each rotifer averaged 0.056 μg C/cell. More research could yield a better understanding of the overall metabolic relationships of rotifers.

Rotifers play an important role in the overall food chain from bacteria and algae to higher organisms. They are widely found in the aquatic environment where there is a suitable environment for growth. Rivers, lakes and reserviors are good sources of rotifers. The environments that favor rotifers tend to favor other higher animal forms.

CRUSTACEANS

Crustaceans are multicellular animals with hard shells to protect their bodies. They also have jointed appendages attached to their bodies. The appendages assist in movement and food gathering. The large size of the crustaceans, 1.5 to 2 mm, makes them visible to the naked eye if one looks very carefully. They also appear quite large in the microscope, requiring low power magnification for good observation. Being more complex than the rotifers, they grow slower and are more sensitive to environmental changes. The crustaceans feed on bacteria, algae, protozoa, and solid organic materials. Figure 6-5 illustrates the *Daphnia* and the

Cyclops, two common crustaceans. They are easily found in freshwater lakes in the warm summer months. They require high levels of DO and a moderate level of nutrients. It has been estimated that *Daphnia* require about 80% of their body

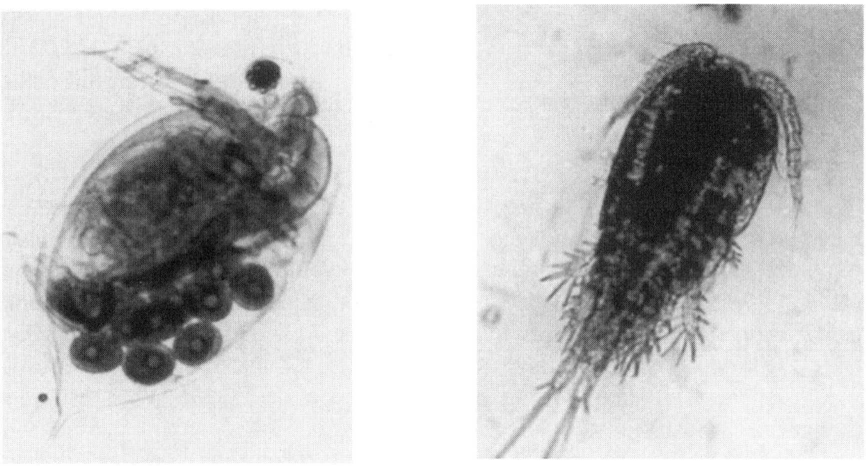

a. *Daphnia* b. *Cyclops*

Figure 6-5 PHOTOMICROGRAPHS OF TWO COMMON CRUSTACEANS

weight each day for maximum growth. Only about 20% of the food consumed ends up as cell mass. The larger mass of the *Daphnia* requires a considerable number of smaller organisms to remain alive and to grow. Since the *Daphnia* are relatively large, they become food for macroscopic organisms in the water environment. The *Daphnia* shown in Figure 6-5 is carrying at least nine eggs. As the microorganisms become larger, they have a harder time obtaining sufficient food to sustain large populations. The crustaceans tend to appear during the warm spring and fall weather, when the algae populations bloom. The crustaceans can completely remove all the algae in a given sector of a lake and then starve for a lack of food. Sustained populations of crustaceans require the continued addition of a suitable food source. The presence or absence of sufficient concentrations of trace metals in the bacteria or algae, used as their food source, also affects the magnitude of growth of the different species of crustaceans. The Federal EPA has proposed the use of *Ceridaphnia* as the indicator organism for effluent toxicity from wastewater treatment plants. Unfortunately, *Ceridaphnia* is a very sensitive crustacean that can be difficult to maintain in the laboratory for routine use. Researchers are currently examining other *Daphnia* in an effort to find a suitable crustacean that is both sensitive to toxic substances and easy to maintain in the laboratory.

NEMATODES AND OTHER WORMS

Microscopic worms are not common in the natural environment. Most of the worms are macroscopic and will not be discussed. The simple microscopic round worms are nematodes. Like other higher animals the nematodes feed primarily on bacteria and small organic particles. A typical nematode is shown in Figure 6-6. Nematodes range in size from 1,000 to 2,000 μm, making them easy to observe

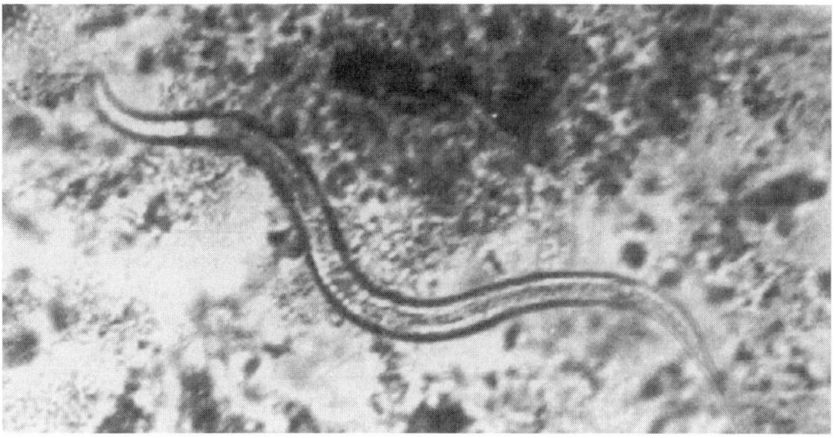

Figure 6-6 PHOTOMICROGRAPH OF A TYPICAL NEMATODE

under the microscope at low power. Nematodes have complex digestive systems that are readily apparent under the microscope. One of the characteristics of nematodes is a constant thrashing motion trying to breakup small particles for use as food. The nematodes require very large quantities of food to survive and will only be found where adequate nutrients and a reasonable environment exists. The female nematodes are responsible for egg production and reproduction of the species. The nematode eggs are well protected by a tough covering of chitinous material. Nematodes require a moderate level of DO to survive. *Diplogasteroides* is one of the common nematodes found in aqueous systems. Nematodes can also be found in soil systems where they tend to be plant parasites. Since nematodes tend to grow best near the soil surface, they are easily washed into natural watercourses. Environmental microbiologists become quite familiar with nematodes.

Other microscopic worms found in the aqueous environment include the bristle worm with large orange spots, making them easy to identify. The bristles projecting along the entire length of the body of the worm makes it recognizable under the microscope. Bristle worms range in size from 3,000 to 7,000 μm in length and are hard to classify as microscopic. They cannot be viewed in their entirety in a single field except at very low power magnification. Bristle worms are not as common as nematodes as they have a more complex metabolic system, requiring more nutrients.

There are many other worms that tend to approach macroscopic size. *Tubifex* is a red worm that is found in sludge deposits. It is definitely macroscopic. There are also many different worm-like larvae that have been observed in trickling filter slimes in wastewater treatment plants. These worms and larvae are important in environmental wastewater treatment systems but are of limited importance in environmental microbiology. Simple recognition of the organisms in samples is usually all that is needed.

ENVIRONMENTAL CONCERNS

The microscopic animals play a dual role in environmental concerns. The major concern lies with the pathogenic protozoa. *Entamoeba histolytica* is no longer a serious problem in the United States, but still remains a problem in many developing nations of the world lacking in modern sanitation. Public health problems have been created in the developed countries by two protozoa, *Giardia lamblia* and *Cryptosporidium parvum.* Pathogenic protozoa tend to be transmitted through drinking water as spores or oocysts. The federal EPA has developed regulations to control transmission of all types of pathogenic microorganisms through public water systems.

Most protozoa play a positive role in wastewater treatment systems by eating the dispersed bacteria and producing a clarified effluent. The success of aerobic wastewater treatment systems depends upon the balanced population of bacteria and protozoa. Protozoa and higher animals contribute to maintaining high water quality in our streams and lakes. The types of protozoa found in various environmental systems are excellent indicators of the current health of those systems. It is essential for environmental microbiologists to have an understanding of the important microscopic animals and their contributions to the success of our environmental pollution control projects.

THINGS TO REMEMBER

1. Protozoa are single cell microorganisms.

2. Protozoa use bacteria as their primary source of food.

3. Protozoa metabolism is similar to bacteria metabolism.

4. Most protozoa are aerobic.

5. Anaerobic protozoa can only grow at high organic food levels.

6. Protozoa form cysts for survival under adverse environments.

7. Rotifers are multicellular microscopic animals.

8. Rotifers are strict aerobes, requiring dissolved oxygen for metabolism.

9. Rotifers reproduce by eggs rather than by simple binary fission.

10. Crustaceans are large, multicellular, microscopic animals with hard shells.

11. Crustaceans can eat bacteria, algae, and protozoa.

12. Nematodes are common microscopic roundworms.

13. Nematodes eat large numbers of bacteria for growth.

14. A few protozoa are pathogenic; but most protozoa help keep the environment clean.

REFERENCES

Anderson, O. R. (1987) *Comparative Protozoology*, Springer-Verlage, Berlin.

Butterfield, C. T., Purdy, W. C. and Theriault, E. J. (1931) Experimental Studies of Natural Purification in Polluted Waters IV. The Influence of the Plankton on the Biochemical Oxidation of Organic Matter, *Public Health Reports*, **46**, 393.

Drake, J. F. and Tsuchiya, H. M. (1977) Growth Kinetics of *Colpoda steinii* on *Escherichia coli*, *Appl. Environ. Microbiol.*, **34**, 18.

Fenchel, T. (1987) *Ecology of Protozoa*, Science Tech, Madison, WI.

Holm, H. W. and Smith, F. A. (1970) *Effects of Protozoa on the Fate of Particulate Carbon*, EPA-660/3-73-007.

Jost, J. L., Drake, J. F., Fredrickson, A. G. and Tsuchiya, H. M. (1973) Interactions of *Tetrahymena pyriformis*, *Escherichia coli*, *Azotobacter vinelandii* and Glucose in a Minimal Medium", *J. Bacteriol.*, **113**, 834.

Lee, J. J., Hutner, S. H. and Bovee, E. C. (1985) *An Illustrated Guide to the Protozoa*, Society of Protozoologists, Lawrence, KS.

Sleigh, M. A. (1989) *Protozoa and Other Protists*, 2nd Ed., Edward Arnold, London, England.

Stone, A. R., Platt, H. M. and Khalil, L. F. (1983) *Concepts in Nematode Systematics*, Academic Press, New York.

Sudo, R. and Aiba, S. (1973) Mass and Monoxenic Culture of *Vorticella microstoma* Isolated From Activated Sludge, *Water Research*, **7**, 615.

Tsuchiya, H. M., Drake, J. F., Jost, J. L. and Fredrickson, A. G. (1972) Predator-Prey Interactions of *Dictyostelium discoideum* and *Escherichia coli* in Continuous Culture, *J. Bacteriol.*, **110**, 1147.

Walz, N. (Editor) (1993) *Plankton Regulation Dynamics*, Springer-Verlag, Berlin.

Chapter 7

SOIL MICROBES

Soil is an interesting medium for growing microorganisms. It contains various nutrients that the microbes need for their metabolism. Unfortunately, the nutrients are not always readily available. Soil is not a *homogeneous* material, but rather is a *heterogeneous* material. Soil consists of ground rock particles, clay particles, and silt particles in various combinations and with varying amounts of water. The solid particles form layers or clumps with open void spaces between the different soil particles. Microorganisms cannot penetrate the solid soil particles. They can live on the surface of the particles, provided there are adequate nutrients and water for their growth. Organic matter discharged from living organisms on the soil surface provides the major nutrients for the microorganisms in the soil. The upper layer of soil provides the best environment for growing microorganisms. Bacteria, fungi, actinomycetes, algae, protozoa and nematodes can be found in organic rich soil. The soil environment determines the specific microbes that can grow and their numbers. The seasons of the year produce the most important environmental variables with both temperature and moisture changes.

SOIL CHARACTERISTICS

Engineering courses dealing with soils are concerned more with the physical characteristics of soils than with their chemical characteristics. Both the physical and the chemical characteristics of soil are important in developing a proper understanding of soil microbiology. Natural forces such as wind, rain, freezing, thawing and pressure help form soil. Rocks are slowly broken into small particles, 50 μ to 1 mm, known as sand. As the particles decrease in size to between 2 μ and

50 μ, they are called silt. Particles less than 2 μ in size are known as clay. All of these particles are essentially quartz, SiO_2, with various chemical contaminants. The primary chemical contaminants include iron, aluminum, calcium, magnesium, potassium and sodium. Since potassium and sodium are monovalent cations that are more soluble than the divalent calcium and magnesium cations or the trivalent iron and aluminum cations, soil particles contain lower sodium and potassium concentrations than the other cations.

As soil particles decrease in size, they are more chemically reactive. The three common clays: kaolinite, montmorillonite and illite have layers of silica joined by aluminum, iron, magnesium and potassium and their hydroxides. Kaolinite clays consist of layers of silica and aluminum hydroxide. Montmorillonite has water layers between sheets of aluminum hydroxide and silica. Loss of the water during dry weather causes the montmorillonite to shrink. When wetted with water, the dehydrated montmorillonite expands. Illite has potassium ions binding the layers of silica. The clays can change properties by ionic displacement of the active cations. Other soil particles include mica (a potassium oxide-aluminum oxide-silicate combination), calcite (calcium carbonate crystals), gypsum (calcium sulfate) and hematite (ferrous oxide).

Organic matter also contributes to the structure of soil. Organic matter deposited on the soil surface undergoes biological decomposition by various groups of microorganisms, eventually forming humus. Humus is complex organic material that is relatively stable, accumulating in the upper layer of soil particles. Humus is dark brown in color and consists of polymers of benzene and oxygen in ether linkages. Humus has carboxyl radicals that allow it to react as an organic acid. The polymers in humus also have some nitrogen links. The various hydrophilic groups allow humus to attract and hold moisture. Humus gives soil some of its structural properties and is important in the cultivation of plants. Overall, soil is a mixture of relatively inert, inorganic particles and organic humus.

The upper layers of soil contain about 50% solid matter and 50% void space. The solid soil particles provide the structure for soil. As forces are applied to the soil surface, the forces are distributed through the solid particles to keep the soil structure from collapsing. Application of concentrated forces on the soil surface causes the soil to compact with a reduction in void space. Moving down through the soil, larger particles are encountered. Eventually, solid rock is reached. There are many places in the world where solid rock is at the ground surface and there is no normal soil. Over time, the environment is continuously wearing down the solid rock and producing smaller particles that eventually become soil.

PRECIPITATION

Precipitation is one of the main forces helping to create and change soil. Rainfall is the most common form of precipitation. Snow, sleet and hail are different forms of precipitation. Precipitation varies widely across the world's landmasses both as to time and intensity. Precipitation determines which areas of the world can sustain life and which areas cannot.

The hydrologic cycle determines how water moves through the environment. Energy from the sun evaporates large quantities of water from the oceans. As the water vapor rises in the atmosphere, it cools and condenses into clouds. Atmospheric pressure differences create winds that move the clouds over land masses. When the clouds move into colder temperatures, the moisture condenses into rain droplets and falls to the ground. Some areas receive considerable rain on an annual basis and produce large quantities of plants that can be used as a source of food. Other areas do not receive significant precipitation, creating desert areas with little vegetation. Desert areas cannot support much life unless water is imported from the wet areas. Figure 7-1 illustrates the various aspects of the hydrologic cycle.

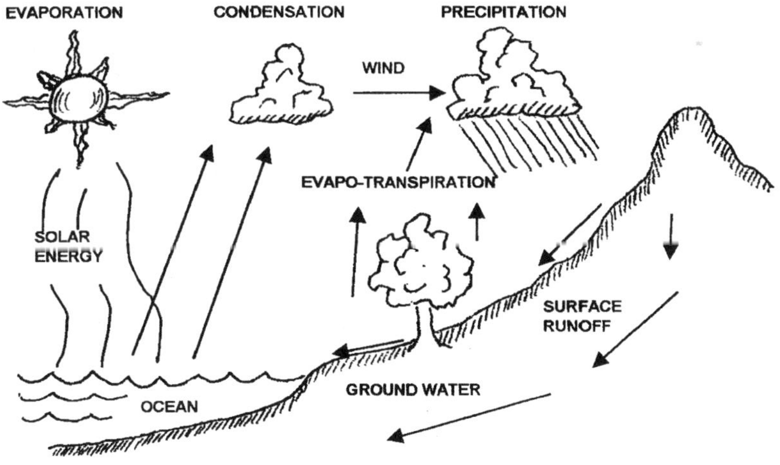

Figure 7-1 SCHEMATIC DIAGRAM OF THE HYDROLOGIC CYCLE

In regions near the north and south poles the seasons vary widely over the course of a year. During the winter period, temperatures drop and precipitation occurs as snow, sleet or hail. If the ground surface has sufficient heat, the frozen precipitation melts and acts the same as rain. If the ground surface temperature is below freezing, the precipitation remains frozen and accumulates on the ground surface until the temperature rises above freezing. Except in areas very close to the north and south poles, the accumulated frozen precipitation melts and becomes fluid, the same as rain.

When precipitation is formed, the water droplets come into contact with the various gases and particulates in the atmosphere. The water absorbs some of the gases and collects the particulates, cleansing the air of contaminants. These contaminants move with the water and react with the particles in the soil as well as with the plants growing in the soil. The initial precipitation is more contaminated than the later precipitation that falls in air already cleansed of contaminants. Precipitation is nature's way of cleaning the air. The lack of alkalinity in pure precipitation allows the absorbed carbon dioxide from the atmosphere to produce carbonic acid and depress its pH. Industrial discharges of nitrous oxides and sulfur dioxide into the atmosphere can cause the pH of precipitation to become quite acidic. Collection of alkaline soil particles in precipitation causes the pH to increase. Chemical analyses of precipitation can provide considerable data as to its contact with industrial pollution and natural contaminants. Once precipitation reaches the soil surface, it collects varying quantities of contaminants, depending on the soil characteristics.

Precipitation that reaches the ground surface is partially evaporated by the heat of the sun, returning back to the atmosphere as water vapor. Part of the precipitation runs into the soil void spaces. Gravity pulls the water down through the interconnecting void spaces in the soil. Plants, growing on the surface of the soil remove some of the water as it moves past their root structures. Plants obtain many of their nutrients from the water in the soil. As the water moves through the plants, it carries nutrients to the various cells. The water taken up by the plants is returned to the atmosphere by transpiration from the plant leaves. The water that cannot enter the soil void spaces moves across the soil surface by gravity to rivers and lakes. If the velocity of water flow is very high, the water can pickup soil particles and move them into the rivers and lakes. Soil movement by rapid surface runoff results in erosion of the soil and loss of plant nutrients. By modifying the soil surface to reduce the velocity of surface runoff, loss of soil can be prevented. This is the basis of most agriculture soil conservation programs.

As rain begins, water runs into the void spaces and fills the voids. Continued rain results in most of the water running off over the soil as *surface water*. When the rain stops, the water drains out of the void spaces into the lower soil layers. The soil particles remain coated with water as a result of surface tension forces. This attached water determines the moisture content of the soil and reduces the void space available for air. As the sun heats the soil surface, the attached water is slowly evaporated, increasing the void space and reducing the soil moisture content. Since the sun's heat cannot penetrate very deep, the lower soil layers retain moisture for a longer period of time than the upper layers of the soil. The soil with attached water and air void spaces is considered *unsaturated*, as far as water is concerned. The water moves downward by gravity until it reaches an impervious layer that prevents it from moving deeper. The water builds up over the impervious layer, creating a *saturated* water zone. The increasing water level above the impervious layer creates a pressure that causes the water to move along the path of least resistance. Accumulation of a large layer of water over the impervious layer results in an *aquifer* that can be used as a source of *groundwater*. Groundwater can be tapped by placing a well into the saturated water zone and pumping the water to the land surface where it can be used. The key to successful use of groundwater is having greater recharge than pumpage. A porous media structure is required for the water to move easily as the hydraulic gradient is created. Dense soil particles create resistance to the flow of water through the aquifer, limiting their value as water supplies except for individual homes.

Some aquifers are located between two impervious layers. The recharge area is located at the soil surface in a coarse media that eventually becomes sandwiched between two impervious layers. The water pressure in the aquifer can increase with the slope of the terrain. Drilling a well into the subsurface aquifer allows the water to rise in the well pipe in proportion to the water pressure and the resistance of the water pipe. When the water pressure is sufficient, the water may actually flow out the top of the pipe without being pumped. In other cases the water may rise up into the pipe, requiring only a little energy to pump the water to the surface. These pressurized aquifers are called *artesian* aquifers. Excessive removal of water from an artesian aquifer will result in loss of pressure, requiring more work to pump the water to the ground surface.

Aquifers have gravel and rock media to minimize the resistance to water flow. In areas with large limestone deposits the water moves through fractures in the limestone. Carbon dioxide in the water can react with the limestone to produce soluble calcium bicarbonate. Over time the loss of calcium carbonate from the limestone leaves an ever-increasing hole. Many large caverns have been formed in limestone regions as a result of water slowly dissolving the limestone. Clay soils do not make good aquifers, as the particles are too small for rapid water movement. Many clay soils swell when wetted, reducing the void spaces between particles. Clay soils can create impervious layers that prevent the passage of water. The groundwater will move across the clay surface until it reaches the end of the clay deposit and moves downward again. In some aquifers the water is trapped in small pockets. Recharging aquifers is essential for continued use of groundwater as water supplies.

Groundwater continues to move slowly until it comes to the soil surface and becomes surface water or it reaches the ocean under the surface. The point where the groundwater comes to the soil surface is called a *spring*. Springs can be classified as *continuous* or *intermittent*, depending on the flow characteristics. Spring water quickly becomes surface water. Surface waters that collect in streams and rivers move more rapidly than groundwater, as it flows back to the ocean. The large volumes of water that are continuously recycled through the environment have created many major rivers that drain large land masses. The major rivers form inland waterways that stimulate economic development to sustain the areas through which they flow. These waterways are used for water supplies as well as for waste disposal. They are used for transportation and recreation. Agriculture and industries find the major rivers very attractive. It is not surprising that these rivers carry large amounts of the suspended soil and dissolved chemicals from the drainage basins into the oceans. The heavy silt particles settle out, creating deltas as they enter the oceans and slow their velocity. The dissolved chemicals accumulate and became part of the ocean salts. With continuous evaporation of water from the oceans, the soluble salts slowly increase their concentrations.

MICROORGANISMS

Moisture, nutrients and particulate surfaces are essential for the growth of microorganisms in the soil. The gaseous atmosphere in the void spaces is also important in microbial growth. The majority of microorganisms in soil are aerobic, requiring unsaturated conditions. Only anaerobic organisms grow in soils saturated with water.

BACTERIA

The wide diversity of metabolism, together with the small size of bacteria permits them to grow in soils that contain significant amounts of nutrients. The surface of the soil screens out most particulates, allowing only small colloidal particles and soluble nutrients to pass with the water into the soil voids. Plowing the soil surface can mix large organic particles into the soil and create an environment for microbial growth to the depth of the plow. Plant roots penetrate the soil, seeking water and nutrients for growth. These plant roots provide additional surfaces for microbial attachment and subsurface organics for growth. Because the environment changes rather drastically in the soil, sporeforming bacteria tend to be most common. When environmental conditions become too difficult for normal growth, the bacteria form spores and remain dormant until the environment returns to proper conditions for normal growth. The diversity of nutrients found in the soil stimulates bacteria that can completely metabolize the organics to carbon dioxide, water and cell mass under aerobic conditions. Alexander found *Pseudomonas, Bacillus, Arthrobacter, Clostridium, Achromobacter, Micrococcus* and *Flavobacterium* as the most common bacteria in organic soils. Nitrifying bacteria are also widely distributed in soils since the decomposition of proteins releases ammonia nitrogen, providing the *Nitroso-* bacteria with food for their metabolism. Conversion of ammonia nitrogen to nitrite nitrogen provides nutrients for the *Nitro-* bacteria, allowing them to oxidize the nitrite nitrogen to nitrate nitrogen. The production of nitrous acid assists in the dissolving of minerals in the soil. If the nitrous acid is not neutralized, the low pH will limit further bacteria metabolism, preventing oxidation of the nitrite ions to nitrate ions. Other inorganic metabolizing bacteria include the sulfur oxidizing bacteria, *Thiobacillus,* and several different iron oxidizing bacteria. The sulfur oxidizing bacteria generate sulfuric acid that also dissolves minerals in the soil. The normal organic metabolizing bacteria and the nitrifying bacteria do not compete for nutrients, but do compete for oxygen. Many of the soil bacteria are facultative with the ability to shift from aerobic metabolism to anaerobic metabolism as the dissolved oxygen is depleted.

The air in the void spaces in the unsaturated zone contains a finite amount of oxygen. This oxygen is in equilibrium with the dissolved oxygen in the water layer attached to the surface of the soil particles. As the bacteria metabolize the organic matter in the soil, the dissolved oxygen in the water is reduced and more oxygen is transferred from the void space into the water. Over time the oxygen in the air is reduced, slowing the rate of oxygen transfer into the water. Near the ground surface the void spaces are in direct contact with atmospheric air that can continuously replenish the oxygen as fast as it is used. Thus, the bacteria near the ground surface are always metabolizing aerobically, unless they are saturated with water. Moving deeper into the soil, gas exchange with the atmosphere becomes more difficult. Depletion of the oxygen results in a pressure decrease; but the production of carbon dioxide results in a pressure increase. The net effect is dependent on the reaction of the carbon dioxide with minerals in the soil. When the carbon dioxide reacts with soil minerals, very little carbon dioxide will be released as a gas to replace the oxygen that was used. With a pressure differential in the void space, gas will move to produce equal pressure. Without a chemical reaction between the carbon dioxide and minerals in the soil, the carbon dioxide will be released as a gas into the void space. In this case, there will be no significant pressure difference to cause the gases to flow in any direction. Molecular forces will cause the gases to move at a very slow rate, not allowing the void spaces at the lower levels to maintain aerobic conditions. Water movement through the void spaces is the primary mechanism for renewing the oxygen in the void spaces. As the water enters the voids and fills the voids, the gases are compressed and form bubbles that are carried downward until they reach pressure equilibrium. As the rain slows and the water pressure decreases, the bubbles may move upward towards the soil surface. Eventually, the bubbles rise to the soil surface. Atmospheric air moves into the new void spaces at the surface, replenishing the oxygen.

Carbon dioxide produced by aerobic metabolism is more soluble in water than oxygen. Once the water attached to soil is saturated with carbon dioxide, the excess carbon dioxide is discharged into the void spaces. Carbon dioxide in water forms carbonic acid that reacts with minerals in the soil to form bicarbonate salts. Carbonic acid reacts with insoluble calcite particles to form calcium bicarbonate that is soluble and moves with the successive flow of water past this area of soil. The insoluble calcite crystals in the soil will slowly disappear over time, leaving larger void spaces. Water moving through the soil is now more mineralized. Having picked up calcium ions, the water is considered to be *harder* than it was before the calcium ions were solubilized. Ammonia nitrogen reacts very readily with carbonic acid, forming ammonium bicarbonate. Magnesium and iron also react to form soluble bicarbonate salts. Thus, microbial metabolism in the soil affects the mineral quality of the groundwater that is produced by successive infiltration of precipitation. The geology of the rocks in any given area determines the potential for minerals to be dissolved by the groundwater. Fortunately, the dissolution of minerals is slow and passes without much notice except for the minerals appearing in the groundwater.

Bacteria metabolism results in the production of new cell mass in proportion to the organic matter metabolized. The new bacteria are retained in the water layers attached to the surface of soil particles. Surface tension forces hold the water layers on the surface of the soil particles. As the bacteria age and lose motility, they use their pili to become attached to the soil particle surfaces. Over time, microbial growth may completely cover the soil particle. The force of water moving through the void space may force some of the dispersed bacteria deeper into the soil environment. Unfortunately, the deeper environment is not as suitable for sustaining biological life as the upper environment. The bacteria die off and become part of the humus material in soil. It takes a long time for the humus material to accumulate in the deeper layers. Humus materials form more rapidly at the soil surface because of the greater amounts of available organics and more rapid metabolism. The bacteria continue to grow as long as the environment permits. Unfortunately, the bacteria located next to the soil particle surface cannot obtain sufficient nutrients and shift entirely to endogenous respiration as other bacteria grow over them. Slowly, the bacteria die and become part of the inert organics in the soil, helping to retain water and hold the particles together. If the bacteria are unable to find suitable electron acceptors for endogenous respiration, they simple remain in an inert state until a suitable electron acceptor becomes available.

Nitrogen fixing bacteria are one of the more interesting groups of bacteria found in soil. These bacteria have the ability to take nitrogen gas from the atmosphere and fix it into ammonia nitrogen in their cell proteins. Unfortunately, nitrogen fixation requires considerable energy and occurs only when ammonia, nitrite, or nitrate nitrogen is not available in the soil. *Rhizobium* is a symbiotic bacteria that fixes atmospheric nitrogen when grown in conjunction with legumes. *Rhizobium* forms nodules on the root structure of the legumes. Air in the soil around the roots supplies the nitrogen gas for the bacteria; while the legume supplies carbohydrate energy for the bacteria. When the *Rhizobium* cells die and undergo lysis, ammonia nitrogen is released and made available to the legumes for their growth. At the end of growing season, the legumes are plowed into the soil where they are decomposed by bacteria and fungi. The protein nitrogen from the legumes is released as ammonia nitrogen for use by crops the next growing season. The use of legumes is an important method for replenishing nitrogen removed by other crops. *Clostridium* and *Azotobacter* are two free-living, nitrogen-fixing bacteria that are found in the soil. *Clostridium* are anaerobic bacteria that fix atmospheric gas in the absence of dissolved oxygen. *Azotobacter* are aerobic bacteria that can fix atmospheric nitrogen in the presence of adequate dissolved oxygen. Nitrogen fixing bacteria are very important microorganisms in nitrogen deficient soils.

Water saturated soil cannot obtain much oxygen from the atmosphere and quickly becomes anaerobic. Swamps tend to be formed when the soil is continuously saturated. Facultative bacteria will be readily found in saturated soils, but their metabolism will be slow under anaerobic conditions. Strict anaerobic bacteria will grow over time. If the environment is satisfactory, sulfate-reducing bacteria will produce sufficient hydrogen sulfide to reduce the O-R-P of the soil to a point where the methane-producing bacteria can grow. The methane produced under these conditions has been called *swamp gas*. Anaerobic environments tend to make many heavy metals soluble as salts of organic acids. On the other hand, the hydrogen sulfide produced by sulfate reducing bacteria can convert the soluble heavy metal salts to insoluble metal sulfides and prevent their movement in the water. It is not surprising that many heavy metals are found in soil as insoluble sulfide salts. The solubilities of the metallic sulfide salts range from 10^{-16} for MnS to 10^{-53} for HgS.

Data on the numbers of active bacteria in soil are highly variable and should be used with care. So many environmental factors enter into the bacteria counts that the data have little validity on a quantitative basis. Part of the problem is many bacteria in soil form spores that allow the bacteria to survive until placed into a good nutrient environment. The various plating media do not distinguish between spores and vegetative cells. Direct microscopic counts using fluorescent dyes have been the developed to measure vegetative cells; but many of the fluorescent dyes are not as specific in the soil environment as in the laboratory environment where they were first developed. One can find bacteria populations from 10^3/g soil to 10^9/g, if the data of Martin and Focht are valid. Chapelle found data that indicated that direct counts gave the highest numbers and that the counts of active bacteria varied with the media used to cultivate the bacteria. Since the soil environment is not uniform like water, adjacent samples from the same soil will not have the same number of living cells. Care should be used with the collection of single samples as well as with the cultivation media. Valid data can only be obtained with large numbers of samples, grown on various media and analyzed statistically. These reasons help explain why soil bacteriology remains an uncharted area of microbiology.

FUNGI

The fungi find the upper layers of soil conducive for growth. As strict aerobes, the fungi are only able to grow when there is adequate oxygen in the water. Fungi are not good at competing with bacteria for nutrients. Under the right conditions they are able to compete quite well. If the pH of soil is acidic, fungi have an advantage over bacteria. If the soil is low in nitrogen, fungi are able to produce more protoplasm than bacteria on the limited nitrogen. Fungi can also grow under less moist conditions than bacteria. Fungi reproduce by spore formation, making it much easier to distribute cell material for future growth than the simple binary fission of bacteria. The large number of spores permits the fungi to grow better than bacteria when soil samples are placed into nutrient media. Fungi will overgrow bacteria in most laboratory media. Fungi mycelium will quickly cover the media surface exposed to air. In soil the fungi filaments grow over the soil particle surface and help to hold soil particles together.

The large quantity of plant residues deposited on the soil surface provides the nutrients needed by certain species of fungi. Lignin is a complex aromatic polymer that combines with cellulose in higher plant tissue to protect the plant tissue from premature biodegradation. A number of fungi have the ability to metabolize the lignin and the cellulose in dead plant tissue, returning these materials back into the environment as basic components. The white-rot fungi, *Phanerochaete chrysoporium*, are common fungi capable of metabolizing both lignin and cellulose. Most of the soil fungi belong to the *Fungi imperfecti*, a large group of diverse fungi whose complete life cycle has not been sufficiently observed to permit proper classification. Yeasts are a special group of fungi that will also be found in soils. Martin and Focht, as well as Ehrlich, indicated that the number of fungi cells range from 10^3/g soil to 10^6/g soil. There is no doubt that fungi are important to the overall soil microbiology.

ACTINOMYCETES

Actinomycetes will also be found in soil. As previously indicated, the *Actinomycetes* are bacteria that have some characteristics of bacteria and some characteristics of fungi. The small diameter of the *Actinomycetes* filaments resembles filamentous bacteria. Yet, the filamentous growths are similar to fungi in their overall characteristics. The *Actinomycetes* filaments can fragment into short segments that act like spores. Each fragment can grow and produce large masses of *Actinomycetes*. Fragmentation occurs mostly under adverse environmental conditions. *Actinomycetes* grow best on dead vegetation, helping to recycle plant nutrients from crop residues. As expected, *Actinomycetes* are widely distributed in agricultural areas. *Streptomycetes* and *Nocardia* were reported by Alexander to be the most prevalent *Actinomycetes* in soil. Normally, the mass of *Actinomycetes* will be greater than the mass of fungi in soil, but less than the mass of the other bacteria. The spore forming ability of the *Actinomycetes* allows them to spread rapidly through soil, the same as fungi. Martin and Focht found that the *Actinomycetes* tend to range from 10^6/g soil to 10^7/g soil.

PHOTOSYNTHETIC MICROORGANISMS

Algae are the most common photosynthetic microorganisms found in soil. They will be found only near the soil surfaces where light is readily available. Since the algae use inorganic compounds for cell protoplasm, they do not compete with the other organisms for organic nutrients. Limited amounts of inorganic nitrogen, phosphates and essential trace metals control the growth of photosynthetic microorganisms in many locations. The blue-green bacteria, previously classified as algae, have the ability to fix atmospheric nitrogen, allowing them to grow in nitrogen deficient soils. Even nitrogen fixing bacteria must have phosphates and trace metals. Growth of photosynthetic organisms results in the creation of organic matter. As the photosynthetic microorganisms die and undergo decomposition, cell nutrients are released for other microorganisms to utilize. Bacteria and fungi will grow in proportion to the available nutrients. Slowly, humus accumulates, creating soil useful for agriculture. Green algae and diatoms are found in soils that have sufficient inorganic nitrogen and other essential elements. The ability to find sufficient nutrients, moisture, and light at the soil surface limits the growth of algae in most soils. Moist garden soils with available nitrogen and phosphorus will show good populations of algae on the soil surface. The availability of sufficient water is essential for algae growth since they use water for their source of hydrogen ions required for cell synthesis. Martin and Focht reported 10^3 to 10^5 algae/g soil.

PROTOZOA

Single cell animals find the bacteria in soil a good source of nutrients. It is not surprising that protozoa are found in the upper layers of the soil where there is adequate moisture, oxygen, and bacteria. *Sarcodina* have the ability to crawl over the soil surfaces and engulf the bacteria attached to the soil particles. *Mastigophora* and *Ciliata* will also be found in soils. The small flagellated protozoa are able to find sufficient nutrients to grow nicely. Small free-swimming ciliated protozoa are also found where the bacteria populations are very active. It takes more energy for the ciliated protozoa to survive than the other protozoa. When the environment becomes unsuitable for protozoa growth, they produce cysts that allow the cells to survive until the environment becomes suitable again. Martin and Focht found 10^3 to 10^5 protozoa/g soil.

HIGHER ANIMALS

Higher animals also find soil a reasonable place for living. Various worms and mites can be found in the soil. Nematodes and other round worms are found in soil. The nematodes live on the bacteria in the soil as well as on various plants. Many nematodes are pathogenic to plants, creating definite problems in certain agricultural areas. Earthworms tend to create large void spaces in the soil and leave their wastes behind for bacteria to stabilize. Earthworms are important aerators for surface soils, creating increased void spaces. Various insects live in burrows in the soil, providing vents for surface air and water to move into the soil. Larger animals also find the soil a suitable environment for living. The waste products of the large animals provide the microscopic organisms with nutrients for growth.

Martin and Focht stated that there could be from 50 to 200 nematodes/g soil. Paul and Clark indicated that each nematode required 5,000 bacteria/min, indicating the nematodes consumed from 0.25 to 1.0 x 10^6 bacteria/min/g soil. These data show that nematodes will only be found in relatively rich soils with high bacteria populations. While all of these organisms can be found in different soils, there are definite environmental constraints that must be given proper consideration. Adequate moisture, oxygen, and nutrients are essential for growth of the higher animals. The presence of specific types and relative numbers of organisms can be used as indicators of the soil environment.

SURFACE METABOLISM

The growth of plants in the surface soil provides many different animals with their source of nutrients. Annuals have a single growing cycle over a one-year period in moderate climates. When the plants die, microorganisms in the soil metabolize the plant residue and return the chemical nutrients back to the soil for reuse the next growing season. Seeds from the growing plants fall to the soil surface and provide for next years crop of annuals. Aerobic bacteria, fungi and actinomycetes are responsible for decomposition of the dead plant tissue in the surface soil layer. The root structures of annual plants are slowly decomposed. Perennial plants tend to survive for several years. The tops of the perennial plants die, but the roots remain viable for several growing seasons. Microorganisms decompose the leaves that drop onto the soil surface by normal metabolism, returning the nutrients to the environment. Because the surface environment is not always favorable for microbial growth, the rate of decomposition of dead plant tissue is slow when compared to the normal rates of microbial metabolism.

Animals deposit waste materials on the soil surface. A few animals bury their wastes just below the soil surface. Either way, microorganisms degrade the waste products under aerobic and anaerobic conditions, returning the nutrients to the soil environment for plant growth the next season. When the animals die, microorganisms slowly decompose the organic material with anaerobic metabolism predominating initially. As stabilization progresses, metabolism shifts from anaerobic to aerobic. Ultimately, the dead animal is degraded with its nutrients being returned to the soil environment for reuse. Only the bones remain to slowly weather away over time.

For centuries soil also received human wastes. In many areas of the world the soil is still used as the primary means for human waste disposal. Microorganisms are able to stabilize human wastes the same as animal wastes as long as the environment is satisfactory. Too much fecal material in a given volume of soil creates an anaerobic environment with depression of the pH from the production of organic acids. Too much acid production stops microbial metabolism in the bulk of the contaminated soil. Only at the edges of the contamination are conditions suitable for aerobic metabolism and complete stabilization of the human wastes. Large amounts of urine decompose with the release of ammonia and an increase in pH. As the pH increases, some of the ammonia is lost to the atmosphere as a gas. With sufficient time, microorganisms can stabilize the entire mass of urine and feces, provided more wastes are not added to create additional problems. Any given mass of soil has the ability to absorb and stabilize a finite amount of organic wastes. If organics are mixed with surface soil at regular intervals and dampened with water, the rate of surface stabilization can be increased with the microorganisms metabolizing aerobically. Pit privies are an excellent example of excess accumulation of human wastes in a small volume of soil. When the human wastes become excessive, the stench of ammonia and anaerobic end products forces people to dig a new pit privy some distance from the old privy. Limited land in urban areas forced the abandonment of pit privies as the readily available soil became saturated with wastes.

Septic tanks were developed to replace privies as indoor plumbing was incorporated into individual houses. Septic tank systems consist of a large concrete, steel or plastic tank, ranging in size from 1,900 L to 5,700 L (500 – 1,500 gallons) followed by a lateral field for soil absorption of the liquid. A schematic diagram of a septic tank system is shown in Figure 7-2. The septic tank is designed to retain the suspended solids from the household wastewaters. The readily biodegradable suspended solids are slowly decomposed anaerobically while the grease accumulates with the other inert materials. The liquid effluent from the septic tank is distributed to a tile drainage field, buried a short distance below the soil surface. The tile drainage field consists of drainage pipes with holes along the bottom of the

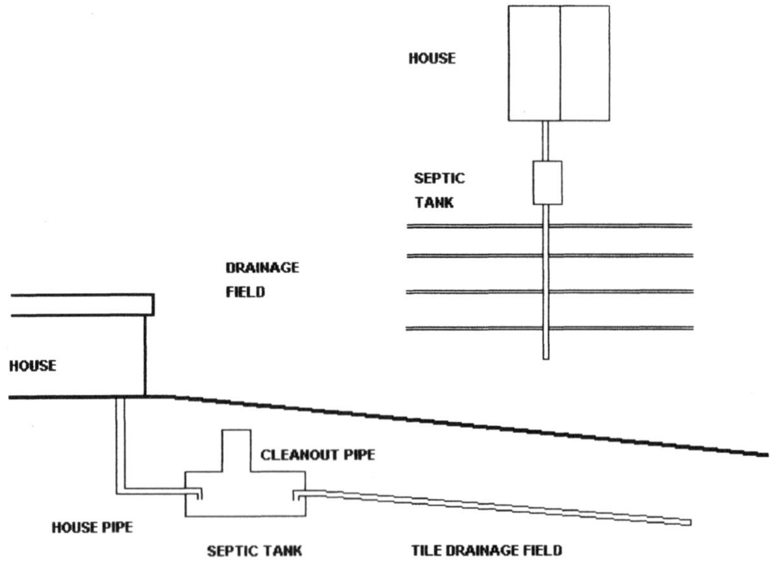

Figure 7-2 SCHEMATIC DIAGRAM OF A SEPTIC TANK SYSTEM FOR A SINGLE FAMILY RESIDENCE

pipes placed in a bed of gravel, completely covering the pipes. The effluent from the septic tank is discharged into the gravel surrounding the pipe and comes into contact with the adjacent soil. As the effluent moves into the void spaces in the adjacent soil, aerobic bacteria in the soil finish the stabilization of the soluble organic compounds and the excess ammonia nitrogen in the effluent. The excess water either evaporates into the atmosphere or seeps through the soil. Septic tanks operate best in dry climates with sandy soil. Clay soils swell when wetted, preventing the water from seeping into the soil. Areas with high rates of precipitation do not lend themselves to the use of septic tanks for the treatment of household wastewaters. The rates of evaporation from the soil surfaces are not great enough to handle the wastewaters and the soils tend to become anaerobic rather than aerobic. The wastewaters from the tile field will flow to the saturated surface of the soil and into the nearest drainage ditch without adequate soil treatment, creating a health hazard.

Over time the wastewater solids accumulate in the septic tank and have to be removed before being discharged into the tile drainage field. Excess suspended solids in the tile drainage field will clog the soil and necessitate digging up and replacing the tile drainage field. Solids should be pumped out of septic tanks every three to five years to prevent problems in the tile drainage field. Care should be made to insure that the entrance to the septic tank is easy to find and is readily accessible for the septic tank pumping truck. Heavy machinery should not be allowed over the tile drainage field as it will compact the soil and adversely affect proper operation of the septic tank. Large trees and shrubs should not be planted over the tile drainage field as the roots will penetrate the tiles and clog them. Grass is best planted in the area over the tile drainage field. The growth of grass will help absorb some of the nutrients and assist in the transpiration of water to the atmosphere. Keeping the tile drainage field area open will allow the heat of the sun to provide the maximum water evaporation from the soil. Septic tank systems are an excellent example of soil processing of household wastewater when properly designed and operated.

As communities grew, soil stabilization of human wastes gave way to collection in sanitary sewers with water used for carrying the wastes to wastewater treatment plants. Commercial development and industrial factories helped to generate community growth. Many of the commercial and industrial facilities processed various chemicals as part of their business operations. Accidental spills and leaks resulted in loss of some of the process chemicals onto the surface of the soil around these facilities. The industrial chemicals in the soil created definite microbial responses, the same as the human wastes did previously. Some of the organic chemicals were easily biodegraded and vanished. Some of the organic chemicals were slowly biodegraded over the years. If the organics were toxic, they did not undergo metabolism by the various microorganisms in the soil. Toxic organic and toxic inorganic compounds remained in the soil and accumulated with time as more spills contributed to the total soil load. The accumulated toxic wastes in the soil were eventually designated as *hazardous wastes*. The importance of hazardous waste is great enough to deserve a separate chapter for their discussion.

Surface metabolism in soil is affected by various environmental conditions. Temperature exerts a major influence on the rate of microbial metabolism in soil. As the temperature decreases, the rates of metabolism of all of the different microorganisms slow. In cold climates the surface moisture can freeze, stopping metabolism completely. Spores and cysts permit the microorganisms to survive the freezing conditions, allowing metabolism to restart when the surface of the soil thaws out with increased temperatures. At the other extreme, warm, wet conditions allows the surface metabolism to proceed at a rapid rate. Thus, moisture combines with temperature as key environmental factors affecting microbial populations in surface soils. The major problem affecting soil metabolism is mixing. The lack of mixing of microbes, water, nutrients, oxygen, and alkalinity in soil limits the rates of metabolism.

DEEP SOILS

Recently, bacteria have been isolated from deep soils, creating speculation as to their metabolism in deep soils. Bacterial isolation from deep soil samples has been limited since conditions do not exist for normal growth. It appears that the bacteria isolated from deep soils were in the soils when geological shifts pushed them from the surface to deep within the ground. Without suitable environmental conditions for normal growth the bacteria were essentially held in a state of suspended animation. Chemicals did not exist to permit the bacteria to continue metabolism. All metabolic reactions reached equilibrium and stopped until the soil was brought back to the surface and a suitable environment was created for growth. Isolation of bacteria from deep soil layers has stimulated additional research in an effort to learn more about these bacteria.

As water moves through soil, the microorganisms in the water tend to become attached to soil particles and to remove nutrients from the water as it passes. Soil attachment is one of the reasons why bacteria are seldom found in groundwater. Most of the bacteria found in groundwater come from contamination created when the well was drilled or from deterioration of the well pipe near the soil surface. Disinfection of the well water initially, destroys all the initial contamination. It has also been found that without proper drainage protection around the water well pipe, contaminated surface water can move along the outside of the pipe to the well screen and back into the well water being pumped from the aquifer. Fecal contamination of well water is entirely related to surface drainage that should not have occurred. Soil particles are excellent media for removing bacteria from contaminated water. Several feet of fine soil particles will remove the bacteria from contaminated water. This relationship forms the basis for sand filtration in water treatment processes. Soil removes all suspended and colloidal particles from water, producing clear water with only soluble chemical contaminants remaining. The geology of the adjacent soil determines the concentrations of soluble chemicals in deep groundwater. Chemical analyses of well waters can be useful in determining the source of the groundwater and possible connections to other aquifers.

Soil characteristics are quite diverse and rapidly change. The variations in chemical and biological characteristics of soil make it difficult to develop a basic understanding of the microbial relationships. Two soil samples taken a few feet apart will result in two unique samples that may or may not yield the same microbial populations. The lack of uniformity in soil makes sampling and analyses a real problem. The best that can be done is to collect large numbers of samples and apply statistical analyses to the results obtained from the samples. Even then, it should be recognized that data simply represent the characteristics at a specific time. The environmental impact on the same area of soil changes with the passage of time, creating new biological relationships. Soil microbiology can be frustrating as well as rewarding. It is constantly changing and there is no such thing as true equilibrium conditions, such as exist in aquatic systems

THINGS TO REMEMBER

1. The upper layer of soil contains about 50% particles and 50% voids.

2. Soil particles are largely disintegrating rock fragments.

3. The hydrologic cycle determines the amount of precipitation that falls on different landmasses.

4. The surface characteristics of soil determine how much water runs off and how much enters the soil.

5. Clay soils often swell when wetted, preventing water from entering the soil.

6. Sandy soils allow surface water to enter the soil because of the large void spaces between particles.

7. Rapidly moving surface water can pick up and move soil particles great distances.

8. Bacteria, fungi, actinomycetes, yeast, protozoa and nematodes can be found in the upper layers of soil.

9. Environmental conditions in the soil determine the distribution of microbes and their survival in soil.

10. Microbial activity in soils is a factor in determining the chemical quality of groundwater.

11. Microbes exist in soil capable of metabolizing all naturally occurring organic materials and related synthetic compounds.

12. Contaminated soils are a good source of bacteria capable of metabolizing the contaminating organics.

13. Bacteria occur in deep soils, but lack the ability to metabolize and grow in that environment.

REFERENCES

Alexander, M. (1961) *Introduction to Soil Microbiology*, John Wiley & Sons, New York.

Chapelle, F. H. (1993) *Ground-Water Microbiology and Geochemistry*, John Wiley & Sons, New York.

Ehrlich, H. L. (1996) *Geomicrobiology*, 3rd Edition, Marcel Dekker, New York.

Martin, J. P. and Focht, D. D. (1977) Biological Properties of Soils, *Soils for Management of Organic Wastes and Wastewaters*, Soil Science Society of America, Madison, WI.

Means, R .E. and Parcher, J. V. (1963) *Physical Properties of Soils*, Charles E. Merrill Books, Columbus, OH.

Paul, E. A. and Clark, F .E. (1989) *Soil Microbiology and Biochemistry*, Academic Press, New York.

Waksman, S. A. (1952) *Soil Microbiology*, John Wiley & Sons, New York.

Chapter 8

WATER MICROBIOLOGY

Water is essential for all biological life. Both plants and animals use water to carry nutrients to their cells and to carry waste products away, permitting them to continue to exist. From the hydrological cycle we learned that as water flows over land surfaces on its journey to the oceans, it picks up soluble minerals from the soil, as well as suspended particles. These contaminants, picked up by the flowing water, provide a good environment for the growth of some microorganisms. Ultimately, the water reaches the ocean. The suspended contaminants slow their velocity and settle out. The soluble contaminants mix with the ocean waters where the continuous evaporation of water from the ocean by the sun concentrates the soluble contaminants to a much greater level over time. The microorganisms in nature have adapted to the highly saline environment in the oceans, creating similar groups as in fresh water. The meeting of fresh waters and saline ocean waters creates a saline gradient that allows the microorganisms to adapt from one environment to the other environment. Contaminant gradients are essential for microbial adaptations from safe environments to harsh environments.

RIVER WATER MICROBIOLOGY

The microbiology in river water starts with the organisms and nutrients picked up as the water flows over the land surface. River water starts high in the mountains. The mountains have sustained centuries of precipitation with small changes. The high altitude and low temperatures in the mountains prevent biological growths except for microorganisms sheltered from the harsh environment. *Cyanobacteria,* which have the ability to fix gaseous nitrogen from the atmosphere, grow very slowly, producing organic matter from sunlight and limited nutrients. The *Cyanobacteria* stimulate the growth of green algae; and together they stimulate fungi growth, creating lichens that form on the rocks. The lichens and atmospheric conditions result in the slow fracturing of the rock, creating pockets of soil that allow a few simple plants to grow. Part of the nutrients released by these slow growing plants are carried by the rain water to lower levels where other green algae are able to grow on the surface of stones. Phosphorus is a limiting element controlling microbial growth in the high altitude streams. Growth and death of the algae will release sufficient organic matter to allow a few bacteria to grow in the water. Bacteria growth also tends to occur on the surface of the rocks. As the water drops to lower altitudes and the temperature increases slightly, a few protozoa will be able to survive on the algae and the bacteria. As a few terrestrial plants grow at the lower altitudes, more nutrients are made available. Higher animals begin to appear, feeding on the plants. Organic waste materials, created by the higher animals, are metabolized by the bacteria in the soil, releasing nutrients and stimulating additional plant growth. Some of the nutrients wash off the land and enter the flowing water, allowing greater growth of bacteria and algae, as well as protozoa. The bacteria are largely aerobic and facultative soil bacteria. *Pseudomonas, Bacillus, Achromobacter, Flavobacterium, Alcaligenes, Nitrosobacter,* and *Nitrobacter* species will be found in varying numbers, depending upon the available nutrients. The clear, cold river water is saturated with dissolved oxygen (DO) and moves quickly down the mountain to meadows where the flow slows and more nutrients are added by animals and decaying vegetation. As the river drops below the timberline, more decaying vegetation is added to the flowing water. Fungi and actinomycetes enter the river; but do not grow significantly since they cannot compete against the bacteria that grow more efficiently. Green algae use the end products of bacterial metabolism and appear on rocks in shallow areas where sunlight furnishes the energy for growth.

The concentration of organic matter in the river water limits the growth of bacteria, fungi, and actinomycetes. As previously indicated in Chapter 2, the organic matter must provide 31.6 J of energy to produce 1.0 mg of VSS bacteria cell mass. With a 1.0 μ^3 volume per cell and a VSS content of 27 percent, it would take 3.7 X 10^9 bacteria to yield 1.0 mg VSS cell mass. The bacteria would have metabolized organic matter containing about 2.2 mg BCOD. From an organic substrate point of view, 1.0 mg BCOD metabolized would yield 1.7×10^9 bacteria. In view of the small size of bacteria, a reasonable number of bacteria will be produced in relatively clean streams. While fungi and actinomycetes cannot be counted the same as bacteria, they produce about the same mass of cells per unit of BCOD metabolized as the bacteria. Algae growth is limited by the availability of phosphates and sunlight, using carbon dioxide from bacteria metabolism as their source of carbon for cell mass. Algae growth is important because of its production of new organic matter in the water. The algae cell mass helps maintain a suitable environment for good growth of higher forms of aquatic organisms. The growth of protozoa and rotifers is limited by the available bacteria and algae. Research by Drake and Tsuchiya in 1977 found the free swimming ciliated protozoa, *Colpoda steini*, produced 0.45 g cell mass for each g *E. coli* metabolized. The ability of the protozoa to continue to grow depends upon the bacteria population. A small ciliated protozoa must consume about 2×10^4 bacteria to obtain sufficient nutrients to reproduce themselves. The problem facing the protozoa is finding sufficient numbers of bacteria. Protozoa grow quite readily if large numbers of bacteria are dispersed in the river water. As the bacteria population decreases from protozoa predation, bacteria become harder to find. The protozoa expend more energy trying to find the bacteria, slowing the growth of protozoa. It is necessary for organic matter to enter the river on a continuous basis to properly stimulate the bacteria that stimulate the protozoa. With limited nutrient supplies the microorganisms grow and die off in a short stretch of the river. Higher microscopic life metabolizes the lower forms of microscopic life at a lower and lower efficiency. Minnows feed on the higher forms of microscopic life and are fed upon by larger fish. The numbers of organisms at each succeeding level decrease as the size of the organisms increases.

****

TYPICAL CALCULATIONS:

1. Determine the number of bacteria/mg VSS if each bacterium has a volume of $1.0\ \mu^3$.

    Bacteria contain 30% dry matter that is 90% volatile solids.
    $1.0\ \mu^3$ water $= 1.0 \times 10^{-12}$ g or 1.0×10^{-9} mg
    Dry weight of one bacterium $= (0.30)(1.0 \times 10^{-9}) = 0.30 \times 10^{-9}$ mg
    VSS weight of one bacterium $= (0.90)(0.30 \times 10^{-9}) = 0.27 \times 10^{-9}$ mg
    Number of bacteria/mg VSS $= 1/(0.27 \times 10^{-9}) = 3.7 \times 10^9$

2. Determine the number of small ciliated protozoa that would be produced by eating 1.0 mg bacteria VSS.

    Number of bacteria/protozoa $= 2 \times 10^4$
    Number of bacteria/mg VSS $= 3.7 \times 10^9$
    Number protozoa/mg bacteria VSS $= (3.7 \times 10^9)/(2 \times 10^4) = 2 \times 10^5$
    (Note: Since the number of bacteria/protozoa has one significant figure and the number of bacteria/mg VSS has two significant figures, the calculated numbers of protozoa/mg bacteria VSS can only have one significant figure.)

****

The potential for pathogenic microorganisms increases as animals begin to encroach on the river water. Water temperature is a primary factor limiting the survival of pathogens discharged into the river. Pathogenic microorganisms grow best at animal body temperature. The cold river water limits the ability of the pathogenic bacteria to compete with the non-pathogenic bacteria for the limited nutrients. Spores and cysts of pathogenic microorganisms are able to survive the adverse environments. Fortunately, the numbers of pathogenic spores and cysts are limited and are greatly diluted in the river environment. Expanding agriculture at the lower altitudes tends to increase the addition of nutrients and stimulates microbial diversity in the river water. Rapidly moving stream water flows over rocks in shallow depths, transferring more oxygen to the water. The demand for oxygen is small compared with the oxygen transfer capacity. As the land surfaces become flatter, the river spreads out and becomes deeper. The velocity of flow slows, allowing some of the larger suspended particles to settle out. Colloidal suspended solids and dissolved organics become the driving forces for dispersed microbial growth. Settled solids accumulate on the bottom of the river and slowly form layers of organic and inorganic solids. The biodegradable fraction of the settled organic solids undergoes slow metabolism. At the water-solid interface bacteria metabolism will be aerobic, as the oxygenated water passes continuously over the settled solids. The aerobic metabolism will only occur in a relatively thin layer of the settled solids. Metabolism shifts from aerobic to anaerobic below the thin aerobic layer of organic solids. The organic rich mud on the bottom of slow moving rivers contains a mixture of facultative bacteria, strict anaerobic bacteria, sulfate-reducing bacteria, and methane bacteria in a common community. The growth of the different groups of bacteria is determined by the chemical characteristics of the anaerobic layer of the bottom mud.

Algae grow as dispersed cells or as attached cells on the wetted surfaces of rocks. Even clear water absorbs light energy, limiting its availability for algae and other aquatic plants below the water surface. Turbidity and nutrients increase in the river water as agriculture expands along the river. The turbidity of the water limits the growth of algae and allows the excess inorganic nutrients to pass unchanged. Bacteria and other non-photosynthetic plants continue to grow. The slower river velocity and deep water retards the transfer of oxygen from the atmosphere to the water. Point source wastewater discharges from small agricultural communities produce additional organic loads on the river water. The bacteria respond by metabolizing the organic waste materials as quickly as possible. The dissolved oxygen in the water begins to drop as the demand for oxygen exceeds the transfer of oxygen from the air into the water. The dissolved oxygen in the water continues to drop until the rate of metabolism of the organic matter balances the rate of oxygen transfer. As metabolism of the organic matter slows even further, the rate of oxygen transfer exceeds the rate of oxygen demand. The net effect is a slow increase in dissolved oxygen in the river water. If no additional organic matter is added to the river, the dissolved oxygen in the river water will eventually return to saturation concentrations. The drop in dissolved oxygen and its return to saturation as the water flows along in the river resulted in the concept of *self-purification* of flowing river water. Unfortunately, self-purification of flowing river water is more related to the stabilization of the organic compounds added to the river water than to the removal of pathogenic microorganisms. Reduction of pathogenic microorganisms in the river water is primarily the result of starvation and predation by protozoa, rotifers and higher animals.

As the size of communities along the river increases, the amount of organic matter discharged to the river increases. The minimum dissolved oxygen level in the river water drops to lower and lower levels. The dissolved oxygen drop and ultimate recovery is termed the *oxygen sag curve*, since the dissolved oxygen tends to sag and rise again along the length of the river. In 1925 Streeter and Phelps developed the first mathematical relationships to express the oxygen sag curve. Over the years the mathematical equations have been modified and adjusted in an effort to develop practical relationships. Unfortunately, the mathematical equations are more useful in textbooks than in practice. Rivers are not uniform microbial reactors. Each river has its own unique characteristics. Most rivers lack sufficient uniformity of hydraulic characteristics to permit accurate mathematical evaluation using the modified Streeter-Phelps equations except for short distances. Milo Churchill recognized that a given stretch of a specific river responded in a similar fashion as the organic load changed. He developed a linear regression analysis to predict changes in the oxygen sag curve over relatively short lengths of a river affected by increasing organic loads. Efforts to make the mathematical prediction equation more precise resulted in loss of its simplicity and its accuracy. Unfortunately, Milo Churchill developed his mathematical techniques before the development of the microcomputer which could have easily handled the complex mathematical equations. The complex biological relationships that exist in rivers follow definite reactions that are difficult to present in a precise mathematical form. There are too many reactions occurring in the microenvironments of streams and rivers to permit development of precise mathematical equations for absolute simulation. Fortunately, absolute predictions are not required in handling environmental problems in streams and rivers. It is far more important to understand the general concepts and to apply a broad approach to the sum of the important reactions than to use imperfect mathematical equations requiring a computer for their solution. Widespread use of modified Streeter-Phelps equations by regulatory agencies to determine various levels of wastewater treatment along major streams has resulted in many, large computer studies of limited value. Real world experience has demonstrated the fallacy of complex computer models in predicting future conditions within a river.

The microbial population at any point in a river is related to the environment that exists at that point. The mixture of organisms is related to the input of nutrients into the river and response of all of the different biological organisms. Each river contains a mixture of bacteria, fungi, actinomycetes, algae, protozoa, rotifers, and crustaceans together with higher plants and animals. Different species of bacteria compete with each other for nutrients. The more rapidly growing bacteria increase rapidly and die off just as quickly. The slower growing bacteria have a hard time competing with the more rapidly growing bacteria; but they die off at a slow rate and predominate downstream over the rapidly growing bacteria. The bacteria also compete with the fungi and the actinomycetes for those nutrients. Since the microscopic plants depend upon absorption of nutrients across the cell wall, they cannot keep other microscopic plants from obtaining some of the nutrients. They all grow in proportion to their ability to obtain and process nutrients from the water. Each small portion of the river will have its own special microcosm. There is no uniformity of microorganisms across the various segments of a river. The microbial populations continuously undergo dynamic changes. The specific microorganisms growing in any section of the river reflect the overall environment that existed when the water samples were collected. Understanding the changing microbial dynamics in streams and rivers is essential for environmental microbiologists.

The growth of bacteria removes dissolved oxygen from the river water while the air above the water surface attempts to replenish the oxygen that has been removed. The surface characteristics of the river determine the rate of oxygen transfer from the air into the water. The driving force for transferring oxygen into the river is related to the difference between the saturation oxygen level and the current oxygen level in the river water, as shown in Equation 8-1.

$$dC/dt = k*(Cs - C) \qquad (8\text{-}1)$$

where: dC/dt is the rate of oxygen transfer from the air, mg/L/hr.
k is the oxygen transfer factor for the river, 1/hr.
Cs is the saturation oxygen concentration in the river, mg/L.
C is the current oxygen concentration in the river, mg/L.

The transfer factor, k, is related to a number of variables, including temperature, water depth, river width, mixing, and specific contaminant concentrations. As long as the microbial demand for oxygen is greater than the rate of oxygen transfer from the air, the dissolved oxygen concentration in the river water will continue to drop. When the dissolved oxygen has been completely removed from the river water, the water becomes septic and anaerobic metabolism is established. Sulfate reducing bacteria, *Desulfovibrio*, will reduce the available sulfates to hydrogen sulfide, creating obnoxious odors. The sulfides will react with iron in the water to form a black, colloidal precipitate, giving the water an unpleasant appearance. With zero dissolved oxygen in the river water, the maximum rate of oxygen transfer will occur at the water-air interface. As oxygen enters the surface water, bacteria and fungi will use the oxygen for aerobic metabolism. Since the fungi are strict aerobes, they tend to grow as filaments attached to submerged objects near the water surface. As the water flows downstream, a point is eventually reached where the rate of oxygen transfer exceeds the rate of oxygen demand. The dissolved oxygen concentration in the river water begins to rise, as the rate of metabolism slows. Protozoa begin to appear and metabolize the dispersed bacteria. As the dissolved oxygen concentration increases, a few scavenger fish appear, followed by the more normal types of fish. Rotifers and crustaceans appear and the turbidity in the water decreases. The fungi no longer grow to any significant extent since the organic compounds have all been metabolized except for a few complex organic compounds that are metabolized at very slow rates. Sulfur oxidizing bacteria, *Thiobaccillus*, begin to appear as soon as the dissolved oxygen rises, oxidizing the soluble sulfides to sulfates. Nitrifying bacteria oxidize the excess ammonia nitrogen to nitrites and then to nitrates. Algae grow on the inorganic compounds in the water and produce additional dissolved oxygen as a major end product. The dissolved oxygen rises to saturation as the river recovers from metabolism of the organic load. Further growth of algae allows the river water to become supersaturated with oxygen. Figure 8-1 is a schematic diagram of the oxygen sag curves for a high organic load. There is no doubt that the microbes living in the river water have the ability to stabilize the biodegradable organics discharged from point sources and non-point sources along the river. The difficulty with using the self-purification capacity of the rivers lies in the loss of the full potential of the river for fish and

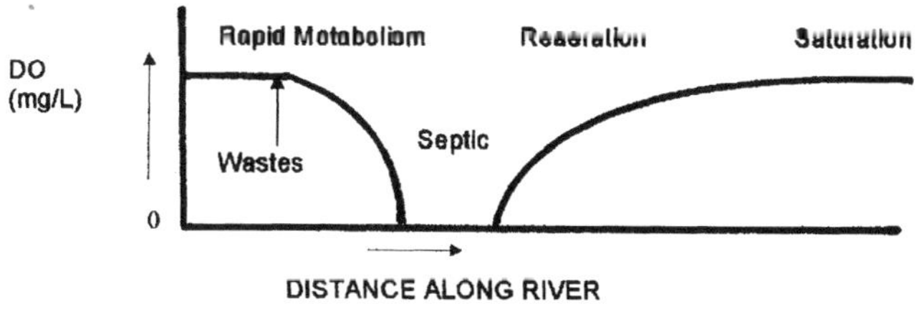

Figure 8-1 SCHEMATIC DIAGRAM OF THE OXYGEN SAG CURVE FOR A HIGH ORGANIC LOAD IN RIVER WATER

recreation in the low oxygen and zero oxygen sections. Beneficial uses of streams and rivers require that the organic loads discharged to these bodies of water be kept below the levels that create low dissolved oxygen conditions. Streams that are used for the highest forms of fish, such as trout and other game fish, require the dissolved oxygen levels to always be above 5.0 mg/l.

Over the years, people discharged their wastewaters into nearby streams and rivers without regard to the impact on water quality or on downstream water users. It was assumed that the natural purification characteristics of the flowing waters would eliminate any harmful impact of the wastewater discharges. No one realized that pathogenic bacteria in the wastewaters were carried downstream and posed a threat to the downstream users of the river water. Industrial discharges added toxic chemicals to the municipal wastewaters in the rivers. The failure of growing communities to accept responsibility for their own wastewater treatment prior to discharge into nearly streams and rivers resulted in increasing river pollution in the United States. Efforts by the States to control wastewater pollution varied from State to State. There were too many different opinions on how rivers should be used and who should benefit. Instead of working together, the States favored positions that maximized the benefits for their own people without regard to the people in neighboring states. It is not surprising that State control of water pollution failed. Ultimately, Congress decided that Federal regulations were required to provide a national policy for water pollution control. The initial efforts at Federal control were almost as fragmented as the States had been. In 1970 President Richard Nixon pulled all of the major environmental pollution components together when he established the Environmental Protection Agency (EPA) under the Office of the President. Senator Edward Muskie from Maine worked hard to insure that the EPA had the proper congressional legislation to accomplish the task at hand. PL 92-500, passed by Congress in 1972, provided the EPA with the authority to set specific effluent criteria for both municipal and industrial wastewater discharges to receiving rivers to maintain suitable water quality. Initially, the EPA set effluent criteria that were attainable with reasonable effort on the part of communities and industries. When suitable progress was made, the effluent criteria were changed to raise the water quality to higher levels. The cost of wastewater treatment increased significantly, as the effluent criteria required additional treatment. Concerns are currently being raised as to whether the cost of improved wastewater treatment effluents exceeds the benefits obtained in improved river water quality. Water resource decisions such as these require social, technical, and political skills to reach the proper decision. While political decisions determine Federal policies, the people ultimately determine the validity of the political decisions. Only a well-informed public can make the best decisions.

Many Third World countries have serious water pollution problems in their streams and rivers. Enteric diseases are endemic in some countries. The future of these Third World countries depends upon creating a safe environment where they can build sound economies with healthy people. They have the opportunity to profit from the experience of the United States and the other advanced nations. Thus far, the efforts at pollution control in some Third World countries have followed the exact footsteps of the United States, making the same mistakes over and over again. The biggest mistake has been the waste of limited financial resources to build complex wastewater treatment plants with all the latest equipment, rather than constructing simple, low cost treatment plants that can produce a suitable effluent with less effort. Conditions in developing countries are not the same as in the United States. They need their own wastewater treatment regulations and wastewater treatment plants rather than simply making copies from the United States or Europe. Time will tell if the developing countries will really learn from the United States or simply follow our footsteps and make all the same mistakes we have made over the years. Knowledge has developed the tools necessary to create economical systems to eliminate environmental pollution in Third World countries, but greed and vanity often combine to prevent the use of that knowledge.

PONDS, LAKES, AND RESERVOIRS

River water tends to remain in a relatively narrow channel until the terrain levels off, creating small ponds or lakes. The river water spreads out in a basin until a point is reached where it forms a small discharge channel and becomes a river again. Water continuously moves downhill by gravity, picking the easiest path of flow. The water velocity slows as it enters a pond or lake. Solids that were kept in suspension by the velocity of flow in the river settle out at the low velocity of flow through the ponds or lakes. The settled solids accumulate at the mouth of the lake until the velocity of flow moves them farther into the lake itself. The retention of water in ponds or lakes can be quite long, especially if the natural terrain allows for some deep areas. The biological responses in the ponds and lakes can be very interesting and quite varied. Organic matter will enter with the river water, depending on the drainage area supplying the water. Several streams may discharge into the same lake, creating a series of different microenvironments. Wind action across the ponds or lakes is very important for mixing the water and for oxygen transfer. Large, flat, shallow lakes are easily mixed by wind, while deep meandering lakes show little wind mixing. It is important to recognize that the water entering from a river will move almost directly to the outlet if there is no significant wind mixing and will remain close to the water surface, no matter how deep the lake is. As shown in Figure 8-2, the incoming water moves across the surface of the lake as a river. The large volume of still water around the moving water channel across the lake acts as a resistance to mixing. The incoming water moves along the path of least resistance. Since wind only contains a limited amount of energy that can be transferred to the water, wind mixing will be limited to the water surface and will not penetrate very deeply. Incoming water is usually near the

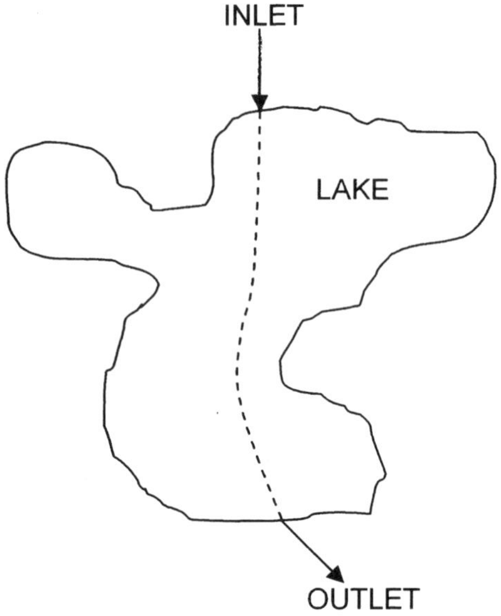

Figure 8-2 A SCHEMATIC DIAGRAM OF THE MOVEMENT OF WATER ACROSS A NATURAL LAKE FROM THE INLET TO THE OUTLET

same temperature as the surface water, eliminating density differences. If the incoming water is from a submerged spring, it will normally be colder than the surface water, giving rise to a density current with the incoming water sinking below the warmer surface waters. In either case the incoming water will move directly towards the lake discharge point. The limited mixing of the lake water allows the microorganisms to develop different populations in the lake, depending upon the water quality at specific locations. Large lakes can show quite diverse microbial populations, making it quite difficult to generalize about the overall microbial characteristics unless large numbers of samples are taken and evaluated at regular time intervals from all significant points in the lake.

The total biological populations in ponds and lakes are relatively small in numbers. The lack of nutrients limits the growth of most microorganisms. Aquatic plant growths in the shallow waters around the edges of lakes stimulate bacteria growth that causes the growth of protozoa and crustaceans. Animals use ponds and lakes as a source of drinking water and deposit their wastes around the edges of the lake. These wastes eventually wash into the lake and stimulate further microbial growth. Since the incoming river water is a continuous source of microbes and nutrients, it is not surprising that the bacteria found in lakes are the same as found in streams and rivers. The same is true for the protozoa. Normally, ponds and lakes will show more crustaceans than streams and rivers. The environment in the lakes and ponds is more favorable for these higher forms of microscopic animals. Algae grow in the shallow areas where the sunlight penetrates the water. Some algae are attached to stones and particles on the lake bottom. Other algae grow as dispersed cells in the upper layer of water across the lake surface. Small fish feed on the dispersed algae and become food for larger fish. The fish discharge their wastes in the lakes and stimulate microbes to further growth. The fish also die and decompose in the lake waters. Nutrients are cycled and recycled through the biological communities in the lakes. Limited organic matter keeps the oxygen demand in the lake at a low level; while wind action across the lake surface causes the dissolved oxygen to reach saturation. Growth of the algae near the lake surface often causes the surface water to become supersaturated with oxygen during the daylight hours. At night algae shift their metabolism with endogenous respiration requiring oxygen the same as bacteria. When daylight appears, the carbon dioxide and ammonia nitrogen released during the night and retained in the water are available for the algae to reuse. Regrowth of algae can restore much of the algae mass lost during the night. With limited input of organic nutrients the algae population will be several times the population of bacteria and fungi in lakes.

Large lakes have long hydraulic retention times and limited nutrient input. Bacteria have more than enough time to metabolize all the readily biodegradable organics. The bacteria not only have time for complete metabolism of the nutrients, they also have time to undergo complete endogenous respiration and die off. Water storage is one of the simplest methods for reducing bacteria populations in contaminated river water entering lakes. Bacteria not only undergo natural growth and death, their numbers are reduced by protozoa predation. Protozoa predation also helps reduce the probability of pathogenic bacteria survival. The microbial dead cell mass slowly settles to the bottom of the lake and accumulates with the other settleable suspended solids. Only the dead cell mass formed near the effluent end of the lake or suspended by wind mixing currents will be discharged in the effluent from the lake.

One of the most interesting aspects of lakes that have depths greater than 20 feet is that they stratify as a result of temperature differentials. Water absorbs heat from the sun, but is not very good at transmitting the absorbed heat. The surface water warms and decreases its viscosity. As winds blow over the water surface, small riffles are formed as the water moves toward the shore. The water moving across the lake surface reaches the shore and returns along the bottom of the lake. The lake bottom has more resistance than the water itself, causing the fluid flow to take the path of least resistance and move back through the water. As the temperature decreases with water depth, the viscosity of the water increases. Since the quantity of water moving back across the lake must balance the quantity of water moving across the surface, a moving upper layer is established with reasonable mixing in the upper layer. The lower volume of water in the deep lake is not affected by normal wind action and remains quiescent. If temperature measurements are taken at regular depths in a deep lake, it can be observed that the upper layer that is mixed by the winds has a slight temperature decrease from the surface downward until the stagnant zone is reached. The temperature will drop quickly in a short distance and remain almost constant to the bottom of the lake. The upper zone has been called the *epilimnion*; while the lower zone has been called the *hypolimnion*. The thin layer where the temperature rapidly declines is called the *thermocline*. Figure 8-3 illustrates the different temperature zones in deep lakes.

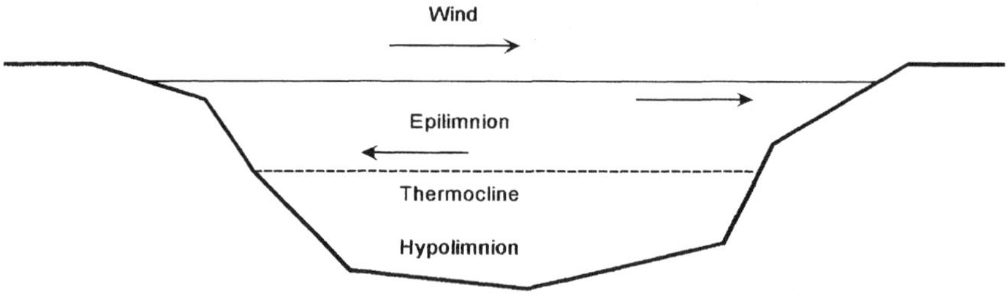

Figure 8-3 A SCHEMATIC DIAGRAM OF WIND MIXING IN A DEEP LAKE WITH A THERMOCLINE

On July 14, 1891, Dr. Thomas M. Drown, Professor of Chemistry at M.I.T. and Consulting Chemist for the Massachusetts State Board of Health, examined the temperature and the percentage oxygen saturation at various depths in Jamaica Pond that was being used for a municipal water supply. The surface water temperature was 24.0° C with 100% oxygen saturation. At 10-ft depth the water temperature had only dropped to 23.8° C and the oxygen saturation was still 100%. At 20-ft depth the water temperature had dropped to 12.3° C with 49% oxygen saturation. At 30-ft depth the water temperature had dropped to 5.8° C with only 29% oxygen saturation. It appeared that the thermocline was located close to the 20-ft water depth. At 35 ft depth the water temperature was 5.6° C with only 4.2% oxygen saturation. At 40-ft depth the water temperature dropped to 5.4° C with 0 mg/l dissolved oxygen. At 48 ft depth, just above the bottom, the water temperature was 5.2° C with 0 mg/l dissolved oxygen. The wind mixing in the epilimnion kept the dissolved oxygen and the temperature high. Aerobic metabolism prevailed near the pond surface. Below the thermocline the temperature was rather uniform and the dissolved oxygen decreased rapidly to zero. The temperature indicated no significant mixing in the hypolimnion; but the thermocline was not a barrier for oxygen diffusion. Dissolved oxygen in the epilimnion slowly diffused into the hypolimnion and was consumed by the microbes in the stagnant zone. Near the bottom of Jamaica Pond the water was completely anaerobic. As suspended solids settled to the bottom and underwent metabolism, the bacteria removed the dissolved oxygen and produced anaerobic end products, creating unpleasant tastes and odors in the water drawn from the lower depths. During the next winter Dr. Drown examined Jamaica Pond when it was covered with an ice layer, 8 to 9 feet thick. Just below the ice, the water temperature was 0° C with 97.5% oxygen saturation. At 10 ft depth the water temperature was 2.0° C with 100% oxygen saturation. The water temperature rose slightly with depth and the oxygen saturation dropped slowly to zero, just above the bottom at 44-ft depth. The ice cover prevented any wind mixing in the pond. The low temperature greatly retarded biological activity and allowed the dissolved oxygen to remain high except at the pond bottom where biological metabolism occurred in the bottom mud.

These early data illustrate the basic concepts of lake limnology. The water temperature in the deep sections of lakes is determined by the adjacent soil temperature. In the summer the wind keeps the epilimnion well mixed; while the hypolimnion is stagnant. As autumn sets in, the water temperature in the epilimnion decreases with no change in water temperature in the hypolimnion. As winter begins, the water temperature in the epilimnion drops below the water temperature in the hypolimnion. The density of the water changes with the surface water becoming slightly heavier than the stagnant water below the thermocline. The heavier surface water suddenly begins to sink, displacing the lighter bottom water, causing a rapid turnover in the lake contents. The anaerobic liquid is brought to the surface and the lake contents are well mixed. The partially decomposed organic compounds on the bottom of the lake are suddenly brought to the surface. This rapid turnover of the lake contents is known as the *fall turnover*. The low temperature keeps the bacteria from rapidly metabolizing the sudden influx of organic matter. The high levels of oxygen saturation at low water temperature keeps the dissolved oxygen high as the temperatures decrease and ice is formed on the surface. In the spring the warm winds help melt the ice and raise the surface temperature. As the temperature of water rises to 4° C, it reaches its maximum density. The heavier surface water again sinks, creating a *spring turnover*. This time the rising surface temperature stimulates the growth of bacteria and algae, which starts the growth of protozoa, rotifers and crustaceans. These microbes, in turn, become food for the fish population. The lake soon returns to normal with the thermocline establishing the mixing barrier between the two zones. Excessive nutrients from the stagnant layer after the spring turnover can stimulate heavy algae growths and greater accumulation of organic matter in the lake. The reuse of nutrients by algae allows the algae populations to increase until equilibrium is reached between nutrient input and nutrient release. Heavy algae growths can be quite detrimental for optimum use of lakes and lake water.

Man-made reservoirs have been constructed on rivers for flood control, water supply, and recreation. Various types of dams have been constructed across rivers having sufficient drainage to provide the desired amounts of water. The dams cause the water to backup, creating artificial lakes. Flood control reservoirs are designed to capture the rapid runoff from heavy rains and to slowly release the collected water after the runoff has stopped. The water level in flood control reservoirs can vary quite widely, reaching a maximum during floods and dropping to a minimum during dry periods. Since floodwaters carry considerable floating and suspended debris, the flood control reservoirs act as collection ponds for the flood debris. The floating debris is often collected on heavy screens near the reservoirs discharge points. The suspended solids tend to settle out in the reservoirs unless the water is released quite quickly. Flood control reservoirs require regular dredging to remove the excess settled silt, sand and rocks. Failure to remove the settled solids will result in the reservoirs filling with settled solids and a loss in storm water holding capacity. Most flood control reservoirs are designed to fill with solids and to lose their effectiveness after many years.

Water supply reservoirs are used where there is inadequate groundwater for meeting the water supply needs. In the northeastern part of the United States water supply reservoirs were placed on watershed basins where public access could be controlled. Fences were erected around the watershed basins to restrict access. The purpose of the restricted access watershed was to protect the quality of the water from wastewater contamination. When the reservoirs were originally purchased, the populations were much smaller and the potential for contamination was small. As populations increased, the uninhabited watershed basins became increasingly valuable. More importantly, animals migrated to the safety of the watersheds. The increased animal populations brought their diseases, some of which could affect people. The most significant of the animal pathogens are the parasitic protozoa carried by small animals. The time may well come when restricted watershed basins are no longer valid. Until then, regular microscopic examination of the biological quality of the water from these water supply reservoirs will help determine the safety of the water along with regular bacteriological analyses.

Recreational reservoirs have been designed for fishing, swimming, and boating. Many of the recreational reservoirs have been designed with houses surrounding most of the shoreline. The problem that these recreational reservoirs are facing is the release of septic tank effluents into the reservoirs by seepage through the soil. The ammonia nitrogen and phosphorus entering the reservoir water from septic tanks stimulate rapid algae growths near the reservoir shores. The recycling of nutrients by algae allows the algae to accumulate slowly over time. The algae result in nuisance growths that eventually have an adverse effect on the recreational uses of the reservoirs. Swimming is affected more than fishing or boating. Efforts have been made to reduce the growth of algae in the swimming areas by applying copper sulfate to the water surface. Copper sulfate is toxic to algae. The application rate is sufficiently low that dilution prevents toxicity to fish in the area. While fish are not affected by the copper sulfate, protozoa are killed. The death of the protozoa allows the bacteria population to increase very rapidly. If water quality samples are being collected from the swimming area to confirm the safety of the water, a rapid increase in total bacteria should be expected after each copper sulfate application until the protozoa population returns to normal levels. The increased bacteria population reflects the importance of predators in keeping the bacteria population under control in reservoirs. Large growths of algae can create distinct odor problems when the algae die and begin to undergo bacteria metabolism. Rapid bacteria growth will quickly exhaust the oxygen in the water, allowing anaerobic metabolism. Use of copper sulfate to control large masses of algae is not a viable solution. The copper tends to precipitate and settle out in the reservoirs. Over time the copper can accumulate to toxic levels in the bottom mud and adversely affect bottom plants and other organisms. Ultimately, fish will disappear from the reservoirs, reducing their recreational value. Algae growths can best be controlled by preventing nutrients from entering the reservoirs. Some housing areas around recreational reservoirs use plastic pipe and small, grinder pumps to remove domestic wastewaters for treatment away from the reservoirs. The use of fertilizers around houses is controlled or prohibited to prevent nutrients from washing off the lawns and gardens into the reservoirs. Minimizing the release of phosphorus to the reservoirs will minimize the growth of algae and will keep the recreational reservoirs operational for a longer period of time.

In the Midwestern part of the United States many of the large reservoirs are multipurpose, rather than single purpose reservoirs. The limited sites required the development of multipurpose reservoirs to meet all the different water demands. Multipurpose reservoirs provide municipal water supply, flood control, and recreation in one reservoir. One of the major problems with multipurpose reservoirs is the varying water level required for flood control. The variations in water levels expected in multipurpose reservoirs must be clearly communicated to adjacent property owners to prevent public disappointment with the different water levels. Historical water records are essential in providing the reservoir operators with data on normal rainfall, floods, and droughts. Although historical water records simply show what has happened in the past, they provide the operator with the best information as to the probability of various events hitting the reservoir in question. The operator normally reduces reservoir volume to a minimum before the flow patterns indicate potential flooding. A specific volume of water is always maintained for water supply. Efforts are made to insure as stable water levels as possible during the recreation periods. Trying to meet the needs of all water uses in multipurpose reservoirs requires continuous monitoring and careful control. Since municipal water supplies have the highest priority from a health standpoint, water quality is important. Thus far, there have been no major health problems from the use of multipurpose reservoirs. This means that as the population increases, many single purpose reservoirs can be converted to multipurpose reservoirs.

WATERBORNE PATHOGENS

Waterborne pathogens have been a major factor in the control of human populations. There are many different types of waterborne pathogens. Viruses, bacteria, protozoa, and worms are part of the important pathogenic microorganisms found in water. Youmans, Paterson and Sommers have presented a good review of infectious diseases that can be used as a reference for additional information. One of the most interesting aspects of pathogenic microbes is the appearance of new pathogens, as old pathogens are brought under control. It is important to recognize the world is a dynamic biological system where every living organism is fighting for survival in a highly competitive environment. Changing the biological populations allows new organisms to take the place of the previous organisms. Not all of the changes are positive from the human point of view. The changes are positive for the new groups of organisms. This is part of the dynamic aspect of biology that has not been fully accepted by our modern society. While the positive benefits of biological change are heralded as advances, the negative aspects are considered as disasters created by modern science. It is important to recognize that the dynamic aspects of biology are part of the natural world in which we live. Positive changes are always matched by negative changes. The elimination of one pathogen allows a new pathogen to grow and take its place.

VIRUSES

Viruses are tiny, parasitic pathogenic microorganisms that are highly specific as to their host. They are simply complex nucleoproteins that utilize the host organism for their source of nutrients. The rapid growth of viruses in the host can cause a failure of the host organ the viruses invaded or the host itself. Unlike higher forms of biological life, the viruses do not respire and simply remain inert outside of the host organisms. Being organic matter, viruses can be hydrolyzed by microbial enzymes. Animals can develop anti-bodies that attack the viruses, giving the animals immunity from those specific viruses. As a net result, viruses are not completely immune from other organisms. They too must work for survival in a highly competitive environment. Viruses are difficult to recognize because they cannot be seen with an optical microscope. Electron microscopes are necessary to show viruses. Special techniques have been developed to isolate and study viruses in pure cultures. There is no general media on which the viruses can be grown. Viruses are specific to host cells. Only when host cells are mass-produced outside of the host can viruses be grown in large numbers and studied in detail in the laboratory. The enteroviruses tend to be the most common viruses transmitted through water. Enteroviruses are produced inside human beings and discharged in their fecal wastes. Polio and hepatitis are two diseases that are caused by enteroviruses. Discharge of human sewage into water without proper treatment is the primary mechanism for transmission of enteroviruses. Contaminated water can transport the enteroviruses considerable distances. Use of contaminated water for irrigation of food crops allows the enteroviruses to become attached to plants. The contaminated water not only poses a threat to the agricultural workers in the agricultural fields, but also poses a threat to people who eat the contaminated crops raw. Vegetables are quite susceptible to surface contamination. Enteric viral pathogens are easily spread when contaminated vegetables are eaten raw. Using the contaminated water as a source of drinking water without proper treatment is a major method for transmitting the viruses to large numbers of people. The people who have developed immunity to the enteroviruses are unaffected; while those individuals with no immunity are adversely affected by the enteroviruses. The very young and the very old are most susceptible to these pathogens. One of the more interesting sources of contamination is shellfish harvested from areas where the contaminated water drains into the ocean. It has been found that bivalve shellfish concentrate the enteroviruses, permitting transmission when the shellfish are consumed raw. Cooking the shellfish destroys the viruses. Treating contaminated water prior to its use as drinking water can also eliminate the pathogens and protect the people using the water. One of the problems facing environmental microbiologists is the "unknown virus". As soon as one group of viral pathogens is brought under control, the opportunity will arise for a new "unknown virus" to take its place. The nature of viruses makes quick identification and treatment difficult. It should be recognized that the fight against viral pathogens is a never-ending fight. Thus far, scientists have demonstrated that even the most difficult viral pathogens can eventually be identified and controlled to a reasonable degree. As long as people exist, there will be an opportunity for enteroviruses to be spread in the environment. With proper care the enteroviruses can be controlled with a minimum of damage.

BACTERIA

Unlike viruses, very few bacteria are pathogenic. Bacterial pathogens can be parasitic or toxin producers. Parasitic pathogens depend upon the host organism for their nutrients. As the parasites increase in numbers, the demand for nutrients adversely affects the nutrient supply required by the host organism. Death soon occurs unless the parasites are destroyed. Some parasitic pathogens limit their growth, allowing both the parasite and the host to survive together. Toxin producing bacteria release protein compounds that are toxic to the host organism. If sufficient toxins are released into the host organism, death soon occurs. There are bacteria that are pathogenic to either plants or animals. From an environmental point of view the enteric pathogens are the most important since they are generated by animals and are carried in their fecal wastes. Discharge of feces into natural waters allows the pathogens to move to other potential sites for infection. Cholera is probably the oldest disease caused by enteric bacteria. Originally, cholera was centered in India and spread to China. Travelers to and from these countries carried the disease to all parts of the world. Cholera is caused by *Vibrio cholerae.* Although cholera is a very old disease, it is not a problem in the developed areas of the world. Cholera is still endemic in the developing areas of the world where sanitation standards are poor to non-existent. Typhoid fever is another enteric disease that is caused by bacteria, *Salmonella typhi*. Typhoid fever was endemic in the United States a little over 100 years ago. Recognition of the fact that enteric bacteria causing typhoid fever were discharged into the environment in untreated wastewaters led to the development of water treatment and wastewater treatment systems that removed the *Salmonella typhi* and other enteric pathogens. Construction and operation of water treatment and wastewater treatment plants in the United States and Europe reduced the transmission of typhoid fever, cholera, and other enteric bacteria diseases. The few cases of typhoid fever that occur today in the developed areas of the world are mostly food related. Like cholera, typhoid fever is still a problem in developing areas of the world that have inadequate sanitation.

Legionella is an interesting bacterial pathogen that is spread through water and air. It obtained its name because the bacteria caused a number of deaths during the American Legion Convention in Philadelphia in 1976. The bacteria were taken in through the lungs and created respiratory infections, primarily in older people who had reduced immune systems. Water droplets circulating through the air conditioning system permitted the *Legionella* to remain viable long enough to infect several people in one of the hotels, housing convention goers. A few cases of Legionaires disease occur each year in various air conditioning systems where *Legionella* have an opportunity to grow and be distributed through the duct systems. With increased emphasis on *Legionella*, it has not been surprising to find that this microorganism exists more widely in the environment than previously recognized. It appears that *Legionella* occurs widely in various bodies of water in the environment. Fortunately, *Legionella* is more of an opportunistic pathogen than a total pathogen. As a net result, it went unrecognized prior to the 1976 outbreak in Philadelphia. Recently, *Legionella* has been found to grow in hot tubs.

Even hot water showers have been implicated in the spread of *Legionella*.

Escherichia coli is primarily a non-pathogenic bacteria that grows extensively in animal intestines. *E. coli* can be easily isolated from the feces of all warm-blooded animals. For these two reasons, *E. coli*, has been used as the primary indicator of fecal contamination in water and foods. Recently, several strains of *E. coli* have been found to be pathogenic, causing deaths from improperly cooked hamburger meat. It appears that the pathogenic strains of *E. coli* are opportunistic pathogens, striking people with reduced immune systems. To date none of the pathogenic *E. coli* have been spread through wastewaters; although, the potential for spreading through contaminated wastewaters always exists.

The development of antibiotics on a large scale during World War II helped to control many of the pathogenic bacteria. Unfortunately, destruction of the most numerous pathogens allowed some strains of the pathogens that were resistant to the antibiotics to find sufficient nutrients to grow to larger numbers. As a net result, the virulent strains of pathogenic bacteria slowly replaced the strains that were easily affected by the antibiotics. It gave the appearance that a new, more virulent pathogen had suddenly appeared that was resistant to the antibiotics. The virulent strains of the pathogens had always existed in the environment; but competition with the normal pathogen had kept their numbers minimized. Once the competition was eliminated, the virulent strains of the pathogenic bacteria predominated. The lack of effectiveness of antibiotics on the surviving pathogenic bacteria has been a cause of concern to the public and has been well publicized by the media. The struggle against pathogenic bacteria is simply part of the continuous struggle to survive. Each new attack on a specific pathogen allows the strongest bacteria to survive. Bacteria that survive antibiotics may modify their gene structure to avoid reaction with the antibiotic, creating a resistant strain of the pathogen. Some non-pathogenic bacteria find antibiotics a source of nutrients and metabolize the antibiotics the same as other organic substrates. The more a specific antibiotic is used, the greater is the opportunity for its metabolism by one or more bacteria in the environment. The battle against pathogenic bacteria is a continuous struggle that has no end and is one that cannot be won completely. Each new victory sets the stage for a more difficult struggle to follow.

FUNGI

It is fortunate that very few fungi are pathogenic to humans. Fungi are very difficult to control once they gain a foothold in an organism. Pathogenic fungi are primarily parasites that overwhelm the host organism. Most of the chemicals used to kill fungi also kill the animal cells being attacked by the fungi. Since fungi are aerobic, it is not surprising that the pathogenic fungi are found in the lungs and in the blood where oxygen is plentiful. It is fortunate that water is not a suitable medium for transmitting fungal pathogens. Fungal pathogens are spread through contact with contaminated soil or spores carried by wind currents. At worst, fungi create a few skin infections from specific surface exposures. When pathogenic fungi cannot be controlled by chemical means, surgery is needed to remove the fungi from the contaminated organism.

Most fungi pathogens attack agricultural plants used for feeding people or animals. The rusts and smuts are typical agricultural fungi pathogens. Some chemicals have been used to control limited fungi infections on plants. More often than not, it is necessary to breed a fungal resistant strain of the plant under attack or to change crops entirely until the fungi and their spores are effectively eliminated. Fungal pathogens have proven to be devastating to local farmers faced with a new epidemic. It takes years to develop sufficient fungi resistant plant seeds to replace the diseased plants.

PROTOZOA

There are very few pathogenic protozoa, when compared to all the non-pathogenic protozoa. The pathogenic protozoa are parasitic pathogens, using the host for nutrients. In most cases the pathogenic protozoa are debilitating, causing a loss of energy in the host organism. Under the right circumstances the pathogenic protozoa can destroy the host organism. *Entamoeba histolytica* is an amoeboid protozoa that was reported by J. Parsonnet to infect about 10 percent of the world's population. It is very common in the developing areas of the world where basic sanitation is not available. The problems *E. histolytica* have already been discussed in Chapter 6. It suffices to say that *E. histolytica* is still a dangerous protozoan pathogen that can be transmitted through contaminated water.

Giardia lamblia is a flagellated, pathogenic protozoa that has recently been found to be transmitted as cysts in improperly treated drinking water. *Giardia lamblia* grows primarily in small, wild animals. The cysts are discharged into streams from the animal waste products. It is not surprising that *Giardia* is found in outdoor sportsmen and naturalists who drink stream water without treating it first. Parsonnet estimated that as much as 3 percent of the U.S. population could be infected with *Giardia*. Most of the people infected with *Giardia* do not show any symptoms of illness.

Recently, *Cryptosporidium* became a major parasitic pathogen transmitted through water. Problems with water treatment in Milwaukee, Wisconsin, resulted in a large number of cases of enteric disease from *Cryptosporidium*. The media helped generate a panic situation without providing a satisfactory understanding of the problem. *Cryptosporidium* is a coccidia which produces small, highly resistant oocyts as a member of the *Sporozoa*. The *Sporozoa* are all parasitic protozoa that form spores as part of their life cycle. It was not recognized that any *Sporozoa* created problems in water until studies on *Cryptosporidium* indicated that these protozoa could create disease in people. While *Cryptosporidium* causes diarrhea, it is not normally life threatening. The main reservoirs of *Cryptosporidium* are dogs and cats, other domestic animals, and small wild animals. Transmission of the oocysts is through animal feces. With the large number of dogs and cats in the United States, it is surprising that the organism has not been a greater threat to public health. It appears that most people develop immunity to *Cryptosporidium* and do not respond to its presence in the environment. The danger from *Cryptosporidium* is in people who have not developed an immunity to this protozoa or who have lost their immune system. This means that the very young, the very old, and people infected with HIV are the most vulnerable to this pathogenic protozoa. Both *Giardia* and *Cryptosporidium* have become of greater importance because of fear generated by the media and the willingness of politicians to take advantage of the situation for their own personal advancement. It is very difficult to detect the oocysts in surface waters. Currently, the Federal EPA is attempting to develop regulations to require all public surface water suppliers to make regular examination of their water to insure that it is safe. The current methods employed for testing for oocysts are not satisfactory and are subject to considerable error. The amount of effort being expended to control *Giardia* and *Cryptosporidium* far exceeds the danger of these organisms in water. Education still has a long way to go before people learn to work together for their common good.

PATHOGEN PROBLEMS

The best defense against all types of disease producing organisms is the natural immunity that develops upon exposure to limited numbers of organisms. When the natural immune system breaks down, the pathogens are able to gain a foothold in the body. Currently, bacterial pathogens and pathogenic protozoa can be controlled by antibiotics and other chemical agents; but viruses and fungi are unaffected by antibiotics at the concentrations normally used. AIDS is a widespread viral disease that adversely affects the natural immune system, making the infected person susceptible to other pathogens. *Giardia* and *Cryptosporidium* become more serious pathogens once the immune system has been damaged by AIDS or other immune destroying diseases. Cancer and natural aging can reduce the immune system, making individuals susceptible to organisms that would normally not affect them.

Natural immunity can protect individuals in the same household from others who develop serious cases of a disease. Personal health is the best protection against most pathogens. Failure to maintain personal health and a suitable lifestyle makes people susceptible to contagious diseases. It is not surprising that the very young and the very old are the groups most affected by pathogenic microorganisms. Over the centuries the pathogenic microorganisms kept the world's population under control. The failure to recognize the true causes of common diseases allowed epidemics to spread quickly and take large tolls in human lives. The discovery that microorganisms were the cause of most of the diseases provided the basis for eliminating the common infectious diseases. Public health was developed over the past 150 years to create a better environment in which people could live longer. There is no doubt that public health has been successful in controlling the most obvious pathogens, allowing the world's population to rapidly increase. It should also be recognized that the additional numbers of people will stimulate the development of new pathogens to help maintain the proper population levels that the world can sustain. Medical science is currently directed towards finding cures for various diseases. For the most part, medical science is directed at the treatment of individuals affected by specific disease organisms. Public health is concerned with preventing the spread of diseases by adjusting the environment to keep the pathogens under control. Together medical science and public health provide the best one-two punch to control the spread of pathogenic microorganisms.

WATER POLLUTION

Water pollution has played a number of different roles throughout history. In areas of the world where water was in short supply, pollution of water supplies was another weapon against the enemy in warfare. As cities began to grow with increasing industrial development, water pollution became more critical. Initially, sewers were constructed to collect storm water from the city streets with discharge into adjacent streams and rivers. When indoor plumbing became a reality, the storm sewers were used to carry the sanitary wastewater, as well as, storm water. The discharge of liquid wastes from the end of the collection pipes into streams and rivers became known as a *point source discharge*. Each pipe discharge presented a separate point source of wastewaters. As populations increased, the intensity of agriculture also increased with more lands being cultivated and more fertilizers and farm chemicals being applied to the agricultural land. Normal rainfall-runoff resulted in some of the soil and chemicals being washed off the fields into surface waters. Some of the fertilizers and chemicals even percolated into groundwater. The runoff from large agricultural areas was considered as being *non-point discharges* since the contaminated water did not enter the streams and rivers through pipes as point sources of wastewaters did. Surface runoff from urban areas that is not collected in pipes is also considered as being non-point source runoff.

Point source discharges carried domestic wastewaters with their potential number of pathogenic organisms into nearby streams and rivers. In those areas where rivers were used for both waste disposal and municipal water supply, enteric diseases became endemic. Once it was recognized that the major diseases were bacterial in origin, efforts were directed towards finding methods to remove and destroy the pathogenic bacteria. Research at the Lawrence Experiment Station in Lawrence, Massachusetts, starting in 1887, quickly found that intermittent sand filters were effective in removing bacteria from polluted waters. Construction of a full-scale intermittent sand filter in Lawrence demonstrated that the use of filtered water throughout the community resulted in a dramatic drop in the cases of typhoid fever. As a net result, many communities using surface water supplies constructed intermittent sand filters to prevent the spread of pathogenic bacteria. Because of the large land areas required for intermittent sand filters, research shifted towards sand filters that could handle larger volumes of water in smaller areas. This new approach produced rapid sand filters. The research at Louisville, Kentucky, starting in 1895, set the stage for rapid sand filter design for treating surface waters in the United States. As more cities built and operated sand filtration plants for treating their water supplies, the rate of enteric diseases began to decrease. Unfortunately, rapid sand filters were not as effective at removing bacteria as intermittent sand filters. In order to insure the safety of the treated water from rapid sand filters, chlorine was added to the treated effluent to destroy any surviving pathogenic bacteria.

Intermittent sand filters were originally designed to treat domestic sewage before it was discharged to streams and rivers. The intermittent sand filters removed organic matter from domestic sewage, as well as, the pathogenic bacteria. One of the characteristics of a well-treated effluent from an intermittent wastewater treatment plant was complete nitrification. While intermittent sand filters were successful in treating wastewaters, the area required for wastewater treatment was also much too great for large cities. The British suffered more serious water pollution problems than the Americans and needed immediate solutions on a large scale. The British built on the research started at the Lawrence Experiment Station and quickly developed the trickling filter by 1897. The trickling filter used rock media rather than sand media. While the degree of treatment in trickling filters was not as great as with intermittent sand filters, the treatment efficiencies were considered as adequate and the economics were much more favorable. Trickling filters quickly became the standard method for treating municipal wastewaters in England and in the United States.

Together, water treatment and wastewater treatment had a profound impact on enteric diseases. The death rates related to waterborne enteric diseases decreased as treatment plants were constructed and placed into operation. The overall effect was a dramatic increase in population and greater longevity. In the United States the population from 1890 to 1990 went from about 63 million to almost 249 million people. Their life expectancy went from about 45 years to 76 years. It is expected that life expectancy in the year 2000 will be around 80 years. Not only has the population increased, but the time each person remains alive has also increased significantly. During this 100-year period, the water treatment and wastewater treatment plants reduced the number of enteric pathogens in the water environment and greatly reduced the potential for the spread of enteric diseases. Even with more people living in larger cities, there was less opportunity for the spread of enteric diseases. While significant progress has been made in the advanced nations of the world, the developing nations have yet to apply the lessons learned from proper water and wastewater treatment. Enteric pathogens have large residual populations in the developing countries with inadequate water and wastewater treatment. When the developing countries construct their water and wastewater treatment plants, the world's population will really explode unless some other pathogens find the increased populations a suitable environment for their growth.

THINGS TO REMEMBER

1. High mountain streams contain little biological life.

2. The lack of available nutrients controls biological growth in streams and rivers.

3. Nutrients primarily wash off the land into streams and rivers.

4. Algae and bacteria grow first and stimulate protozoa and higher animals.

5. Point source organic discharges stimulate oxygen sag curves in streams and rivers.

6. Non-point source organic discharges tend to lower DO levels over long stretches of streams and rivers.

7. Excess organic matter in rivers can produce bacteria metabolism, using all the available dissolved oxygen and creating septic conditions in the river water.

8. Wastewater treatment reduces organic loads on streams and rivers, helping to

keep them aerobic.

9. Ponds, lakes, and reservoirs have similar microbiology as streams and rivers.

10. Long retention periods in lakes and reservoirs allow complete metabolism of incoming organic matter.

11. Deep lakes and reservoirs undergo temperature stratification.

12. Spring and fall turnovers are normal in deep lakes and reservoirs.

13. Primary concern with protected watersheds around water supply reservoirs is from animal wastes.

14. Multipurpose reservoirs have proven as successful as single purpose reservoirs in meeting the water resource needs of the public.

15. Waterborne pathogens will always be a concern for environmental microbiologists.

16. Water treatment and wastewater treatment have proven effective in controlling waterborne pathogens.

REFERENCES

Churchill, M. A. (1954) Analysis of a Stream's Capacity for Assimilating Pollution, *Sew. & Ind. Wastes*. **26**, 887.

Churchill, M. A. and Buckingham, R. A. (1956) Statistical Method for Analysis of Stream Purification Capacity, *Sew. & Ind. Wastes*, **28**, 517.

Churchill, M. A., Elmore, H. L. and Buckingham, R. A. (1962) The Prediction of Stream Reaeration Rates, *Jour. San. Engr. Div., ASCE*, **88**, SA-4, 1.

Edwards, D. D. (1993) Troubled Waters in Milwaukee, *ASM News*, **59**, 342.

Fuller, G. W. (1898) *The Purification of the Ohio River Water at Louisville, Kentucky*, D.Van Nostrand, New York.

Knight, P. (1993) The Hemorrhagic *E. coli*: The Danger Increases, *ASM News*, **59**, 247.

Mass. State Board of Health (1892) *Twenty-Third Annual Report of the State Board of Health of Massachusetts*, Boston, MA.

Mass. Sate Board of Health (1893) *Twenty-Fourth Annual Report of the State Board of Health of Massachusetts*, Boston, MA.

Mass. State Board of Health (1894) *Twenty-Fifth Annual Report of the State Board of Health of Massachusetts*, Boston, MA.

Parsonnet, J. (1992) Gastrointestinal Microbiology, *Encyclopedia of Microbiology*, **2**, 245, Academic Press, San Diego, CA.

Pontius, F. (1994) What You Should Know About *Cryptosporidium*, *Opflow, AWWA*, **20**, No. 6, 5 and No. 7, 6.

Staff (1992) Peru's Cholera Epidemic: Health System's Baptism of Fire, *ASM News*, **58**, 178.

Streeter, H. W. and Phelps, E. B. (1925) A Study of the Pollution and Natural Purification of the Ohio River III. Factors Concerned in the Phenomena of Oxidation and Reaeration, *Public Health Bulletin No. 146*, U.S.P.H.S., Washington, D. C. (1925).

Youmans, G. P., Paterson, P. Y. and Sommers, H. M. (1975) *The Biological and Clinical Basis of Infectious Diseases*, W. B. Saunders, Philadelphia, PA.

Chapter 9

WATER SUPPLY AND TREATMENT

Water supply is one of the most critical factors in the world's rapid population growth. Water is required to meet the physiological needs of people and their domestic animals, as well as, for personal hygiene. Water is also required for industrial plants and agriculture. Concern over the microbiological quality of water began in the 1880s when bacteriology was in its infancy. The discovery that typhoid fever and cholera were being spread through polluted surface waters had a very pronounced effect on the water industry. Protection of watersheds from contamination and the development of water treatment systems became priority items. Since groundwater was not contaminated, efforts were directed to using groundwater as the primary water supplies where possible.

Initially, the water supply industry in the United States was privately owned and operated. As cities grew, local governments took over the responsibility for water supply and distribution within their jurisdiction. Most state health departments developed engineering sections to help local communities in the planning and design of water systems. Private consulting engineering firms took responsibility to provide the detailed design of water treatment plants and water distribution systems. Various equipment manufacturers were established to provide the mechanical equipment required in the water systems. The municipalities took responsibility for the operation of the water treatment plants and the maintenance of the water distribution systems after construction. As the number and size of public water systems increased, the state health departments developed minimum standards for water quality, water treatment plant designs, and water distribution system requirements to protect the public. The engineering sections of state health departments slowly shifted their efforts from assisting local communities with their water system designs to approving water treatment plant plans and specifications produced by qualified consulting engineers. Chlorination was developed to insure that the water reaching the public was safe from pathogenic bacteria. The transmission of enteric diseases through public drinking water steadily declined to zero in the United States. The success of water treatment in public health was so great that the public began to take clean water for granted and the water industry became complacent with the status quo.

After World War II, the rapid growth of industry and the total population created increased pollution of both groundwater and surface waters. While bacterial contamination had been brought under control, environmental activists began to raise concerns about viruses, chlorine resistant pathogens, and a host of toxic chemicals in public water supplies. By 1970 the environmental movement and the media had convinced the American public that the Federal government, rather than the State governments, should control the public water systems to insure the safety of drinking water. In 1974 Congress passed the Safe Drinking Water Act, giving the EPA the power to set water quality criteria for public water supplies. Since the EPA assumed the responsibility for developing water quality criteria,, the number of problems with water quality has increased at a more rapid rate than when the States set water quality criteria. The difficulties with water quality appear to be as much related to increased governmental bureaucracy and politics as to public health. As our population has increased, the need for Federal guidance and coordination in the development of water supplies and water quality has become more apparent. There is no doubt that the Federal government has more resources than State governments to set water policies for the nation as a whole and should be setting national water policies. The basic problem lies in the diversity of this country. The United States is not a uniform geographic region that allows application of a single set of water quality criteria over the entire country. The Federal water quality criteria should set broad objectives with the details for achieving the water quality left to the States and the local governments. Local government has the ultimate responsibility for obtaining the water supply for their citizens, for treating that water supply to a safe level, and for distributing the water to everyone residing in their jurisdiction. Water is a local issue, as well as a regional issue. The State and Federal governments have very little control over actual operation of the water systems at the present time. They may set policies, but implementation of those policies only occurs at the local level. The time may come when either the State governments or the Federal government take over and operate all public water systems. That time is still in the future. It is unfortunate that the public and national politicians do not understand how water systems developed in the United States and what made them the safest public water systems in the world.

GROUNDWATER SUPPLY

As previously indicated, groundwater is the primary water supply of choice with excellent microbial characteristics. Figure 9-1 is a schematic diagram illustrating the production of groundwater. Precipitation from clouds falls onto the surface soil. Sandy loam soils allow some of the surface water to seep into the soil. The water moves by gravity through the void spaces between the soil particles. The tiny soil particles near the ground surface filter out the suspended particles, including bacteria and other microorganisms, carried by the surface water. As the water move

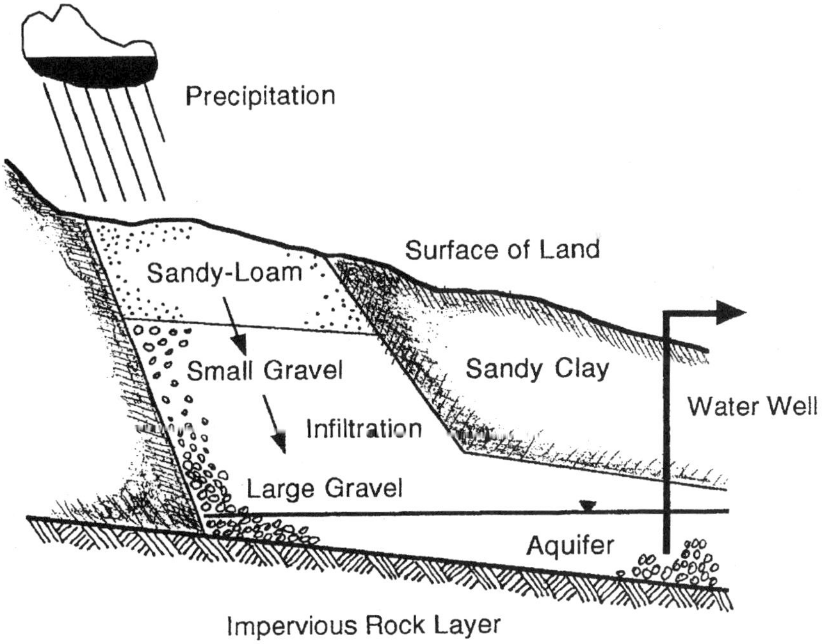

Figure 9-1 SCHEMATIC DIAGRAM OF GROUNDWATER PRODUCTION FROM PRECIPITATION

moves through the porous soil, various chemicals are removed by adsorption onto soil particles. A few chemicals from the soil particles dissolve into the moving water. The soil soon changes to pure sand and small gravel, allowing the water to move quickly downward, creating an *unsaturated zone* until the water becomes part of the *saturated aquifer*. The aquifer moves through the large gravel bed that lies over an impervious rock layer. The impervious rock layer stops the water from moving downward, allowing the water to accumulate in the gravel layer. As the water accumulates, it builds up a pressure head that pushes the water through the porous aquifer. The friction, produced by the water flowing around the gravel particles, determines how fast the water flows away from the recharge area. The lack of available nutrients in the water limits the growth of microbes as the water moves deeper into the ground. The primary path for microbes to reach groundwater supplies is around the well pipes from the ground surface to the well screens. By properly protecting the well pipes, surface contamination can be minimized. The chemical quality of groundwater is determined by the geology around the groundwater. Some groundwater has very little dissolved mineral materials; while other groundwater will have considerable amounts of dissolved salts. If the chemical quality of groundwater is satisfactory for all desired uses, the water is pumped out of the ground to a covered storage tank where it is chlorinated to prevent the growth of pathogens that might accidentally enter the water from an unknown source. If the groundwater has too much mineral material for satisfactory use, the groundwater must be treated to remove the excess minerals. Treatment can range from simple chemical precipitation to demineralization. Highly mineralized groundwater is used only as a last resort when no other suitable water is available.

Normally, groundwater in the covered storage tank is pumped into the distribution system to meet the varying demands of all of the users. The greatest danger of contamination lies in the distribution system. A large number of varying size pipes provides a complex, network from the source of the water to every user. Since water pipes and wastewater drainage pipes are located near each other in the ground, there is always a risk of contamination from the drainage pipes. The risk is small, but is there. The greatest danger lies in *cross-connections* that occur accidentally between water lines and sewer lines. The water pressure in water distribution lines acts as a positive factor minimizing contamination from low-pressure sewage lines. Negative pressure can occur in water distribution systems from time to time, creating a condition where contamination could occur. If there is a cross connection between the water line and a wastewater drainage line, contaminated water can be drawn by the negative pressure into the water line and ultimately distributed by the water system. Water distribution maintenance crews make special efforts to prevent cross-connections and the potential for contamination in the distribution system. When new lines are added to the distribution system, the lines are carefully flushed and disinfected prior to becoming part of the distribution system. Water in the distribution system has a chlorine residual to kill any pathogenic bacteria that might accidentally find access into the water pipes. In large water distribution systems it is necessary to add chlorine at various points in the distribution system to insure that the chlorine residual reaches the farthest points in the distribution pipes. The success of any water system depends upon the effectiveness of the water system employees.

Unfortunately, groundwater supplies are limited. Some areas of the country have good groundwater resources and some areas have none. The key to groundwater supplies lies in recharging the aquifers from surface waters at a rate equal to or greater than the rate of removal by pumping. Most groundwater sources are recharged by precipitation some distance from the water wells with the rate of recharge depending upon the rate of precipitation in the recharge area. A few communities are treating their wastewaters to a high degree and recharging local aquifers with the treated wastewater to keep from overdrawing the groundwater supply. Many communities have seen their groundwater supplies drop as the demands for groundwater exceed recharging in their aquifer. Excessive pumping of groundwater will ultimately remove the available water supply and will require the development of a new water supply. Establishing equitable guidelines for withdrawing groundwater is an important function for State governments with assistance from the Federal government. As future water demands increase, communities that previously had only groundwater supplies will have to develop surface water supplies. Reuse of properly treated wastewaters will help insure adequate water for increasing populations. A key part of wastewater treatment for reuse will be in the removal of potentially pathogenic microorganisms. It will also be necessary to remove the increased minerals and residual organic compounds remaining in the treated wastewaters to prevent the buildup of salts and toxic compounds in the reuse water. Reuse of wastewaters will be one of the major challenges facing the water supply industry as the demand for water increases.

SURFACE WATER SUPPLY

When there are not adequate ground water sources to meet the water needs of a community, it is necessary to use surface water to meet the community water needs. Surface waters require more treatment than groundwater. Part of the problem lies in the fact that no two surface waters have exactly the same chemical and microbial characteristics. Even two surface water supplies, located on the same river, will have slightly different characteristics as a result of different drainage between the two water intakes. Surface waters have both suspended solids and dissolved solids. The suspended solids are normally organic solids combined with inorganic solids, but may be entirely inorganic solids. Some of the suspended solids consist of readily biodegradable organic compounds together with some organic compounds that are non-biodegradable in the aqueous environment. Microorganisms are part of the suspended solids. The higher plants and animals found in surface waters are not considered as part of the suspended solids. The dissolved solids are mostly inorganic with some soluble organic compounds. The soluble organic compounds may be biodegradable, non-biodegradable, or a combination of the two relationships. Colloidal suspended solids are also found in surface waters. Colloids are very tiny particles that act like soluble solids and are often measured as part of the soluble solids. Time, temperature and other environmental factors determine microbial responses in the surface water supplies.

Because of the recognition of the impact waterborne pathogens have had on society, much of the effort on surface water quality is directed towards control of these previously important pathogens, even though they have little current impact in the United States. Direct isolation and rapid identification of microbial pathogens have always been problems with surface waters. The non-pathogenic bacteria tend to overgrow the pathogens in most isolation media, making it hard to recognize growth of the pathogens. As the number of pathogens have decreased in water, isolation of pathogens has become even more difficult. The difficulty in isolating and identifying pathogenic bacteria stimulated a search to find suitable *indicator organisms*, non-pathogenic organisms that are easily cultivated and identified and grow under the same conditions as pathogens. Current research is being directed towards the development of sensitive gene probes that will allow rapid detection of specific pathogens in mixtures of natural bacteria.

The primary source of human pathogens in water has been from human wastes generated by people affected by the disease-producing organisms. Animal wastes also carry pathogens that affect people as well as other animals. Originally, bacteria were the pathogens of concern. Widespread epidemics of cholera and typhoid fever stimulated research on the microorganisms responsible for these two diseases. Both cholera and typhoid fever were caused by specific pathogenic bacteria contained in human feces. Discharge of domestic wastes into surface waters allowed these pathogenic bacteria to be dispersed downstream. Other research indicated that the coliform bacteria were found in large numbers in feces and in water polluted with human wastes. Since the coliform bacteria grew in large numbers in the intestines of warm-blooded animals and were non-pathogenic, the coliform bacteria were used as indicators of water polluted with human wastes.

The ability of the coliform bacteria to metabolize lactose sugar with the production of acid and gas provided the simple test to demonstrate the presence of coliform bacteria in water samples. The presumptive test used lactose broth with submerged, inverted fermentation tubes to capture some of the gas released during metabolism. Small samples of water were added to sterilized tubes of lactose broth. The presence of gas in the inverted fermentation tubes after 24 to 48 hrs incubation at 37° C was considered as presumptive evidence of coliform bacteria in the sample.

Since some soil bacteria also fermented lactose broth with the production of gas, it was necessary to transfer a small portion from the gas producing tube into a second media for the confirmed test. Brilliant-green lactose bile (BGB) broth or eosin methylene-blue (EMB) agar plates could be used for the confirmed test. BGB broth consisted of a mixture of the dye, brilliant-green, with ox bile, lactose, and peptone. The (BGB) media inhibited the growth of the soil bacteria while allowing the coliform bacteria to grow. The production of gas in the fermentation tubes of the BGB media indicated the presence of coliform bacteria. The EMB agar plates could be used instead of the liquid BGB media. Normally, a small loop of sample from the gas producing lactose broth tube was streaked on the EMB agar plate and incubated for 24 hours to determine the nature of the growth. EMB agar allowed differentiation between *Escherichia coli* and other coliform bacteria. *E. coli* produced a flat, dry growth with a green metallic sheen on the surface of EMB agar. The other coliform bacteria produced large, rounded growths that had a wet appearance. To insure that the bacteria were enteric coliforms, it was necessary to continue through the completed test. The completed test consisted of transferring a drop of the BGB media that showed gas back to lactose broth and incubating for 24 hours to demonstrate gas production. With growth on the EMB agar, a part of a single colony was transferred to lactose broth for incubation. The final step in the completed test was taking a sample from the lactose broth tubes showing gas and making a smear on a glass slide, followed by Gram staining. Microscopic examination should show only small Gram-negative rods, indicative of coliform bacteria. The presence of Gram positive, spore-forming bacteria indicated a soil contaminant. The use of the presumptive test followed by the confirmed test, the completed test, and Gram staining provided the desired information on the presence or absence of coliform bacteria. The basic problem with the coliform test was that it required almost a week to reach the final conclusion. If the water had been contaminated, the damage would have been done to the consumers before the test had been completed. The coliform data were simply historical data that had no value unless a problem occurred and an investigation was made to determine the possible cause of the problem. Coliform testing at water treatment plants became a simple routine that had little significance since the water treatment processes effectively removed the coliform bacteria and pathogenic bacteria from the surface waters being treated.

The "total coliform bacteria test" often measured soil coliform bacteria, as well as, enteric coliform bacteria, raising the question as to the validity of using the total coliform test for water quality. The good thing about the coliform test was it indicated contamination in the water sample, even though it was supposed to indicate fecal contamination. The debate over coliform testing continued for years and stimulated considerable research on improving the test. Eijkman's research on coliform bacteria in 1904 found that coliform growth at 46°C differentiated between fecal coliform bacteria and non-fecal, coliform bacteria. In the United States, many years later, E. E. Geldreich and others found that the optimum temperature for growth of fecal coliform bacteria was 44.5°C. Fecal coliform bacteria grow normally at 44.5°C, while the non-fecal coliform bacteria are killed at this temperature.

Another major advance in coliform technology came with the development of the membrane filter. The membrane filter technique was developed after World War II to assist in the rapid detection of bacteria that might be spread as part of biological warfare. Ultimately, the membrane filter technique was adapted for use in the water field to measure fecal coliform bacteria. The key to using the membrane filter in water testing was in the development of a media that provided positive differentiation of the fecal coliform bacteria from the other bacteria normally found in water. The membrane filter consisted of a small, circular, cellulose acetate membrane that retained bacteria on its surface while allowing the water to pass through the tiny pores. The membrane and bacteria were then transferred to a media pad in a sterile plastic petri dish. Surface tension allowed the media to travel up through the tiny pores to the membrane filter surface and to stimulate bacteria growth. Incubation at 44.5°C allowed growth of the fecal coliform bacteria in a lactose broth type media. The major shortcoming of the membrane filter technique was that it was only useful with water samples having low suspended solids.

Coliform bacteria counts on raw waters containing high suspended solids required use of the Most Probable Number (MPN) technique. The MPN test used five tubes of lactose broth in each of three serial dilutions to cover the expected count. Incubation at 44.5°C allowed the fecal coliform bacteria to predominate over the other bacteria. The MPN counting technique required a statistical evaluation of the most probable number of fecal coliform bacteria, based on the number of positive and negative gas producing tubes. Tables in *Standard Methods* provide the most probable number of coliform bacteria from tests using three serial dilutions of five tubes in each dilution. The fecal coliform bacteria test has come under criticism since it does not measure viruses or protozoa pathogens. It is important to recognize that the fecal coliform test is only indicative of a relative level of fecal contamination in the sample of water being tested. There is no easy method to test directly for all the different pathogenic microorganisms. In the United States the concentration of pathogenic microorganisms to the concentration of fecal coliform bacteria in the environment has been steadily reduced over the years. The potential danger from polluted water with the same number of fecal coliform bacteria is less today than it was 50 years ago. Yet, public health officials still use the same numbers of fecal coliform bacteria as indicators of pathogens as they have for almost 100 years. There is no doubt that minimizing fecal coliforms in surface water reduces the probability of pathogens creating a health hazard. It is important to recognize that a zero health hazard cannot exist in the real world environment. It is possible to reduce health hazards to a low level, but never eliminate them completely. The problem lies in the susceptibility of the human beings living in the environment. Not all people have the same immune system or the same resistance to microbial pathogens.

Current concerns about pathogens in surface water center on pathogenic protozoa, primarily *Giardia* and *Cryptosporidium*. These protozoa have become the focus of new EPA regulations. The basic problem with the protozoa pathogens is the difficulty in identifying the cysts and the oocysts. The ability of water treatment operators to identify cysts and oocysts from all the other suspended matter in surface waters is quite limited and is the source of poor results to date. Much research is being carried out on developing rapid techniques for identification of *Cryptosporidium* oocysts. A study by Pittsburgh water treatment plant personnel with 8 quality control samples supplied by the EPA showed 34% mean identification of *Giardia* and 29% mean identification of *Cryptosporidium*. These results clearly show that the best identification techniques available, when this study was made, fell short of the desired results.

Increased urban development and high intensity agriculture resulted in the discharge of additional quantities of complex chemicals in both wastewater effluents and storm water runoff. Improved analytical techniques allowed for greater identification of various chemical pollutants in surface waters. Removal of the hazardous chemicals in surface water has created a host of new water treatment techniques to go along with the older water treatment techniques that have been in use for years. Treatment is essential for the proper use of surface waters for municipal water supplies.

STORAGE

It has been recognized for many years that storage is the simplest method for improving the quality of surface water. Long-term storage in a lake or reservoir can improve the microbial quality of the surface water by sedimentation, starvation, and microbial predation. Gravity settling will remove many suspended particles, including spores, cysts, and oocysts. Wind generated mixing is the primary factor affecting gravity sedimentation of small particles. Lakes and reservoirs that are large, flat and subject to strong wind currents will not show as much settling as lakes and reservoirs protected from strong wind currents. To minimize wind mixing in reservoirs, they should be constructed in long, narrow valleys where the prevailing wind blows across the narrow width of the reservoir. Growing a barrier of trees around the reservoir can help deflect the wind up and over the water surface, reducing wind mixing. While gravity settling removes some of the microbes from the water, starvation and predation by microscopic animals are much more effective in bacteria reduction. The limited nutrients and the low water temperatures in the water prevent significant growth of pathogenic organisms. With a long storage time, the pathogenic organisms and the non-pathogenic organisms both starve to death. If starvation is not adequate, protozoa, rotifers, and crustaceans combine to remove many of the bacteria, pathogens as well as non-pathogens. The dispersed bacteria are easily found as a source of food for the microscopic animals. Even bacteria on the surface of particles are fair game for these predators. Only the bacteria inside of flocculated particles are safe from the predators. From a microbial point of view, the photosynthetic algae find the reservoirs a suitable environment for growth and long-term survival. It is not surprising to see the numbers of algae increase in the surface water in reservoirs and lakes.

Understanding the basic metabolism of algae provides the knowledge of how algae respond in water reservoirs. Since the algae use sunlight for their energy, they grow best near the water surface in proportion to the limiting nutrient in the water. Normally, phosphorus limits the growth of algae in natural waters. Once the algae grow, most of the nutrients incorporated into the algae cells are released back into the water by endogenous respiration. In sunlight the nutrients released by endogenous respiration are synthesized back into cell mass. Since about 20% of the nutrients remain tied up as dead cell mass, regrowth of algae is slowly diminished. The dead algae cell mass settles and accumulates as stable organic solids on the bottom of the reservoir. The growth and death of algae result in removal of nutrients from the water, improving the chemical quality of the water. In some reservoirs the growth of algae can raise the pH of the surface water above pH 8.5 with the formation of insoluble calcium hydroxylphosphate and calcium carbonate. The calcium hydroxylphosphate and calcium carbonate particles settle to the bottom of the reservoir along with the dead algae. In effect, the algae soften the water by precipitating the calcium ions. Multistage reservoirs are more effective in increasing the water retention time and improving water quality than single stage reservoirs of the same capacity. Algae growth occurs to the maximum extent in the first reservoir, leaving fewer nutrients for algae growth in the following reservoirs.

In some parts of the world reservoir water is quite turbid. Wind action across the surface of the reserviors provides adequate mixing to keep colloidal clay particles in suspension. The high turbidity in the reservoir water prevents algae growth, even in the presence of high concentrations of nitrogen and phosphorus. In shallow reserviors the aquatic plant growths on the surface of the water obtains sufficient light, but blocks the light from penetrating the plant leaves into the water below. With heavy coverage of aquatic plants little light penetrates into the water layer, limiting the algae growth.

Algae and actinomycetes can contribute to unpleasant tastes and odors in water stored in large reservoirs. Excessive growths of some algae and actinomycetes create specific organic compounds that impart foul tastes and odors to the water. Taste and odor producing algae grow during the warm summer months and die off in the autumn, releasing volatile oils that create the tastes and odors in the water. The addition of copper sulfate to the reservoirs before the algae grow to critical levels has proven effective in preventing tastes and odors. Control of the algae requires regular microscopic examination and counting of the major species. By plotting a graph of the algae populations against time, the rate of change in numbers can be visualized, allowing the copper sulfate to be applied only when necessary to keep the algae population under control. Copper sulfate is applied from a boat travelling over the surface of the reservoir. Normal wind mixing in the reservoir will disperse the copper and reduce its concentration below toxic levels. The death of algae and microscopic animals will release some nutrients that will stimulate the growth of bacteria in the upper layer of water. The microscopic animals will quickly return when the copper concentration decreases and the bacteria population increases. These reactions will take several days to occur and may be observed by anyone making daily observations. Environmental conditions in the reservoir will allow the algae to regrow. Multiple applications of copper sulfate may be required in reservoirs with high nutrient concentrations and warm climates. The problem with copper sulfate lies in the fact that the copper accumulates on the bottom of the reservoir. As long as the copper is attached to particles or forms an insoluble chemical and remains on the reservoir floor, it will not have a toxic effect on plants or higher aquatic organisms. In the presence of bottom scavenging organisms the ingestion of precipitated copper could allow the copper to accumulate to toxic levels over time. For this reason some aquatic biologists have opposed the use of copper sulfate in reservoirs. Normally, the copper accumulates at low concentrations in layers of sediment at the bottom of the reservoir and does not create secondary problems. Recently, efforts have been made to develop organic algicides that do not accumulate in the reservoir, but are slowly biodegraded by bacteria.

The actinomycetes are washed from agricultural fields in the autumn when crop residues begin to decompose. The water carrying the actinomycetes often discharges into water reservoirs. Like the algae, some actinomycetes release taste and odor producing organics into the water when they die. Actinomycetes normally do not grow in sufficient numbers in non-agricultural areas to produce tastes and odors. The taste and odor producing organics are not toxic. They are simply nuisances that result in large numbers of customer complaints and provide a negative image for the water department. Actinomycetes cannot be controlled by the low copper sulfate dosages used to control algae. Fortunately, the taste and odor compounds from both the algae and the actinomycetes can be controlled by activated carbon. Powdered activated carbon is often used as part of the water treatment process, adsorbing the offensive organic compounds. Chemical precipitation is then used to remove the activated carbon. The negative aspect of powdered activated carbon is the one-time use of the carbon. Fortunately, tastes and odor compounds from algae and actinomycetes occur only over a short period of time and require limited amounts of activated carbon. If the taste and odor-producing organisms occur continuously over time, economics would favor the use of granular activated carbon over powdered activated carbon. The granular activated carbon would be regenerated by heat and reused many times over, making it more economical than powdered activated carbon.

Storage reservoirs have also been examined for pathogenic protozoa. Since wild animals are a source of pathogenic protozoa, it is not surprising that researchers are finding *Giardia* cysts and *Cryptosporidium* oocysts in reservoir waters. Fortunately, cysts and oocysts settle out slowly and are removed to varying degrees in water storage reservoirs. Wind mixing and short-circuiting flow though the reservoir are two critical factors in keeping protozoa cysts and oocysts in suspension.

PRE-SEDIMENTATION TANKS

Surface waters that are taken directly from turbid rivers, rather than from lakes and reservoirs, are passed through pre-sedimentation tanks to remove the settleable suspended solids. Pre-sedimentation tanks are used when space does not permit the use of large reservoirs for both storage and pre-sedimentation. Pre-sedimentation tanks range in size from a low of 4 hours to over 24 hours capacity. Mechanical cleaning equipment is used for removing the settled solids on a continuous or semi-continuous basis. Small water treatment plants may have pre-sedimentation tanks that use simple gravity sedimentation and periodic flushing to remove the settled solids. Mechanical cleaning is preferable in new systems. Gravity cleaning is confined to old tanks that are converted to pre-sedimentation tanks as new sedimentation tanks are constructed. Pre-sedimentation tanks produce minimal changes in the microbial water quality. Only the microbes attached to the large suspended particles will be removed. A schematic diagram of a typical water treatment system is shown in Figure 9-2, starting with pre-sedimentation and ending with chlorination. Chemical precipitation of tiny suspended solids with alum utilizes a rapid mix tank, followed by a slow mixing tank and a quiescent sedimentation tank. The clarified effluent is then passed through a rapid sand filter to remove the residual chemical precipitate. Disinfection with chlorine normally

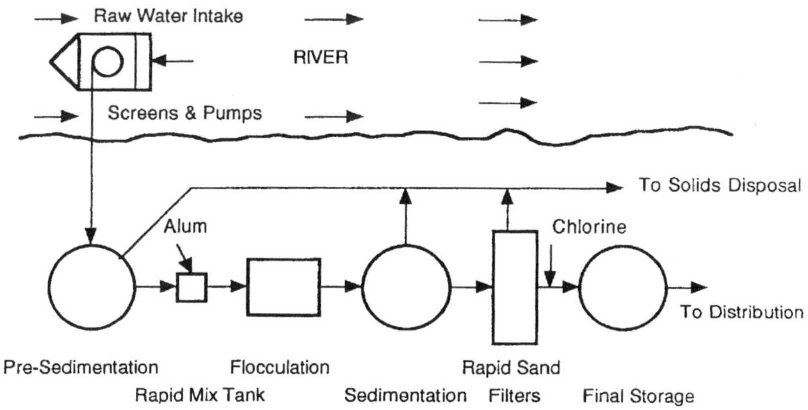

Figure 9-2 SCHEMATIC DIAGRAM OF A SURFACE WATER TREATMENT PLANT

follows sand filtration. Storage allows plenty of time for the chlorine to kill any remaining pathogenic bacteria before the water is pumped into the distribution system.

PRE-CHLORINATION

Surface waters high in fecal coliform bacteria have been pre-chlorinated to reduce the bacteria population prior to treatment. Agricultural areas with large, unconfined, domestic animal populations often experience high fecal coliform bacteria counts after heavy rains and rapid runoff. The runoff carries some of the animal manure off the fields into the surface streams and rivers, creating increases in bacterial counts. Even forested areas with high animal populations will show high coliform bacteria counts in storm water runoff. The surges in fecal coliform bacteria have been controlled by pre-chlorination. It is expected that any pathogens that might be carried with the fecal coliform bacteria would also be reduced. The development of better chemical instrumentation in the 1970s produced equipment that measured trihalomethanes in μg/l concentrations. Research on chloroform and its potential for causing cancer produced a positive correlation in some animal experiments, causing environmentalists to insist on limits for chloroform in drinking water. It took a while before water chemists realized that the chlorination of natural organics in surface waters was producing a complex mixture of chlorinated organic compounds. The chlorine added to kill the fecal coliform bacteria and other pathogenic organisms reacted with some organic compounds the same as it reacted with the microorganisms. The chlorine concentrations were not sufficient to completely oxidize the organic compounds in the water. The chlorine was just sufficient to reduce the fecal coliform bacteria to normal levels prior to water treatment. The chlorine formed addition compounds, depending upon the reactivity of the organic compounds in the water. The stable humic acids formed from the microbial metabolism of organics in the soil reacted with chlorine to form chloroform and other chlorinated methane compounds that were designated as trihalomethanes (THM). Although surface waters had been pre-chlorinated for years without any evidence of increased cancer rates, the EPA established very low THM concentrations for finished water, 100 μg/l. In many surface waters pre-chlorination made it impossible to meet the EPA criteria without additional treatment at greatly increased costs. The net effect was to require very careful control of pre-chlorination in surface waters showing excessive THM formation. Each water system had to develop its own special treatment procedures to meet all the new water criteria mandated by the EPA as a result of pressure from environmentalists and Congress. There has been no evidence generated to indicate that the EPA THM standards have produced a decrease in cancer in the United States. There has not been any additional evidence that THM at the concentrations found in drinking water causes cancer in people drinking the water. On the other hand, it is not possible to conclude that everyone drinking water with 100 μg/l for 70 years would not get cancer. One of the problems that modern society is facing today is too much technical information without adequate knowledge on how to use that information. It is also faced with a media system that thrives on creating scare tactics without proper investigation.

Concerns over THM concentrations soon led to other methods of disinfection than chlorine. As other disinfectants were being promoted as being superior to chlorine, new disinfection by-products (DBP) were discovered. The net effect has been the creation of an entirely new area of research in the water treatment field. The researchers, the equipment manufacturers, the consulting engineers and the Federal EPA have had a field day with DBP formation. In 1996 the Federal EPA published its "Monitoring Requirements for Public Drinking Water Supplies, Final Rule" in the *Federal Register*. This regulation is known as the "Information Collection Rule" (ICR) and is designed to provide information on DBP and microbial contamination in large water treatment plants. It is also hoped that small-scale studies will be made on the removal of these contaminants. This rule was developed because neither the State Regulatory Agencies nor the Federal EPA had sufficient data to determine the magnitude of problems created by DBP and microbiological contamination in large water treatment plants. The failure of water treatment plant administrators to recognize the need for these data caused the Federal EPA to require its collection. It will be interesting to see what the results of this rule are in the next few years. Clearly, it is easy to see why the environmentalists feel that the water industry has failed to provide the leadership expected and needs strong federal guidance to insure satisfactory water quality.

CHEMICAL PRECIPITATION AND SAND FILTRATION

Microorganisms are small, suspended particles that are removed the same as other small suspended particles. Small, suspended particles tend to act as colloidal particles, remaining suspended for very long time periods under completely quiescent conditions. The addition of aluminum sulfate (alum) to the water results in the formation of an aluminum hydroxide precipitate that incorporates the suspended solids into large flocs that settle under quiescent conditions. Initially, alum is added to a rapid mix tank having a liquid retention time of 60 seconds. Mixing is carried out by high-speed mechanical mixers or by hydraulic turbulence. The primary reaction in the rapid mix tank is neutralization of the negative charges on the suspended particles by the aluminum ions. As the alum dissolves in water, trivalent aluminum ions are formed along with divalent sulfate ions. The aluminum ions with their three positive charges neutralize the negatively charged microorganisms and other negatively charged suspended particles before the aluminum ions reacts with the alkalinity in the water to form an aluminum hydroxide precipitate. The reduced surface charge on the suspended particles makes them easier to be incorporated into floc particles.

After the rapid mix, the water flows into a slow mix, flocculation tank. The purpose of the flocculation tank is to disperse the water throughout the tank and to create a gentle mixing so that the small particles will collide and increase in size. As the particles collide, their surface charges are low enough that they will stick together. Van der Waal's surface forces control the growth of the floc particles until the particles grow so large that the hydraulic shear forces break the large floc particles into several smaller flocs that try to reform. The flocculation tank normally has a 20 to 30 minute retention time. The key for good floc tank operation is to keep the floc in suspension while not tearing the floc particles apart. A critical part of the design of the flocculation chamber is the transfer to the settling tank. Care should be taken to minimize water turbulence that tends to break up the newly formed floc particles. Many treatment plants use an entry wall of large planks spaced at regular intervals to create a wooden baffle wall to absorb the energy of motion in the flocculation tank while allowing the water and floc to enter the settling tank at a low velocity. A typical wooden plank baffle wall between the flocculation basin and the sedimentation tank is shown in Figure 9-3. The photograph shows the wooden baffle wall from the flocculation basin side with two slow moving paddle flocculators. The water moves from the left side of the picture to the right side during flocculation. The alum floc,

Figure 9-3 PHOTOGRAPH OF A WOODEN BAFFLE WALL BETWEEN THE FLOCCULATION BASIN AND THE SEDIMENTATION BASIN

sets of vertical planks that dampen the mixing velocity when the floc moves into the sedimentation tank. The alum floc settles immediately on entering the sedimentation tank. The settled sludge is continuously scraped to a sludge collection sump for removal. Heavy-duty sludge collection equipment is used to move the settled sludge to the sludge collection hopper. Periodically sludge is pumped out of the sludge collection hopper for disposal. The sedimentation tanks at this plant are rectangular with 2 to 4 hours hydraulic retention time. The keys to success for the sedimentation tanks lie in the efficiency of floc formation, the rate of floc settling, and the collection of the settled sludge. The reduction in bacteria from flocculation and sedimentation ranges from 99.9 percent to 99.99 percent and higher.

Solids contact tanks have become popular in recent years as replacements for the separate rapid mix-flocculation-sedimentation tanks. Solids contact tanks are circular tanks that provide the rapid mix-flocculation-sedimentation functions in baffled tanks. Although the different equipment manufacturers provide their own special baffle arrangements, the solid contact tanks are all quite similar. The raw water enters a small chamber in the center of the tank where the alum is added. Small mechanical mixers provide the rapid mixing. Sheet metal baffles slow the flow down as it moves to the lower floc chamber that widens out. Paddles on the center shaft provide the slow mixing necessary for floc formation. The lack of turbulence in the lower section allows the heavy floc to settle out. The light floc particles pass under the baffle skirt into the settling section. A series of radial weirs creates a fairly uniform vertical flow. The expanding section of the settling portion of the tank allows the floc velocity to slow significantly. The floc quickly forms a blanket of floc particles that rapidly increases its density as more floc is trapped in the blanket. Settled sludge is removed from the bottom of the tank as well as from the dense floc blanket. Some solids contact systems employ a small solids recirculation system to maintain a definite quantity of floc particles in the flocculation section at all times. The floc blanket is maintained at a specific height by balancing the floc density with the rate of hydraulic flow. One of the functions of the floc blanket is to remove the tiny floc particles by simple filtration as the water passes through the blanket. The simplicity of design and construction of solids contact tanks has made them popular with small water systems.

Rapid sand filters normally follow the sedimentation tanks. These beds of sand or sand and anthracite coal are used to remove most of the fine suspended solids not removed by chemical coagulation and sedimentation. By loading the sand beds at relatively low hydraulic rates, 2 gpm/sf surface area, it is possible to obtain effluents with less than 0.1 NTU and very low bacteria counts. Protozoa cysts and enteroviruses are also removed by chemical coagulation, sedimentation, and rapid sand filtration. The water is applied uniformly over the surface of the sand filter. The suspended solids are removed by the sand particles at or near the surface of the filter. The head of water over the sand filter bed increases to maintain the constant rate of filtration through the filter, as the sand filter accumulates suspended solids on the surface of the sand filter. Periodically, it is necessary to backwash the sand filter to remove the accumulated suspended particles. Backwashing the sand filter is accomplished by forcing clean water up through the sand bed at a high enough velocity to cause the bed volume to expand at least 50 percent. Some sand beds use a secondary water spray in the expanded portion of the bed to force the floc particles from the sand grains. After the accumulated suspended solids have been washed from the sand bed, the backwash water is turned off and the sand settles. The sand stratifies with the heaviest sand settling to the bottom of the bed and the smallest sand grains settling on top of the sand bed. The raw water is then reintroduced to the top of the sand bed, starting normal filter operations. Often, the first water passing through the filter is wasted to insure that the filter media has settled and is working properly. With very high quality water after filtration, a low chlorine concentration easily kills the remaining pathogenic bacteria with a short contact time. A moderate chlorine dose will inactivate enteroviruses. It is important to recognize that chlorine will not destroy all the bacteria unless sufficient chlorine is added for *breakpoint* chlorination. With breakpoint chlorination, the chlorine will oxidize any organic matter in the filtered water. For water low in soluble organic compounds, breakpoint chlorination can be employed at a reasonable cost.

The introduction of organic polymers to improve alum coagulation has produced better water quality for application to the sand filters. The organic polymers have been combined with multimedia filters to permit increased hydraulic loading rates at 4 gpm/sf and higher. The higher loading rates tend to produce greater stress on the floc particles at the media surfaces and increase the number of bacteria passing through the filters. Backwashing must be done on a more frequent interval and must be more thorough than at the 2 gpm/sf rate. The production of satisfactory water at the higher hydraulic loading rates depends upon the effectiveness of disinfection. More water is treated with the same filter area; but operational controls are more sensitive.

SOFTENING

In hard water areas many municipal water systems soften the water to a limited extent. Calcium and magnesium salts in water create hardness. Simple softening consists of reducing the calcium concentration by precipitation as calcium carbonate and settling the precipitate prior to filtration. The addition of lime to the water raises the pH and causes the soluble calcium bicarbonate to shift to insoluble calcium carbonate. After softening the treated water is saturated with calcium carbonate, giving about 40 mg/l hardness. It is necessary to add carbon dioxide to convert the excess calcium carbonate back to calcium bicarbonate prior to sand filtration to prevent the calcium carbonate from cementing the sand grains together. Calcium carbonate is a crystalline structure rather than a flocculent structure and is not as effective as alum in removing suspended solids. If it is desired to reduce the magnesium, as well as, the calcium in the water, sufficient lime must be added to raise the pH close to 12 and convert the magnesium bicarbonate to magnesium carbonate and then to magnesium hydroxide. The magnesium hydroxide has a flocculent structure and is useful in removing suspended particles. Raising the pH to 12 also kills the vegetative microbial cells, producing high quality water from a microbial point of view. Simple lime softening will raise the pH to around 10 and kill the vegetative bacteria in the incoming water. If the water does not have sufficient alkalinity for simple lime softening, soda ash is added to provide the necessary alkalinity. Softening normally follows alum treatment in highly turbid waters. Solids contact tanks are commonly used for water softening. The heavy calcium carbonate crystals produce a good blanket of suspended solids that can handle high fluid flow rates. The excess softening sludge can be mixed with acidic soil to raise the soil pH and provide alkalinity to neutralize acids produced by microbial metabolism of organic matter in the soil.

ACTIVATED CARBON

Powdered activated carbon is often used on a temporary basis to remove taste and odor causing organic compounds from surface waters. Powdered activated carbon provides very large surface areas for rapid removal of the low concentrations of organic compounds. The organic compounds released by algae and actinomycetes are adsorbed by powdered activated carbon that can be removed by alum coagulation and sedimentation. Powdered activated carbon is normally used only once and is not regenerated. Powdered activated carbon is also being used to remove THM compounds formed by pre-chlorination and the organic compounds that could produce THM compounds in the treated effluent. Granular activated carbon is used where water contains soluble organic compounds that must be removed on a continuous basis. Since organic compounds accumulate on the surface of the carbon, granular activated carbon filters are normally used after chemical coagulation, sedimentation, and rapid sand filtration. Application of filtered water allows the activated carbon to adsorb the maximum amount of organic compounds before regeneration. Periodically, the granular activated carbon is removed and regenerated by heat with the destruction of the adsorbed organic compounds. About 10 percent of the activated carbon is lost during heat regeneration, requiring the addition of new activated carbon on a semi-continuous basis. Adsorption of organic matter on the surface of activated carbon particles will stimulate the growth of non-pathogenic bacteria. Bacterial metabolism of the adsorbed organics can lengthen the operational periods between heat regeneration. Extended bacteria metabolism will cover the activated carbon particles with living and dead cell mass, slowing further adsorption of organic matter. For these reasons, granular activated carbon is used for treating waters having low concentrations of organic compounds.

DISINFECTION

Disinfection of the treated water is designed to kill any pathogens that remain in the water after sand filtration and to prevent any pathogens from surviving in the distribution system following treatment. As previously indicated, chlorine has been the disinfectant of choice. Large water treatment systems employ gaseous chlorine (Cl_2). Small water treatment plants find sodium hypochlorite ($NaOCl$) and calcium hypochlorite ($Ca(OCl)_2$) easier and safer to use than gaseous chlorine. Both sodium hypochlorite and calcium hypochlorite are as effective disinfectants as chlorine, when properly used. Chlorine gas reacts with water to form hypochlorous acid and hydrochloric acid as shown in Equation 9-1.

$$Cl_2 + HOH \rightarrow H^+ + Cl^- + HOCl \tag{9-1}$$

While hypochlorous acid undergoes ionization the same as hydrochloric acid, at acid pH values the hypochlorous acid is largely unionized. Sodium hypochlorite and calcium hypochlorite dissolve in water, creating alkaline end products that raise the pH. The net effect is for the hypochlorous acid to ionize and not be as reactive as the unionized hypochlorous acid. Once the hypochlorous acid has been created, it reacts with unsaturated carbon-carbon bonds to form addition products. The hydrogen ion reacts at one carbon atom and the hypochlorite ion reacts at the other carbon atom. The addition reaction blocks further chemical reactions at this location. Hypochlorous acid reacts with each and every unsaturated carbon-carbon bond that it comes into contact with. Since bacteria metabolism utilizes unsaturated carbon-carbon bonds for normal chemical reactions, reaction with hypochlorous acid stops further bacteria metabolism. Once the bacteria cannot metabolize, growth is stopped. If the hypochlorous acid is removed from the double bonds in bacteria, bacteria growth can continue. If more hypochlorous acid is added to the system, the chlorine atoms are removed as chloride ions. The carbon-carbon double bond is broken by oxidation, effectively destroying the molecule required for normal bacteria metabolism. Thus, chlorine causes disinfection by a two step reaction. The first step is an addition reaction; and the second step is an oxidation reaction. The addition reaction does not kill the bacteria, simply blocking metabolism. It is not surprising that the addition reaction is reversible in the right environment. Fortunately, reversing the addition reaction does not occur in the normal water treatment plant. If sufficient chlorine is added to the water, all of the remaining organic compounds are oxidized and a residual of hypochlorous acid remains.

In recent years concerns have been raised about the potential toxicity of the chlorinated addition products and partially oxidized end products. Some environmental activists have indicated a potential for cancer formation with prolonged consumption of chlorinated drinking water. Efforts have been made to indicate the number of cases of bladder cancer and the number of cases of rectal cancer caused on a yearly basis in the United States from chlorinated drinking water. Most of these studies depend upon the use of statistical techniques of limited value. The basic problem in trying to determine a direct cause and effect relationship between chlorinated drinking water and cancer lies in the complexity of modern living and the multitude of factors that affect the health of each individual in the United States. Water chlorination was started on a large scale in Boonton, New Jersey, in 1908 and quickly was adapted across the country. Typhoid fever and general enteric diseases were endemic in the United States until water treatment and wastewater treatment plants began to reach sufficient numbers. The data shows a steady drop in deaths related to enteric diseases in proportion to the number of people served by safe water. The drop in number of cases and deaths correlates quite well with water treatment plant operations. While chlorination was part of the water treatment process, it is not possible to determine the magnitude of the reduction that was the direct result of chlorination. Improvements in modern medicine along with improved sanitation have both contributed to the reduction of enteric diseases in both North America and Europe. The data over the past 70 years do not indicate that the use of chlorine in water treatment has caused any adverse health effects for the general public. Yet, concerns have been raised about the safety of chlorinated drinking water.

Ozone (O_3) has been used for disinfection in areas having low cost hydropower. The basic problem with ozone is the low efficiency in the conversion of oxygen to ozone. Only about 5 percent of the total energy used to generate ozone is converted to ozone. If energy is readily available, ozonation can be a viable alternative to chlorination. The value of ozone is its strong oxidizing reaction. Ozone reacts very quickly with unsaturated chemical bonds. Reaction with saturated chemical bonds depends upon the energy required to break the bonds and the available energy from the ozone. While ozone kills the pathogenic bacteria, it also forms various oxidized intermediate chemicals from the oxidation of organic compounds in the water. Thus, ozone creates its own set of disinfection by-products (DBP). With increased interest in ozonation, equipment manufacturers are developing more efficient ozonators. The use of pure oxygen in ozone production is more efficient than the use of air. One of the drawbacks to the use of ozone comes from its rapid breakdown to O_2. Another drawback is the lack of residual protection in the distribution system. Thus, ozone cannot be used beyond the treatment plant.

Ultra-violet (UV) light has long been known for its disinfection properties. Sunlight is the best source of UV light. Since water quickly absorbs UV light energy, UV radiation only works in water that is clear and in a thin layer. Currently, banks of UV lamps are placed in the stream of clarified water to create disinfection. As the water passes the lamps, the microorganisms absorb the UV energy, disrupting normal metabolism. Like ozone, UV light does not produce a residual effect in the treated water. The high cost of electric energy has limited the use of UV light for disinfection. The increased interest in UV disinfection has stimulated equipment manufacturers to develop more economical UV systems.

Ammonia nitrogen is not a practical disinfectant; but it has value when applied with chlorine. Ammonia reacts with chlorine to form chloramines, Equation 9-2.

$$NH_3 + Cl_2 \rightarrow NH_2Cl + H^+ + Cl^- \qquad (9\text{-}2)$$

The initial reaction produces monochloramines and hydrochloric acid. While chloramines are toxic to bacteria, they are not as effective as hypochlorous acid for disinfection. Part of their value lies in the slower rate of reaction. The chloramines persist for a longer period of time in water than chlorine, providing longer contact periods in the distribution systems. There have been reports of nitrification occurring in the distribution system when chloramines were being used. The nitrifying bacteria find the distribution system a good environment for growth, using ammonia nitrogen for their energy source, bicarbonate alkalinity as their carbon source, adequate phosphates, and dissolved oxygen for cell synthesis. The limited nutrient concentrations in the treated water favors the formation of bacteria films on the surface of the water pipes by the nitrifying bacteria, making it difficult to remove the bacteria films once they have become established. One of the major reasons for using ammonia with chlorine is the rapid reaction of chlorine with the ammonia, minimizing the formation of trihalomethanes and other disinfection by-products (DBP). If it is desired to provide breakpoint chlorination, all of the ammonia nitrogen will have to be oxidized in addition to the organic compounds in the water. For this reason ammonia is normally not added during the treatment process when breakpoint chlorination is used.

Ultrafiltration is the newest form of disinfection to be tried in the water field. Ultrafiltration uses membranes having very small pore sizes to permit retention of all particles above 0.01μ in size. Ultrafiltration is designed to retain all viable microorganisms, as well as, cysts and spores. The basic problem with ultrafiltration is clogging of the membrane pores and cleaning the contaminants from the membranes. Membrane failure is one of the major concerns with using ultrafiltration. Its use has been primarily in small water systems. More improvements in membrane design could provide for increased use of ultrafiltration for disinfection. The drawback in the use of ultrafiltration is the same as with UV disinfection and ozone disinfection. There is no protection from contamination in the distribution system. Like it or not, chlorine or chloramines provide the best protection from cross-connection contamination in water distribution systems.

ENVIRONMENTAL REGULATIONS

Prior to 1970 the responsibility for water treatment plant design, operations, and water distribution systems belonged to the various State Health Departments. The U. S. Public Health Service (USPHS) had a modest program related to drinking water quality to assist the State regulatory agencies as problems arose. With the formation of the federal EPA in 1970 the clean water programs were moved from the USPHS to the EPA. Most of the water programs at this time were related to wastewaters with clean water programs remaining in the background. As more funds were allocated for wastewater programs, the clean water program administrators recognized that Congress was reacting to public concerns over potential problems from polluted water. By indicating that water pollution also threatened the safety of public water supplies, concerns were raised as to the safety of public drinking water. Suddenly, the clean water programs were pushed into the national spotlight, the same as wastewater had been a few years earlier. In 1974 Congress passed the Safe Drinking Water Act, expanding the clean water programs and giving control of drinking water quality to the federal EPA. With pressure from various environmental organizations the federal drinking water regulations have undergone continuous change. In 1986 Congress passed the Amendments to the Safe Drinking Water Act. The latest legislation resulted in the Safe Drinking Water Amendments of 1996. One can expect modifications of the Safe Drinking Water legislation to occur at regular intervals. Congress has been persuaded that it needs to micromanage the environmental field rather than setting goals and allowing the professional experts determine the methods for achieving those goals. The political impact of the environmental issues has become too large to leave to technically trained professionals.

A number of the federal regulations have been designed to improve the microbiological quality of drinking water. One of the regulations is the Total Coliform Rule (TCR), setting the maximum contaminant level (MCL) for total coliform bacteria, fecal coliform bacteria, and *E. coli* in treated drinking water. Essentially, the maximum contaminant level goal (MCLG) has been set at zero. This goal permits the use of relatively simple testing which shows a positive (+) or negative (-) reaction rather than the more complex counting techniques that have been used in the past. The simpler techniques permit more sampling to insure good quality water reaching the customers. The regulations require that samples showing positive tests must be tested for either fecal coliform bacteria or *E. coli*. It has been recognized that statistically, a sample could show a positive test when the coliform bacteria were not present in the sample originally. When a water treatment plant operator tests more than 40 water samples/month, up to 5 percent of the water samples can show positive results before a violation occurs. For small water systems that take less than 40 water samples/month, only one positive sample is allowed before a violation occurs. There is no doubt that a zero coliform count is an admirable goal; but in view of the relative number of pathogens in water as compared to the number of coliform bacteria, it is an overly stringent regulation. The Colilert test has been approved for the initial screening for total coliform bacteria. If the Colilert sample shows positive for coliform bacteria, it can be viewed under ultra-violet light. If the sample fluoresces, it is considered as being positive for *E.coli.* If the sample does not fluoresce, a sample must be transferred to other media to determine the presence of fecal coliform bacteria.

The Surface Water Treatment Rule (SWTR) focuses on *Legionella, Giardia*, enteric viruses, and heterotrophic bacteria plate count. In order to achieve the desired reductions in these organisms, the EPA has proposed specific treatment technology, including disinfection. The goal for all of these organisms is zero. The initial target for *Giardia* is 99.9% reduction or inactivation. The target for enteric viruses is 99.99% reduction or inactivation. The sudden increase in disease transmission by *Cryptosporidium* just after the SWTR was introduced led to the proposed Enhanced Surface Water Treatment Rule (ESWTR). The problem in getting the ESWTR adopted lies in the inability to accurately identify and count *Crytosporidium* oospores. The current techniques are so poor that the errors far exceed the desired limits. In an effort to overcome its lack of data on the two pathogenic protozoa, the EPA has proposed the Information Collection Rule (ICR) to require the various water systems to submit data on a regular basis to provide the EPA with some idea of what is happening. This is a classic example of government attempting to regulate something without having any idea what they are doing. The EPA plans to use the submitted data to develop final ESWTR regulations. Without an accurate measure of these parameters, the validity of the data is suspect. It is apparent that the EPA is under pressure for immediate action by politicians and environmental activists who want absolute answers that are not available.

Considerable progress has been made to improve the microbiological quality of drinking water during the past 100 years. The United States has the best water systems in the world. These water systems are not perfect; and they never will be perfect. The water profession will continue to strive to make the finished water as safe as possible. As our population continues to expand and as we reduce existing pathogens, new microbiological pathogens will appear on the scene as we have seen in the past two decades. These new pathogens will pose new challenges and will require greater technical advancement to bring them under control. The key to success lies in the creation of a real partnership between the public, the government, and the water industry. Continuation of the current adversarial relationships will simply postpone progress and will result in increased costs to the public for the benefit of a few individuals. Unfortunately, the media tends to focus on the negative aspects of American life rather than on the positive accomplishments that have been achieved. People have come to expect technological solutions and breakthroughs on a daily basis. Our modern world is the culmination of efforts by many people who have worked diligently without the recognition heaped upon fleeting stars that accomplish nothing of lasting value.

THINGS TO REMEMBER

1. Groundwater has few microorganisms and is generally the best source for a water supply.

2. Groundwater supplies are limited by the rate of recharge of fresh water into the aquifer.

3. Most microbial pollution of groundwater comes from surface water entering around well pipes.

4. Surface waters are being increasingly used to meet the water demands of expanding populations.

5. Surface waters contain microorganisms that must be removed prior to being used as drinking water.

6. The simplest method for removing microorganisms from surface waters is long-term storage.

7. Microbial removal in long-term storage reservoirs is by starvation, predation, and settling.

8. Algae can create taste and odor problems in storage reserviors.

9. Controlling algae populations with various algicides before they reach critical levels is the easiest way to prevent tastes and odors from dying algae.

10. Pre-sedimentation tanks can remove gross settleable solids from surface waters, but have little effect on removing microorganisms.

11. Pre-chlorination, previously used to control excessive fecal coliform bacteria in raw waters, has fallen out of favor because it causes THM from natural organics in the water.

12. Alum is the preferred chemical for precipitation in water treatment plants.

13. Alum is added to rapid mix tanks to lower the surface charges on particles, including bacteria.

14. Slow mixing allows the floc particles to increase to their maximum size.

15. Care must be taken between the slow mix tanks and the sedimentation tanks to prevent floc breakup.

16. Most of the suspended contaminants and alum flocs are removed in the sedimentation tanks.

17. Rapid sand filters remove the fine suspended particles remaining after sedimentation.

18. Disinfection is used after rapid sand filtration to insure the destruction of all pathogenic bacteria.

19. Ozone and UV lights are being tested as replacement for chlorination.

20. As old pathogens are destroyed, new pathogens arise to take their place.

21. Water treatment has made drinking water in the United States the safest in the world.

22. Water treatment technology is constantly changing to meet new challenges.

REFERENCES

Crockett, C. S. and Haas, C. N. (1995) Protozoan Monitoring: from the ICR to the ESWTR, *JAWWA,* **87:** 8, 50.

Crockett, C. S. and Haas, C. N. (1997) Understanding Protozoa in Your Watershed, *JAWWA*, **89:** 9, 62.

Frost, F., Craun, G. F., Calderon, R., and Hubbs, S. A. (1997) So Many Oocysts, So Few Outbreaks, *JAWWA,* **89:** 12, 8.

Geldreich, E. E. (1966) *Sanitary Significance of Fecal Coliforms in the Environment*, Federal Water Pollution Control Admin. Pub. WP-20-3.

Hale, F. E. (1942) *The Use of Copper Sulfate in Control of Microscopic Organisms*, Phelps Dodge Refining Corp., New York.

Hudson, H. E. (1962) High-Quality Water Production and Viral Disease, *JAWWA*, **54**, 1265.

Keswick, B. H., Gerba, C. P., DuPont, H. L. and Rose, J. B. (1984) Detection of Enteric Viruses in Treated Drinking Water, *Appl. Environ. Microbiol.*, **47**, 1290.

Klonicki, P. T., Hancock, C. M., Straub, T. M., Harris, S. I., Hancock, K. W., Alyaseri, A. N., Meyer, C. J., and Sturbaum, G. D. (1997) Crypto Research: Are Fundamental Data Missing ?, *JAWWA*, **89:** 9, 97.

LeChevallier, M. W., Norton, W. D., Siegel, J. E. and Abbaszadegan, M. (1995) Evaluation of the Immunofluorescence Procedure for the Detection of *Giardia* Cysts and *Cryptosporidium* Oocysts in Water, *Appl. Environ. Microbiol.,* **61**, 690.

LeChevallier, M. W., Norton, W. D., and Atherholt, T. B. (1997) Protozoa in Open Reservoirs, *JAWWA*, **89:** 9, 84.

McGhee, T. J. (1991) *Water Supply and Sewerage*, 6th Ed., McGraw-Hill, New York.

Melnick, J. L., Safferman, R., Rao, V. C., Goyal, S., Berg, G., Dahling, D. R., Wright, B. A., Akin, E., Stetler, R., Sorber, C., Moore, B., Sobsey, M.D., Moore, R., Lewis, A. L. and Wellings, F. M. (1984) Round Robin Investigation of Methods for the Recovery of Poliovirus from Drinking Water, *Appl. Environ. Microbiol.*, **47**, 144.

Metcalf, T. G., Wallis, C. and Melnick, J. L. (1974) Environmental Factors Influencing Isolation of Enteroviruses from Polluted Surface Waters, *Appl. Microbiol.,* **27**, 920.

Miller, K. J. (1994) Protecting Consumers from Cryptosporidiosis, *JAWWA*, **86:** 12, 8.

Moore, B. E. (1993) Survival of Human Immunodeficiency Virus (HIV), HIV-Infected Lymphocytes, and Poliovirus in Water, *Appl. Environ. Microbiol.*, **59**, 1437.

Nieminski, E. C. and Schaefer, F. W. and Ongerth, J. E. (1995) Comparison of Two Methods for Detection of *Giardia* Cysts and *Cryptosporidium* Oocysts in Water, *Appl. Environ. Microbiol.*, **61**, 1714.

Pontius, F. W. (1993) Protecting the Public Against *Cryptosporidium*, *JAWWA*, **85:** 8, 18.

Pontius, F. W. (1993) Reg-Neg Process Draws to a Close, *JAWWA,* **85:** 9, 18.

Pontius, F. W. (1993) Information Collection Rule to Gather Critical Data, *JAWWA*, **85:** 10, 16.

Pontius, F. W. (1996) Inside the Information Collection Rule, *JAWWA*, **88:** 8, 16.

Pontius, F. W. (1997) Expedited Microbial/Disinfection By-product Rules Defined, *JWWA*, **89:** 9, 20.

Robeck, G. G., Clarke, N. A. and Dostal, K. A. (1962) Effectiveness of Water Treatment Processes in Viral Removal, *JAWWA*, **54**, 1275.

Sobsey, M.D., Wallis, C., Henderson, M. and Melnick, J. L. (1973) Concentration of Enteroviruses from Large Volumes of Water, *Appl. Microbiol.,* **26**, 529.

Sorber, C.A., Sagik, B. P. and Malina, J. F. (1971) Monitoring of Low Level Virus in Natural Waters, *Appl. Microbiol.*, **22**, 334.

States, S., Stadterman, K., Ammon, L., Wright, D., Conley, L., and Sykora, J. (1997) Protozoa in River Water: Sources, Occurrence, and Treatment, *JAWWA*, **89:** 9, 74.

Stetler, R. E., Ward, R. L. and Waltrip, S. C. (1984) Enteric Virus and Indicator Bacteria Levels in a Water Treatment System Modified to Reduce Trihalomethane Production, *Appl. Environ. Microbiol.,* **47**, 319.

Swartz, S.G. (1953) Algae Control and Methods of Enumeration, *Taste and Odor Control Jour.*, **19**: 11, 1.

Van Olphen, M., Kapsenberg, J. G., Van de Baan, E. and Kroon, W. A. (1984) Removal of Enteric Viruses from Surface Water at Eight Waterworks in the Netherlands, *Appl. Environ. Microbiol.*, **47**, 927.

White, G. C. (1981) Disinfection by Chlorine, *The Quest for Pure Water*, **2**, 2nd Edition, 1 - 23, American Water Works Association, Denver, CO.

White, G. C. (1981) Other Methods of Disinfection, *Ibid, 24 - 46.*

Chapter 10

WASTEWATER CHARACTERISTICS AND COLLECTION

Once water has been used, it becomes wastewater. Ever since people began to live in cities, collection and return of wastewater back into the environment has been a problem. There are three basic sources of wastewater to be handled. One source of wastewater is from precipitation and is called *storm water.* The second source of wastewater is generated in each house and is called *domestic sewage or domestic wastewater.* The third source of wastewater is from the manufacture of industrial products and is called *industrial wastewater.* Each of these three wastewaters has its own characteristics and impact on the environment. Over the years concern for the pollution potential of wastewater has produced numerous methods for processing wastewaters prior to their discharge back into the environment.

The growth of cities during the 19th century created a number of positive benefits and a number of negative problems. Construction of houses and buildings close together created large areas of impervious surfaces and a reduction in the area of land that absorbed water during rainfall events. Streets were constructed to permit easy traffic movement through the city. The streets went from dirt to gravel that was further crushed and compacted by use, also increasing the impervious area. As the cities grew, the storm water runoff volume increased. For the most part, storm water runoff was a nuisance for everyone in the cities. Eventually, private citizens began constructing drainage ditches along the edges of the roads to handle the storm water runoff. Over time the storm water runoff problems increased, creating problems at an increasing frequency. Eventually, municipal government stepped in and assumed responsibility for the drainage ditches. The major streets soon became paved with bricks or granite blocks and the storm water ditches were eliminated. The storm water collection system moved from surface ditches to pipes buried beneath the ground surface. Surface inlets were placed at regular intervals along the paved roads to collect the storm water. With everyone using horses for personal transportation and for commercial transportation, it is not surprising that horse manure on the streets was a major urban problem. While most cities attempted to collect the manure from the major streets on a daily basis, manure was something everyone had to deal with on a personal basis. Heavy rains were always welcomed in cities since the dust was cleared from the air and the residual manure was washed from the streets. The underground storm water sewers discharged into the nearest natural drainage ditch that carried the storm water into the adjacent stream or river. Local industries normally dumped all solid waste materials on the ground where the wastes accumulated with time. Liquid wastes were dumped into the storm water drainage ditches or directly into the stream or river adjacent to the industrial plant. Storm waters often carried some of the solid wastes from the industrial dumps into the natural drainage ditches. Water pollution was considered a normal part of urban growth. Since the water pollution had little immediate effect on local citizens, municipal government ignored this growing problem.

The development of improved water supplies by municipal governments provided sufficient water for all the community needs. It is not surprising that with more water people began to construct houses with indoor plumbing fixtures. As the popularity of bathrooms increased, a new problem arose. Cesspools in the urban areas were no longer able to handle the increased volumes of wastewaters generated on a daily basis. Even though municipalities had regulations against the discharge of household wastes into storm sewers, it was not long before people were connecting their household drains to the storm sewers. As the volume of domestic wastewaters increased, storm sewers became combined sewers, handling sanitary sewage as well as storm sewage. Municipalities changed their regulations to permit household connections to storm sewers. The pollution load on the receiving streams and rivers increased significantly. By 1870 inland cities were beginning to feel the effects of increasing pollution. Most of the scientific community believed that the self-purification within streams and rivers was more than adequate to handle the pollution loads. Yet, a few individuals were concerned that downstream water users could be faced with future problems. The Massachusetts State Board of Health (MSBH) authorized a study of the chemical quality of a surface water supply, Mystic Lake, which was potentially affected by upstream tannery wastes. Professor William Ripley Nichols of MIT collected and analyzed water samples between the tanneries in Woburn, MA, and Mystic Lake. Professor Nichols demonstrated that the organic matter discharged into the river was oxidized as it moved downstream. While there was no apparent pollution from the tanneries in Mystic Lake, it was recognized that increased industrial activities at the tanneries could create a problem for the people using Mystic Lake as their water supply. Professor Nichols' report stimulated the Massachusetts legislature to authorize the MSBH in 1872 to study the water quality of all public water supplies along the major river basins in Massachusetts and to determine their potential for being polluted by upstream discharges.

By 1886 the Massachusetts legislature recognized that there was a real need for treatment systems to remove the ever-increasing levels of contaminants from polluted river water. The net result was the MSBH establishing the first research center in the United States that was directed entirely towards understanding the problems associated with removing pollutants from surface water at Lawrence, MA. The success of the research at the Lawrence Experiment Station over the next decade demonstrated the value of biological treatment for the removal of pollutants from both contaminated river water and municipal wastewater. British research responded to the research results from the Lawrence Experiment Station, developing the first trickling filter to treat large volumes of municipal wastewater. The trickling filter was followed in 1914 by the development of the activated sludge process at Manchester, England. Over the years research has led to newer biological wastewater treatment processes and to a better understanding of all biological treatment processes. While we have gained much knowledge, there is still much more to learn. We have come a long way in the past 130 years, but the path stretches out in front of us, showing us that even more knowledge lies ahead.

WASTEWATER CHARACTERISTICS

The first step in the design of successful wastewater collection and treatment systems is concerned with an accurate measurement of the wastewater characteristics being treated. It is essential to know the actual wastewater characteristics if the best treatment systems for processing the wastewaters are to be developed. Normal procedures divide the wastewater characteristics into physical characteristics, chemical characteristics, and biological characteristics. The primary physical characteristic of wastewater is the fluid flow rate, normally measured by engineers. Wastewater flow rates are measured as fluid volume over a period of time. English units for wastewater flow rates, gallons/day, are still widely used in the United States. The metric units for wastewater flow rates, liters/second or cubic meters/second, are used in the rest of the world. Although the U.S. Congress adopted the metric system over 100 years ago, the general public has yet to accept the change. Other physical characteristics of wastewater include viscosity and temperature. The chemical characteristics of wastewaters range from a few simple chemical parameters to a large number of chemical parameters. Samples of wastewaters are normally collected in the field and carried to the chemical laboratory where chemists make the various chemical analyses and report the data to the person requesting the analyses. A few chemical analyses are made in the field by chemical technicians. The biological characterization of wastewaters started slowly with a few parameters. In recent years the biological characteristics have increased at a rapid rate and will continue to increase in the future. Biologists are the latest addition to the wastewater characterization team. Unfortunately, chemists, biologists, and engineers do not have the same educational backgrounds. Their basic fields have separate technical languages that tend to create barriers, rather than producing a uniform set of wastewater characteristics that everyone can understand. The chemists, the biologists, and the engineers have tended to generate their own wastewater characteristics without regard to the other technical specialists. The lack of communications between these technical groups has been a real handicap over the years. The engineers were the first to bridge the communication gap by taking more chemistry courses and learning the language used by chemists. The engineers also had to bridge the gap with the biologists. It took the engineers longer to learn biology than it took them to learn chemistry. Some engineers have yet to learn the biology they need to use biology properly in solving environmental pollution problems. By learning more about chemistry and biology, engineers have helped the chemists and the biologists communicate better between themselves. Slowly, but surely, the chemists, the biologists, and the engineers are learning to work together to understand the important wastewater characteristics needed to protect the environment.

Sampling is the most important factor in every wastewater analyses. Normally, there are two types of samples, grab samples and composite samples. As the name implies, a grab sample is a single sample collected at a specific instance of time at a definite location. Analyses of grab samples have limited value by themselves unless the wastewater flows are essentially constant over time. Composite samples are composed of a series of grab samples taken at finite flow intervals over a desired time period. A 24 hour composite sample is often used as a convenient sample period. It is also possible to construct a 24 hour composite sample by collecting a series of 24 grab samples at one hour intervals and then apportioning the size of each hourly sample in direction proportion to the wastewater flow at that time. Thus, the 24 hour composite sample represents a series of grab samples weighted for the flow. The problems with mechanical samplers lie in size of the composite sample, the length of the line from the sample intake to the sample collector, the size of the individual samples, and the wastewater characteristics. The size of the composite sample storage container limits the size of each composite sample. Composite samples are refrigerated to minimize biological activity between sample collections. Efforts to collect more representative samples of widely varying wastewaters have resulted in the use of small samples taken at frequent intervals. Small samples are fine for soluble and colloidal waste materials, but are not suited to wastewaters having large suspended solids. Small sample lines tend to screen out large suspended solids. Failure to clean the influent lines and the sample chamber at frequent intervals will result in microbial growths that will change the waste characteristics being sampled. Since the wastewater analyses will be dependent upon the composite samples, care must be taken to insure that the composite sample is a valid sample.

STORM WATER

Storm water characterization began with flow rate and flow rate variation measurements. The initial problem with storm water was the proper sizing of drainage ditches and collection pipes. Observations showed that storm water runoff was a function of the rate of rainfall, the imperviousness of the drainage area, and the surface area being drained. Rainfall data collected by the local weather bureaus were used to determine the magnitudes of rainfall events over several years time. Initially, engineers determined the quantity of storm water that occurred at a reasonable frequency and designed the sewers to handle the anticipated flow. Once sewers were constructed, engineers began to collect additional flow data over time to provide more accurate information for future designs. It did not take engineers long to realize that their initial designs were too large. Time of flow for the wastewater to reach the sewer inlet and the time of flow within the sewers allowed engineers to design smaller storm sewers. As engineers began to collect data on storm water, people began connecting sanitary drains to the storm sewers. Although sanitary sewer connections to storm sewers were illegal in most cities, the connections increased rapidly as indoor plumbing became accepted in urban areas. Combined sewers became standard engineering practice in the United States. As the wastewater flows in combined sewers increased, the capacities of many sewers were exceeded during heavy rainfalls, creating surcharges in the sewers. Initially, surcharged sewers backed up into nearby houses or overflowed from manholes into the streets. It did not take engineers long to design overflows to take the excess wastewater flows to nearby streams and rivers. When the federal EPA required all municipalities to construct secondary sewage treatment systems in 1972, problems surfaced in cities with combined sewers. The combined sewers hydraulically overloaded new wastewater treatment plants, causing violations of effluent discharge permits. The combined sewers also allowed sewage overflows to continue, creating serious pollution problems in adjacent streams and rivers. It became apparent that the combined sewer concept, which had been extensively used in the United States from 1880 to 1970, was no longer valid and had to be replaced with separate sewers for storm water and for sanitary wastewater. While the federal EPA mandated separate sewers in all new construction, it had a major problem in all of the old sewer systems. Complex sewer systems had been covered with streets and buildings. Many cities lacked detailed records of sewer locations and overflow locations. The federal EPA responded by establishing a combined sewer overflow (CSO) program. The CSO program was designed to establish the wastewater characteristics and the required treatment for the combined sewer overflows before they were returned to the environment. The ultimate goal of the CSO program is elimination of all combined sewers and the CSO program. Unfortunately, the combined sewer overflows have entirely different chemical and biological characteristics than discharges from separate storm water sewers. It is not surprising that data presented in the literature have not always been as clear as they should be to prevent misunderstandings of storm water characteristics. In addition, the federal EPA has a SSO program, sanitary sewer overflow, to reduce all overflows from surcharged sanitary sewers. The SSO program has resulted from builders overloading existing sanitary sewers before the local public works department can upgrade the existing sewers. The SSO program will probably continue *ad infinitum.*

Separate Sewers

The discharge from separate storm water sewers consists of surface runoff following a precipitation event. Normally, the separate storm sewers have no discharge during dry periods unless groundwater leaks into the sewers or illegal connections exist, allowing domestic wastewaters or industrial wastewaters to enter the storm water sewers. The normal discharge flow pattern from separate storm water sewers following a storm event is shown in Figure 10-1. The storm water runoff begins after the start of the precipitation event. It takes time for the water to move across the surface of the drainage area and reach the storm sewer inlet. The flow in the storm sewer requires additional time before the flow reaches the sewer outlet and is discharged into the receiving body of water. The discharge flow rises quickly as the runoff collects in the sewer, reaches a peak and then decreases, rapidly at first and then, slowly for a long period of time. The exact shape of the flow curves and the magnitude of the peak flows will vary considerably, depending on the magnitude and direction of the precipitation events. The same storm water system can show widely varying flow patterns, making it difficult to predict the magnitude of the peak discharge for different storm events, as well as, the length of the discharge event. Since the storm patterns, affecting a given collection system, tend to come at specific times of the year and from the same general direction, most storm water flow patterns will be predictable within a specific degree of error. Unusual storms will produce different flow patterns. Statistical evaluations of the measured data are used to determine the probability of storms of various magnitudes. It is possible to determine the frequency of precipitation events and the expected magnitudes of the storms that occur at different time periods. Like all statistical measurements, the predictions are not absolute and are subject to the generation of additional data

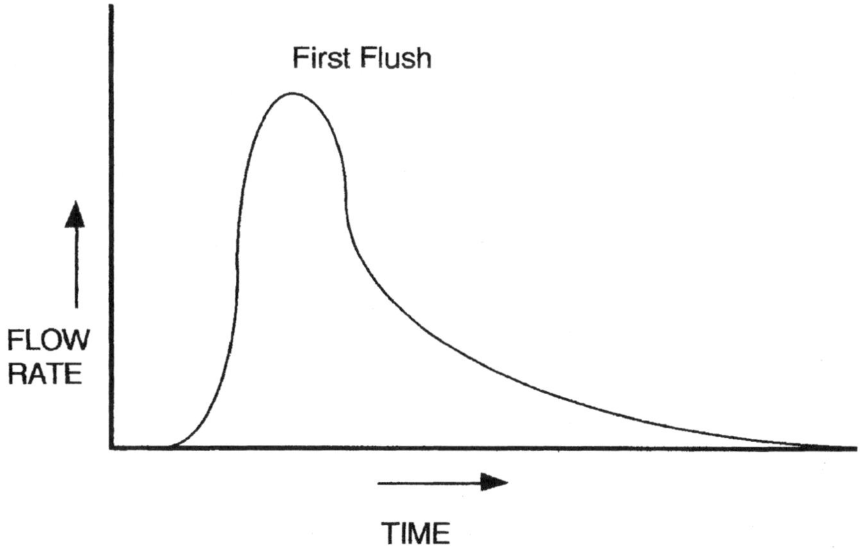

Figure 10-1 A PLOT OF STORM WATER FLOW RATE OVER TIME AFTER A TYPICAL STORM EVENT

The initial rise in storm water flow has been termed the *first flush*. The first flush is the storm water that initially moves over the environmental surfaces, picking up most of the contaminants. The suspended contaminant concentrations in the storm water tend to follow the flow pattern. The velocity of the storm water flow over the ground surface determines what suspended matter can be picked up and carried with the storm water. Small particles are easily collected first and larger particles follow. Once the easily removable particles have been collected, the magnitude of suspended particles in the storm water runoff decreases. The first flush also picks up the readily soluble materials. The concentration of soluble materials quickly rises and then begins to decrease. The overall shape of the various contaminant concentration curves follow the same general shape as the storm water flow curve with the soluble contaminant concentrations peaking first, followed by the suspended contaminant concentrations. If the majority of contaminants come from the farthermost part of the collection system, the contaminant concentrations will peak after the storm water flow peaks. Normally, contaminant concentrations peak before the discharge flow peaks and drop to low concentrations as long as storm water discharges from the sewer.

The area under the flow discharge curve is the total runoff from the precipitation event. The area under the contaminant concentration curve is the total amount of contaminant washed off the drainage basin surface. Since storm water pollution was not considered as being significant until recently, limited data have been collected on contaminant concentrations from separate storm water sewers. It was not until the federal EPA decided to evaluate storm water runoff from urban areas in the late 1960s that adequate storm water quality data were collected. Both batch sampling and composite sampling have been used to collect representative samples for analysis. The initial results of storm water analyses were highly variable. After considerable research, it has been possible to develop some general characteristics for storm water collected in separate sewers. Table 10-1 presents the data reported in the EPA 1983 National Urban Runoff Program Report.

Table 10-1 MEDIAN CONCENTRATIONS OF STORM WATER POLLUTANTS FROM URBAN AREAS (mg/L)

1. Total Suspended Solids (TSS)	100
2. Biochemical Oxygen Demand (BOD)	9
3. Chemical Oxygen Demand (COD)	65
4. Total Phosphorus (TP)	0.33
5. Soluble Phosphorus (SP)	0.12
6. Total Kjeldahl Nitrogen (TKN)	1.5
7. Nitrites and Nitrates (NO_{2+3})	0.68
8. Copper (Cu)	0.034
9. Lead (Pb)	0.14
10. Zinc (Zn)	0.16

The data in Table 10-1 were collected from 81 sites in 22 cities during more than 2,300 storm events. The median concentrations represent the middle values for the collected data. Half of the data had values greater than the median and half of the data had values less than the median. Suspended solids were the major pollutants contained in storm water, as would be expected. Tiny soil particles tend to be carried by the wind during dry periods and deposited on urban surfaces. Precipitation events remove the small, suspended particles from the surfaces of buildings, sidewalks, parking lots and streets. Most of these small, suspended solids are insoluble, inert particles. Very few chemicals dissolve into the storm water. The storm water that falls on impervious surfaces moves rapidly across the impervious surfaces by gravity towards the storm water inlets. Some of the storm water falls onto soil surfaces that tend to be permeable to water. Initially, the soil surfaces are dry, allowing some of the precipitation to move into the void spaces in the surface soil. The soil particles filter out the suspended particles and allow the soluble contaminants to move deeper into the soil. It does not take long for the surface soil to become saturated with water, allowing the soil surface to act as an impervious surface. As more water accumulates, surface runoff is generated on both the soil surfaces and the impervious surfaces, creating the total runoff flow. Since limited surface contamination exists, the storm water flows tend to dilute the contaminants that are picked up. The data on storm water characteristics show little biodegradable organic matter. The 9 mg/l BOD_5 indicates a total BCOD of about 16 mg/l. The non-biodegradable chemical oxygen demand (NBCOD) can be determined by subtracting the biodegradable chemical oxygen demand (BCOD) from the total chemical oxygen demand (COD). These data indicate that the storm water averages about 49 mg/l NBCOD. Examination of the current database on storm water characteristics indicates that urban storm water runoff does not normally represent a major source of environmental pollution in streams and rivers in the United States.

Urban storm water also picks up soil microorganisms. Concerns over coliform bacteria have shown that most of the coliform bacteria in storm water are soil coliform bacteria. The small numbers of fecal coliform bacteria in urban storm water are from the feces of cats, dogs, and rodents. Household pets can be a source of pathogenic protozoa, primarily *Cryptosporidium*. While separate storm sewers carry some pathogenic microorganisms, the dilution effect of the large volumes of storm water minimizes the potential hazard of pathogenic microorganisms from storm water. Illegal connections or illegal dumping of concentrated wastes into storm sewer systems can be detected by measuring high concentrations of bacteria or other contaminants in the effluent discharged from storm sewers.

Combined Sewers

The overflow from combined sewers has different characteristics than the discharge from separate storm water sewers. Combined sewer overflows are simply dilute municipal wastewater. The relative volumes of municipal wastewater and storm water runoff determine the characteristics of combined sewer overflows when the overflow events occur. Originally, combined sewers were designed to carry the domestic wastewaters from a given population of residents plus the runoff from storms of a specific magnitude. As long as the municipal wastewater flow and the storm water flow stay below the design limits, there should be no discharges from the combined sewers except at the terminal end of the sewer. Since unusual storms occur from time to time and local public works departments do not expand their collection systems as often as they should, design engineers tend to place overflows at convenient points along the combined sewers to allow surcharged flows to discharge directly into adjacent waterways before the combined wastewaters back up into adjacent houses or streets. As the municipal wastewater loads on combined sewers increase, the combined sewer overflows will occur at increasing frequencies and their characteristics approach the characteristics of municipal wastewater. The discharge of large volumes of untreated municipal wastewaters from overflowing combined sewers constitutes a definite pollution hazard for downstream environments. It is not surprising that the EPA CSO program was established to focus on controlling the combined sewer overflows. Because of the large number of older cities in the United States with combined sewers, it will take years to eliminate the combined sewer overflow pollution. Unfortunately, citizens tend to mix up information on combined sewer discharges with separate storm water sewer discharges, creating considerable confusion. It is important to recognize the characteristics of both types of storm water discharges and which one represents the most serious threat to the environment.

RURAL RUNOFF

Rural runoff consists of the storm water runoff from uninhabited areas, as well as, from agricultural areas. Because of the differences in chemical characteristics of storm water runoff from uninhabited areas and from agricultural area, agricultural runoff will be examined separately. While most people think that storm water runoff from uninhabited areas is free of pollutants, it is not. Decaying leaves and vegetation, animal waste products, and soil form the basic pollutants from uninhabited areas. The local topography determines the characteristics of the pollutants found in rural runoff. Rapid runoff will usually carry suspended solids into adjacent rivers and lakes. Rapid runoff will usually have little soluble contaminants. Flat topography often provides an environment for trees, bushes, and other forms of vegetation. Forested areas provide suitable habitats for various animals. Runoff from flat land areas is slow, allowing time for various materials to dissolve in the runoff water. The forest litter can undergo both aerobic and anaerobic metabolism. Organic acids together with tannins can be found in the runoff water. In soft water regions the organic acids can depress the pH below 6, producing acid waters that limit normal biological development in affected streams and lakes. The tannins impart a brown color to the water. Animal wastes can provide organic pollutants, as well as, nitrogen and phosphorus. Animal wastes can also be a source of pathogenic protozoa, bacteria, and viruses. While natural vegetation helps to hold the soil and minimize erosion, heavy storms and steep slopes combine to provide considerable loss of soil from uninhabited areas. Nature is constantly changing the rural topography, creating positive and negative impacts. The variability of pollution from uninhabited areas prevents the development of specific wastewater characteristics that can be used as a general guide. Each area has its own waste characteristics that must be determined from field measurements and evaluated separately.

AGRICULTURAL RUNOFF

As more and more land has been developed for agricultural purposes to feed the expanding populations of the world, agricultural runoff has created its share of pollutants in storm water runoff. Decomposing crop residues, fertilizers, pesticides, herbicides, animal manure, and soil are all common pollutants from agricultural land. The quantities of pollutants carried in the runoff water depend upon the characteristics of the agricultural region. It is interesting to note that efforts to increase agricultural yields have generated greater pollution loads on the environment. Initial efforts at controlling farm pollutants were directed towards soil conservation techniques. The loss of soil from plowed fields after rapid rains resulted in increased suspended solids in streams and lakes. Soil conservation procedures produced contour plowing and grass strips on steep slopes to minimize runoff water velocity and the carrying capacity of the runoff flows. Although the soil conservation techniques reduced the loss of soil, soluble pollutants were still carried by the runoff water. In the 19th century agricultural fertilizers were primarily human wastes from urban areas and animal manure. Human wastes were collected in carts from cesspools in or near each house and carried outside the urban areas to agricultural land. Manure was collected from urban areas, as well as, from dairies and cattle farms for use as fertilizer. Edwin Chadwick in England championed the concept of sanitary sewers in urban areas to collect human wastes and to distribute them to agricultural areas outside the urban areas. He felt that domestic sewage had an economic value that should be recovered. The development of new water supplies in the 1850s provided more water to urban dwellers, stimulating the use of bathtubs and water closets. Suddenly, the sanitary sewers had more water to carry human wastes away from all the houses. The diluted sewage had less value as a useful fertilizer than the concentrated cesspool wastes. The development of modern plumbing eliminated the application of untreated human wastes on agricultural lands as a fertilizer in the United States and Europe.

As the automobile replaced the horse as the primary mode of transportation, manure supplies from urban areas decreased and helped to stimulate the development of low cost, chemical fertilizers. Chemical fertilizers made large scale farming a reality and changed the social characteristics of the industrial countries of the world. The less developed countries continued to use animal manure and human wastes for their primary source of fertilizer. As the agricultural chemical industry developed, herbicides and pesticides were created to improve crop yields. Unfortunately, the success of these chemical agents was matched by a negative reaction. The herbicides and pesticides produced serious water pollution problems. The pesticide, DDT, was used to save millions of lives after World War II; but it had serious pollution characteristics that did not become apparent until it had been widely used over the entire world. DDT was not biodegradable in the natural environment, accumulating as more DDT was used. Being strongly hydrophobic, DDT was not very soluble in water. Unfortunately, the soluble DDT was ingested by aquatic organisms and concentrated in fatty tissue. As the organic matter moved up the food chain, DDT accumulated in the fatty tissues of higher animals. The success of DDT as a pesticide led to other chlorinated compounds that were even more toxic than DDT. Rachel Carson focused attention on the damaging effects of DDT and the chlorinated pesticides in 1962 with the publication of her book, *Silent Spring.* The failure of the chemical industry to respond positively to the questions posed by Rachel Carson led to increased militancy of the environmental movement and helped in the formation of the federal EPA in 1970. Congress eventually passed legislation allowing government control of the production and use of toxic chemicals. Although DDT was banned in the United States, it is still being used in many of the developing countries of the world. Its positive value in some parts of the world outweighs its negative aspects. Increasing federal regulations in the United States has forced the chemical industry to develop biodegradable chemicals that do not accumulate in the environment, but break down to non-toxic components over time. Strong efforts are being made to phase out the manufacture and use of all non-biodegradable pesticides and herbicides.

Currently, agricultural runoff contains organic nitrogen, ammonium nitrogen, nitrate nitrogen, and phosphates from fertilizers and crop residues. The nutrient elements can stimulate excessive growth of algae in rivers, streams and lakes. The high concentrations of algae can adversely affect normal biological development in the water, as well as, create organic compounds capable of causing tastes and odors if the water is used for domestic consumption. Organic nitrogen compounds stimulate bacteria to metabolize the organic matter and release nitrogen back into the water as ammonium compounds. Nitrifying bacteria will use the ammonium nitrogen under aerobic environments to create nitrites and nitrates. Since both groups of nitrifying bacteria use dissolved oxygen in their metabolism, they can put a stress on the oxygen resources in the receiving streams and lakes. At night the algae have a high oxygen demand rate for endogenous respiration. The combined metabolism of the algae and bacteria at night can often remove the available DO, creating anaerobic conditions that have an adverse impact on the high-level plants and animals in the water. Suspended solids carried by agricultural runoff include crop residues, soil particles, manure, and microorganisms. After the harvest of crops, the crop residues slowly decay and are plowed back into the surface of the soil to prevent nutrient loss from the soil. Storm water can carry off crop residues at various stages of decay before the residual crop materials are plowed back into the soil. Fortunately, most of the crop residues are retained in the agricultural soil. Like urban storm water, it is not possible to predict agricultural storm water runoff characteristics. Each farm and each farmer create their own storm water runoff characteristics.

The one agricultural waste that can be measured on a semi-quantitative basis is animal manure. The creation of feedlots and confined animal buildings produced a serious problem for animal manure processing for return to the environment. The characteristics of animal manure are directly related to their feed characteristics. Animals fed silage will produce more manure than the same animals fed grain.

Animals kept in confinement are normally fed high-energy grains to minimize manure production and maximize weight gain. Since feed requirements are associated with the size of animals being grown, manure production data are often presented in terms of the weight of pollutants/day/animal for different sizes of animals. Cattle tend to be grown in feedlots with pigs and chickens being grown in confined buildings. In feedlots the manure is subjected to natural environmental conditions before the manure is collected. Wind, rain, heat, cold, hooves, and time all affect the microbial reactions in the manure and determine the final manure characteristics. It suffices to look at the fresh manure characteristics of cattle and to recognize that the characteristics change with the environmental conditions in the specific feedlot. Animals, grown in confined buildings, discharge their wastes through slotted floors into pits. Periodically, the pits below the animal floors are flushed out with water and collected in large, open lagoons. The microbial reactions in animal waste lagoons are largely anaerobic with the production of nuisance odors if the lagoon is not properly designed and operated. When conditions are suitable, wastes are pumped from the lagoons and applied to land for use as a fertilizer and for additional biological stabilization. Care must be taken not to overload the land with either too much nitrogen or too much biodegradable organic matter. It is also important to minimize the water used to prevent the applied wastes from moving over the land surface into nearby surface waterways. Minimum water consumption means a concentrated wastewater that must be properly handled if pollution is to be prevented. A good starting point for determining the wastewater characteristics is determining the total feed used and the total water consumption. All feed and water used by the animals will appear as either weight gain or as waste products. This simple mass relationship can be very useful in evaluating the overall wastewater characteristics in confined animal operations. For the most part, animal wastewater characteristics are evaluated on the same basis as domestic wastewater. Detailed waste characteristics for various types of animal wastes can be found in the US Dept. of Agriculture, *Agricultural Waste Management Field Handbook.*

DOMESTIC WASTEWATER

Domestic wastewater is all of the wastewater from residences, public buildings and commercial establishments required to maintain the residential community. Residential wastewater comes from single and multi-family residences and apartments. Wash waters and toilet discharges make up the residential wastewaters. Domestic wastewater from public buildings is largely washroom discharges and floor wash water. Commercial establishments create washroom discharges and some process wastewater from food preparation activities. The characteristics of domestic wastewater are related to the habits of the people living in the community and the number of people in the community. People are creatures of habit and follow definite activity patterns on a regular basis. Work schedules determine the primary activities of the people. In the United States people tend to work a 40-hour week over 5 days at 8 hours per day. Additional work hours are required at home to maintain their lifestyle.

People produce about the same amount of personal waste materials each and every day. Feces consist of unmetabolized food and bacteria produced during the digestion of food. The primary unmetabolized food components are cellulose, hemicellulose, and lignin with some lipid materials and complex nitrogen compounds. Washing is the other activity that produces wastewater. Washing dishes, clothes, and people uses considerable water and produces various contaminants together with soaps and detergents. These wastes together with toilet paper constitute domestic wastewater. It is not surprising that domestic wastewater characteristics show little variation on a day-to-day basis.

Most of the organic contaminants in domestic wastewaters are suspended solids. In the United States the average suspended solids production for domestic wastewater is 76 g/person/day (0.17 lbs/person/day). Approximately, 80 percent of the suspended solids are volatile solids, giving 61 g VSS/person/day (0.13 lbs/person/day) on the average. It has been found that about 65 percent of the VSS in domestic wastewater are biodegradable in a water environment. The soluble organic solids are about 95 percent biodegradable in a water environment. Most of the soluble solids in domestic wastewater are inorganic solids from the water used to carry the contaminants to the wastewater treatment plant. The oxygen demanding potential of domestic wastewater is measured by the 5-day biochemical oxygen demand test (BOD_5 or just BOD). The average BOD_5 for domestic wastewater is 67 g/person/day (0.15 lbs/person/day). Approximately 40 g BOD_5/person/day (0.09 lbs/person/day) is suspended BOD_5 with about 27 g BOD_5/person/day (0.06 lbs/person/day) as soluble BOD_5. Other significant contaminants include nitrogen and phosphorus. The total nitrogen in domestic wastewater is measured as Kjeldahl nitrogen (TKN). Kjeldahl nitrogen measures both the organic nitrogen (Org-N) and ammonia nitrogen (NH_3-N). The suspended solids contain approximately 2.5 g Org-N/person/day (0.006 lbs/person/day). The soluble TKN averages 12.5 g/person/day (0.028 lbs/person/day). The NH_3-N ranges from 4 to 8 g/person/day (0.009 to 0.018 lbs/person/day) with the soluble Org-N varying between 8.5 and 4.5 g/person/day (0.019 and 0.01 lbs/person/day). Most of the soluble Org-N is urea that is quickly hydrolyzed to NH_3-N by bacteria in the wastewater. The phosphorus in domestic wastewater ranges from 2 to 4 g/person/day (0.004 to 0.009 lbs/person/day), depending on the phosphorus limitations in the commercial synthetic detergents. Some areas of the country limit the amount of phosphates that can be used in commercial synthetic detergents to minimize the amount of phosphorus in municipal wastewaters. It is just a matter of time before the phosphate content of synthetic detergents is controlled nationwide. The pH of domestic wastewater ranges between 6.0 and 9.0 with most wastewater having a pH around 7.0. The pH of domestic wastewater is largely related to the alkalinity of the carriage water and the time of travel in the sewerage collection system. Wastewater in areas of the country having soft water has a pH around 6.0 to 6.5 and an alkalinity between 50 and 100 mg/l as $CaCO_3$. Areas of the country having moderately hard water will have domestic wastewater with a pH between 7.0 and 8.0 and alkalinity values between 100 mg/l and 300 mg/l as $CaCO_3$. Hard water areas will have wastewater with a pH between 7.5 and 9.0 and alkalinity values from 250 mg/l to 500 mg/l and higher as $CaCO_3$. Alkalinity measurements should always be made on centrifuged samples of wastewaters containing suspended solids to eliminate the error generated by the reaction of the sulfuric acid with the suspended solids.

Many of the bacteria in domestic sewage come from feces. Soaps, detergents and high temperatures used in washing clothes and dishes minimize additional bacteria. It has been estimated that bacteria make up 25 percent of the weight of feces. The fecal bacteria are highly variable with many different species growing according to their ability to metabolize the organic compounds. The large intestine provides the environment for the growth and survival of fecal bacteria. The majority of bacteria in feces are facultative bacteria with some strict anaerobic bacteria. Fecal coliforms and fecal streptococcus are the most unique bacteria in domestic wastewater. They are used as the primary indicator bacteria for fecal contamination. In spite of their uniqueness the fecal coliforms and fecal streptococcus do not appear to be more than 1.0 percent of the total fecal bacteria population. The limited data on bacteria in domestic sewage indicate that there are between 1,000,000 and 10,000,000 bacteria/ml by the time the domestic wastewater reaches the treatment plant. Total coliform bacteria counts average about 30,000/ml with 10,000 fecal coliforms/ml and about 2,200 fecal streptococcus/ml. The majority of fecal bacteria are common soil bacteria that grow under anaerobic conditions. Methane bacteria and sulfate reducing bacteria can also be found in feces. The relatively short retention time for the feces in the large intestine limits the growth of highly specific bacteria. It is interesting that cellulose-degrading bacteria do not grow in human intestines, allowing the ingested cellulosic materials to be discharged in the feces. Once domestic wastewater enters the sanitary sewers, soil bacteria, actinomycetes, fungi, and protozoa find a suitable environment for growth. Most of the microorganisms in municipal wastewater are dispersed in the liquid or contained within suspended solids that are carried by the wastewaters. A few microorganisms become attached to the sewer surfaces as the wastewaters are collected and flow to the treatment plant. The microbial attachments occur at pipe joints, house connections, and cracked pipes. Suspended solids tend to accumulate at these points during low flow periods, providing a source of nutrients for the microbes over a longer time period than indicated by the fluid retention time. The joints, cracks, and junctions in the sewer pipes provide points for filamentous microorganisms to grow in sanitary sewers. Many of the bacteria continue to grow anaerobically inside fecal particles. The fluid surface allows some oxygen to move from the air above the wastewaters into the liquid, where the microbes quickly use the dissolved oxygen in aerobic metabolism. Fungi and protozoa grow at the wastewater surface under aerobic conditions. Many of the microbes produce spores or cysts to allow their survival under adverse environmental conditions in the sewers. When the wastewater reaches the treatment plant it is usually septic since the overall demand for oxygen exceeds the oxygen transfer in the sewer. At the same time, the metabolism by the various groups of microorganisms during their travel in the wastewater collection system has changed the characteristics of the wastewaters and has even produced some stabilization of the organic matter.

5-Day Biochemical Oxygen Demand

The BOD_5 test is widely used to measure the pollutional characteristics of wastewater. The primary value of the BOD_5 test lies in its measurement of the oxygen demand required to stabilize the biodegradable pollutants in wastewater. The BOD_5 test began as a measurement of the oxygen demand in streams and rivers after the addition of municipal wastewater or industrial wastewater. The test was then applied to domestic wastewaters and treated effluents in an effort to predict the potential oxygen demand load on the receiving streams and rivers. Over time, the BOD_5 test became accepted as one of the primary measures of wastewater pollution characteristics. Unfortunately, the BOD_5 test is not a perfect test and yields highly variable results. It is one of the few biological tests applied to wastewater characteristics. Most wastewater tests are either physical tests or chemical tests. Because of the 5-day incubation period, the BOD_5 results are primarily of historical value in evaluating wastewater treatment plant efficiencies. In spite of efforts to develop other tests to replace the BOD_5 test, none have been successful. As long as the limitations of the BOD_5 test are understood and properly applied to the data, the BOD_5 results can be useful in evaluating the oxygen demand potential of wastewater.

Basic Test - The BOD_5 test employs 300-ml glass bottles with glass stoppers. Each BOD bottle has a flared lip around the mouth of the bottle for a water seal. A series of 3 to 5 BOD bottles is used for each sample. *Standard Methods* indicates that the oxygen demand should not remove the last 1.0 mg/L DO for valid BOD results. When the DO drops below 1.0 mg/L, the DO can become a limiting factor in the rate of microbial metabolism. The solubility of oxygen from the air into water at 20° C and 1.0 atmosphere pressure is only 9.1 mg/L. As a net result, the maximum available DO for a valid test will be 8.1 mg/L. Municipal wastewater samples normally have BOD_5 values many times greater than 8.1 mg/L, making it is necessary to dilute the wastewater samples to yield BOD_5 values less than 8.1 mg/L. At the other end of the spectrum, the diluted sample should exert a 5-day oxygen demand of at least 1.0 mg/L to be valid. *Standard Methods* recommends that the diluted samples have a minimum 5-day oxygen demand of 2.0 mg/L with 1.0 mg/L of the 2.0 mg/L oxygen demand coming from the microbial seed added to the BOD bottle. Municipal wastewaters normally do not need additional microbial seed. The use of several different dilutions for each sample increases the probability of having at least one or more bottles with at least 2.0 mg/L oxygen demand and having at least 1.0 mg/L DO remaining. Using several dilutions within the desired range of DO greatly increases the validity of the data. All valid data are averaged together to determine the sample BOD_5. In recent years, several commercial companies have offered dehydrated bacteria seed for the BOD test. While the commercial seeds give good results with the G-GA standards, it is important to make periodic checks using normal seeds to insure that the results with the wastewaters are adequate. Good BOD seed should have a mixture of bacteria and protozoa with the bacteria acclimated to the major organic compounds in the wastewaters being tested.

Currently, dilution water for the BOD_5 test is prepared from distilled water or demineralized water. The dilution water is normally stored for several days at 20° C prior to use to allow the entire bottle of dilution water to reach the desired temperature. A phosphate buffer is added to the dilution water to keep the pH around 7.2 for good microbial growth. Ammonium chloride, magnesium sulfate, calcium chloride and ferric chloride are added to insure sufficient nitrogen and key trace metals are available to the bacteria. The dilution water is prepared and aerated for about 30 minutes to insure saturation of the dilution water with DO. After the air has been turned off, the dilution water is allowed to sit for 5 to 10 minutes to allow all the tiny bubbles to rise to the water surface. The water is carefully siphoned into two BOD bottles that do not have any samples. One Blank will be used to determine the initial DO of the dilution water. The other Blank will be incubated with the BOD samples for 5 days before measuring its DO. The change in DO from the initial Blank to the incubated Blank should not exceed 0.2 mg/L. Changes greater than 0.2 mg/L in the Blank DO indicates some problems with the dilution water. The DO of the wastewater to be placed in the BOD bottles is usually measured to permit calculation of the initial DO in the BOD bottles. The microbial seed is added first, if required. Then, the test samples are pipetted into each bottle. Finally, the BOD bottles are filled with dilution water that is siphoned from the dilution water bottle, taking care to keep the discharge tip just below the liquid surface to prevent adding additional DO. The glass stoppers are put into each BOD bottle and water placed around the flared lip to create a water seal. Sample sizes for the BOD bottles are determined from the estimated BOD_5 values of the wastewater samples. Raw municipal wastewaters without major industrial waste sources have BOD_5 values close to 200 mg/L. Primary sedimentation removes about 30% of the BOD_5; and biological treatment removes 85% to 95% of the raw wastewater BOD_5. Sample sizes of 3, 5 and 10 ml should provide good results for raw domestic sewage. When sampling wastewater containing suspended solids, volumetric pipettes with large tips should be used for measuring the samples directly into the BOD bottles. The large tip on the pipette permits suspended solids to be properly sampled. The BOD bottles are then placed in a darkened incubator for 5 days at 20° C. A darkened incubator is necessary to prevent the growth of algae during incubation. Algae generate DO during growth and would adversely affect the BOD data. At the end of the 5-day period, the BOD bottles are removed from the incubator and the DO remaining in each bottle is measured. Currently, direct reading DO probes with mechanical mixers are used to determine the DO values in the BOD bottles.

Laboratories located at altitudes greater than sea level will normally have oxygen saturation values in their dilution water less than 9.1 mg/L at the start of the BOD tests. The saturation oxygen concentration in laboratories located above sea level is directly proportional to the fraction of the atmospheric pressure at the test location times the saturation DO at 1.0 atmosphere pressure. At 1,000 ft altitude, the atmospheric pressure is 0.97 atmospheres. The oxygen saturation can be calculated as shown in Equation 10-1.

$$\text{Oxygen saturation} = (0.97)(9.1) = 8.8 \text{ mg/L} \qquad (10\text{-}1)$$

As the atmospheric pressure decreases, the saturation DO in the BOD bottles decreases. Smaller size wastewater samples will be required for the same wastewater BOD concentrations than at lower elevations.

Because of problems with the microbial seed in BOD tests, a glucose-glutamic acid (G-GA) standard was recommended to confirm the validity of the BOD test. The glucose-glutamic acid standard is a mixture of 150 mg/L glucose and 150 mg/L glutamic acid. *Standard Methods* indicates that the BOD_5 value of the G-GA standard should be 198 mg/L with a one standard deviation (1σ) of $\pm$ 30.5 mg/L. Testing the G-GA standard is similar to testing the raw wastewater samples. A major difference in the results lies with the microbial seed used for the BOD test. While domestic wastewater contains sufficient numbers and types of microorganisms for the BOD test, the G-GA samples do not contain any microorganisms and must be seeded prior to testing. Domestic wastewater normally contains all the microorganisms needed to metabolize the G-GA standard and is used for the seed. Settled domestic wastewater has a BOD_5 value close to 140 mg/L. A two ml sample of settled domestic wastewater added to each BOD bottle will create about 0.9 mg/L oxygen demand. Adding 3 ml, 5 ml, and 8 ml of the G-GA standard to three separate BOD bottles would provide the additional organics for metabolism by the microorganisms in the settled domestic wastewater seed. Dilution water should be added to fill the BOD bottles for the normal tests. The 3 ml G-GA sample should generate 2.0 mg/L DO depletion for a total of 2.9 mg/L oxygen demand. The 5 ml sample should have a total DO depletion of 4.2 mg/L; and the 8 ml sample should exert 6.2 mg/L DO depletion. The DO depletion of the G-GA samples can be obtained by subtracting the initial DO of the dilution water from the DO remaining in the samples after 5 days incubation. The accuracy of the DO depletion data for the wastewater seed sample can be improved by using three or more BOD samples. A plot of the total DO depletion data for each G-GA sample bottle against the sample size on rectangular graph paper can be used as a check on the validity of the BOD_5 data, Figure 10-2.

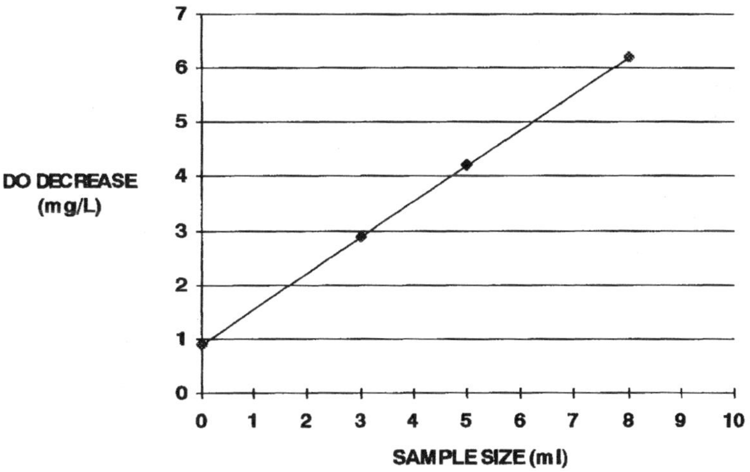

Figure 10-2 DO DEPLETION DATA PLOTTED AGAINST G-GA SAMPLE SIZE

Good BOD data will fall on a straight line with the vertical intercept confirming the DO depletion of the wastewater seed sample. The G-GA samples should give better data than municipal wastewater samples since all the organic compounds in the G-GA samples are soluble and readily biodegraded. With the G-GA data having a $\pm 1\sigma$ of 15%, normal municipal wastewater data could easily show a $\pm 1\sigma$ between 20% and 25%. This level of data variation indicates that routine wastewater BOD_5 data should only be reported to two significant figures. It should also be noted that wastewater BOD_5 values less than 2.0 mg/L have no real significance. The G-GA results should produce a BOD_5 value between 170 mg/L and 230 mg/L. If the G-GA BOD_5 results are less than 170 mg/L, it indicates that the seed did not have sufficient numbers of microorganisms or the dilution water contained some toxic materials. If the G-GA BOD_5 results are greater than 230 mg/L, it means that the seed had too many microorganisms or nitrification occurred during the test. Overall, the G-GA test has helped laboratory analysts improve their BOD techniques and provide more reliable data on wastewater samples.

The key to understanding the BOD test lies in understanding the fundamental concepts of aerobic metabolism. The BOD bottle at the start of the incubation period contains a definite quantity of biodegradable organic compounds, a pH buffer, nitrogen and phosphorus for maximum cell synthesis, an unknown mixture of microorganisms, up to 9.1 mg/L DO, and a temperature of 20°C. Most wastewaters are composed of an unknown mixture of chemical compounds. Municipal wastewaters contain everything dumped down the sewers. The major compounds in municipal wastewater are carbohydrates, proteins, and lipids. Various synthetic organic compounds can be found if one looks very carefully. Synthetic detergents are the largest group of synthetic organic compounds in municipal wastewater. Starches and proteins are the most readily biodegradable organic compounds in municipal wastewaters. Cellulose and lipids are slowly metabolized. If municipal wastewaters contain 1,000,000 bacteria/ml, a 300 ml BOD bottle containing 3 ml of raw wastewater should have an initial bacteria population of 10,000/ml. If the bacteria double their population every 3 hrs, it would only take 20 hrs for the bacteria to metabolize all the biodegradable organic compounds in the BOD bottle. The bacteria would have reached their maximum concentration at 1,000,000 bacteria/ml, and would have used 1.1 mg/L DO. The remaining 100 hours to complete the BOD test would have been used for endogenous respiration. Bacteria have an endogenous respiration rate close to 0.01/hr at 20°C. Approximately 50% of the cell mass would have been oxidized by the end of 5 days with an additional oxygen demand of 1.2 mg/L. The total oxygen demand at the end of 5 days would have been 2.3 mg/L instead of the expected 2.0 mg/L. A number of factors could have contributed to the difference in the oxygen uptake. The production of internal storage products in the bacteria or increased external slime could have reduced the oxygen uptake. The accumulation of waste products around the bacteria could have reduced the endogenous respiration rate below 0.01/hr. Protozoa metabolism of some of the bacteria would have reduced the bacteria endogenous respiration and increased the protozoa synthesis oxygen demand. Statistically, the variations in the BOD_5 data are sufficient that there is no significant difference in the two oxygen uptake values.

Previous Studies - In 1931 C. T. Butterfield, W. C. Purdy, and E. J. Theriault of the U.S. Public Health Service published the results of their research on the biochemistry of the BOD test. Their studies dealt with pure cultures and mixed cultures of bacteria and protozoa on the growth of the microorganisms and the oxygen uptake in the BOD test. Using a dilute nutrient solution of glucose and peptone, they demonstrated that bacteria were responsible for the aerobic stabilization of the organic matter. The protozoa only grew to significant levels after the bacteria had grown, confirming that the bacteria concentrated nutrients for the protozoa in the dilute environment of the BOD test. Their studies with pure cultures and mixed cultures of bacteria indicated that the mixtures of bacteria metabolized the substrate faster and to a greater extent than the pure cultures. In effect, the most efficient bacteria predominated in the BOD bottle. It was not until 1953 that S. R. Hoover, L. Jasewicz, and N. Porges published their paper indicating that the BOD test was a two-phase test. The first phase was the stabilization of organic matter by growth of the bacteria and the second phase was endogenous respiration. The first phase was complete in 24 hrs.

A.W. Busch began his study on the BOD test in the 1950s, publishing papers on the subject over a 10-year period. Busch found that there was a plateau in the rate of oxygen demand around 24 to 48 hours when the bacteria finished metabolizing the organic matter in the BOD bottle. Essentially, Busch confirmed the earlier research on the rapid growth of bacteria and the slow growth of protozoa. A. F. Gaudy was stimulated by Busch and examined the microbiological changes in the BOD5 test. His research showed that the plateau was related to the types and numbers of organisms at the start of the incubation period. A small population of protozoa in the BOD bottle produced the plateau that Busch noted. A larger population of protozoa allowed metabolism to produce a continuum in oxygen utilization without producing a plateau. Both Busch and Gaudy extended their studies on the BOD5 test into microbiological reactions for WWTP operations. .

Suspended Solids - Suspended solids in municipal wastewaters have been found to be about 50% biodegradable. There has been concern that suspended solids do not exert any oxygen demand in the BOD_5 test. Most BOD data have shown that biodegradable suspended solids do exert a BOD_5 in proportion to the bacteria's ability to come into contact with the suspended solids. A study by J. L. Balmat, published in 1957, showed settleable solids being metabolized much slower than colloidal or soluble solids. The problem with settleable solids in the BOD bottle is the creation of a layer of settled suspended solids that limits the transfer of oxygen to the bacteria in the settled solids and the transfer of waste products of metabolism back into the liquid. Bacteria must come into direct contact with the biodegradable parts of the settled solids and hydrolyze the biodegradable suspended solids to soluble solids before the bacteria can metabolize them. Balmat found that settleable solids were metabolized at about 1/5 the rate of soluble organic solids. This means that the biodegradable suspended solids should be metabolized in about 5 days. Daily shaking of the BOD bottles to disperse the settled solids can insure optimum metabolism of the biodegradable suspended solids. Homogenization of the sample can reduce the size of suspended particles, allowing easier metabolism of the biodegradable suspended solids.

Nitrification - With excess ammonia nitrogen in municipal wastewaters, nitrification can occur in the BOD bottles under the right circumstances. Normally, municipal wastewaters contain small quantities of nitrifying bacteria. In the aerobic environment of the BOD bottle with excess NH_3-N, the nitrifying bacteria will grow the same as the carbonaceous bacteria. The low energy yield from the metabolism of NH_3-N results in a slow increase in the number of nitrifying bacteria. It takes about 7 days to produce sufficient numbers of nitrifying bacteria to exert a measurable oxygen demand. For this reason the BOD_5 data have been considered as being the result of carbonaceous metabolism. Unfortunately, aerobic biological wastewater treatment systems increase the number of nitrifying bacteria to a point that the treated effluent samples will show partial nitrification in the BOD_5 test. This means that the BOD_5 data for the raw wastewaters and the primary effluent show only carbonaceous oxygen demand, while the treated effluent shows both carbonaceous and nitrogenous oxygen demands. Calculation of treatment plant efficiencies with these data will result in an error that indicates the treatment plant is not as efficient as it really is from a carbonaceous organic removal point of view. Research by J. C. Young in 1973 led to the use of 2-chloro-6-(trichloromethyl) pyridine as an inhibitor for effluent nitrification in the BOD bottles. It is also possible to pasteurize the wastewater samples by heating to 70°C for 5 minutes and reseeding the BOD bottles with a non-nitrifying seed. It is important that the treated effluent BOD data be measured as carbonaceous BOD. The maximum nitrogenous BOD can be calculated from the NH_3-N and the NO_2-N data in the effluent sample. It should also be recognized that using dilution water several days after preparation allows nitrifying bacteria to increase and adversely affect the data. Failure to clean the dilution water bottle at least once a week can result in nitrifying bacteria becoming attached to the walls of the dilution water bottle and seeding all the samples. Always make dilution water fresh each day and only make enough dilution water for the daily tests.

In spite of its faults, the BOD_5 test is one of the primary measures of environmental pollution and treatment efficiency. Many efforts have been made to replace the BOD_5 test with other tests, but none have succeeded. Mathematically, the carbonaceous BOD_5 values are about 2/3 of the ultimate carbonaceous BOD (BOD_u). The BOD_u value occurs after about 20 days incubation. At one time, it was believed that the BOD_u represented the theoretical carbonaceous oxygen demand (TBOD). Limited research on pure organic compounds indicates that the BOD_u results are lower than the TBOD. The difference between the BOD_u and the TBOD is the chemical oxygen demand of the dead cell mass that remains un-oxidized after all the biodegradable organic matter has been metabolized. The BOD_5 has been estimated to be 0.58 times the TBOD. This relationship allows the calculation of the TBOD from the BOD_5 results. The calculated TBOD results are no better than the original BOD_5 data.

Chemical Oxygen Demand

A major effort was made to replace the BOD_5 test with the dichromate chemical oxygen demand test about 50 years ago. Although the COD test did not replace the BOD_5 test, it found a place in the analyses of wastewater contaminants. The COD test uses potassium dichromate with concentrated sulfuric acid, heat, and time to oxidize organic matter to carbon dioxide, water, and ammonium sulfate. The COD results measure the total carbonaceous oxygen demand. It also can oxidize reduced inorganic compounds that exist in the wastewater. The COD test uses a 2-hour refluxing with silver sulfate and mercuric sulfate as catalysts for the oxidation reactions. The silver sulfate catalyst is essential for complete oxidation of acetate and long chain aliphatic compounds. The mercuric sulfate prevents the chemical oxidation of chlorides that occur in domestic wastewater. Only a few complex aromatic compounds are not oxidized by the COD test. The variability of the COD is less than the BOD_5 test, ± 5 to 10 percent from the average COD value. Unlike the BOD_5 test, the COD test yields results in about 3 hours. The COD results measure both the biological oxygen demand and the non-biodegradable oxygen demand of the wastes. By combining the COD results and the TBOD results, it is possible to estimate both the biodegradable oxygen demand (BCOD) and the non-biodegradable oxygen demand (NBCOD). Equation 10-2 gives the basic equation for calculating the NBCOD from the BOD_5 and the COD data.

$$\text{NBCOD} = \text{COD} - (\text{BOD}_5/0.58) \quad (10\text{-}2)$$

A better understanding of the domestic sewage characteristics can be obtained by measuring the COD and the BOD_5 on both the total sample and the filtrate from the suspended solids measurements. The total sample analyses yield the suspended and the soluble COD and BOD_5 results. The filtrate samples yield the soluble COD and BOD_5 results. The suspended COD and BOD_5 results can be obtained by subtracting the soluble results from the total results as shown in Equation 10-3.

$$\text{SS COD} = \text{COD} - \text{SCOD} \quad (10\text{-}3)$$

Equations 10-2 and 10-3 can be combined to determine the biodegradable fractions of the suspended solids and the soluble organic matter in domestic sewage and industrial wastewaters. Data on domestic wastewater has shown that about 65% of the volatile suspended solids (VSS) are biodegradable (BVSS) and 35% are non-biodegradable (NBVSS). The NBVSS are only non-biodegradable in the liquid environment. If the NBVSS are placed on soil, fungi and actinomycetes will slowly degrade some of the NBVSS, producing humus-like materials. The soluble organic matter in fresh domestic wastewater is about 95% biodegradable. The non-biodegradable, soluble organic compounds in domestic wastewaters appear to be complex polysaccharides from bacteria in the wastewater and other complex organic compounds that result from current personal activities of each and every person in the community served by the sanitary sewerage system. Recent analyses of municipal wastewater indicate that some of the medicines that people use pass through the digestive system into the wastewater.

Wastewater Flow

The flow pattern for municipal wastewater arriving at small to medium sized WWTP during a normal workday is shown in Figure 10-3. In the early morning hours, the wastewater flow drops to a minimum in residential communities. As people awake and start their day's activities, the wastewater flow increases rapidly. Breakfast, lunch, and dinner related activities form the pattern for most residential wastewater flow. Wastewater flow peaks after lunch, drops somewhat, and rises around dinner before falling as bedtime returns. Wastewater flow continues to drop during the early morning hours back to the minimum flow. This 24-hour flow cycle is repeated consistently during the workweek. Although there is a random pattern for each day's activities, the flow pattern for the community wastewater produces limited variations on a day-to-day basis when it arrives at the

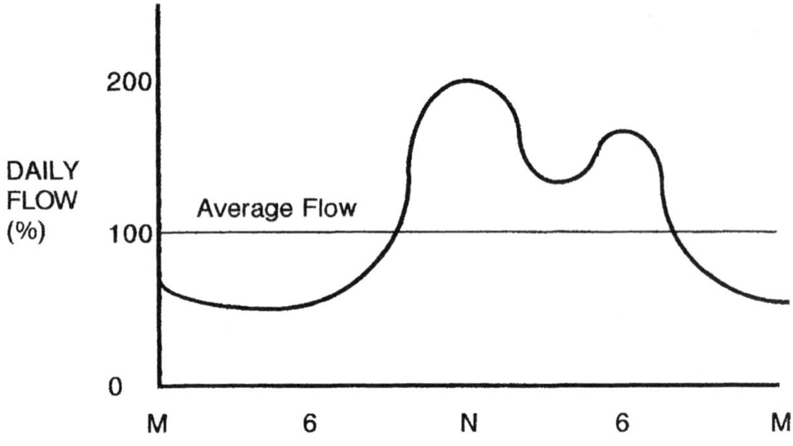

Figure 10-3 GENERALIZED DOMESTIC WASTEWATER FLOW PATTERN OVER 24 HOURS

WWTP. As the size of communities expand, the sewerage systems expand and provide increased wastewater retention in the collection system that levels out variations in both flow and contaminant concentrations. The collection system acts as a large surge tank. The daily domestic wastewater flow also varies with the days of the week. The daily flow variations over each week depend upon the overall work patterns of the people in the community. In the United States people work about 40 hours a week over a five-day period. Most offices and factories operate from Monday through Friday. Commercial businesses are open Saturday with some open for part of the day on Sunday. As a net result, municipal wastewater flows tend to drop slightly on Saturday and drop a little more on Sunday. The five-day workweek pattern results in slightly lower wastewater flows on Monday and Fridays than on the three middle days. The wastewater flow pattern occurring on Monday reflects starting the industrial operations with the wastewater flow pattern on Friday reflecting the shutting down of industrial operations. In areas of the world where the same activities occur seven days a week, the domestic wastewater characteristics will not show any variation from day to day. The chemical characteristics of domestic wastewaters are related to the wastewater flow with the concentrations of the various pollutants being slightly lower in the early morning hours and slightly higher in the afternoon. On a mass balance basis the hourly pollutant loadings show a wide variation from early morning until afternoon. It is important for design engineers and WWTP operators to recognize the impact of the variations in hourly loading rates at the WWTP. The hourly loading variations are a function of the size of the wastewater collection system. In small WWTP, having flows under 4.4 L/s (0.1 mgd), the hourly mass loadings can vary by a factor of 10 or more. As the WWTP flow rises to 44 L/s (1.0 mgd), the mass loading variations drop to a factor of about 6. At a WWTP flow of 440 L/s (10 mgd), the mass loading variations can drop to a factor of 3 or 4. As the size of the WWTP increases both the hourly flow variations and the hourly mass loading variations almost disappear. This relationship in hourly load variations makes a large WWTP easier to design and operate than a small WWTP. The large WWTP has little hourly wastewater flow variations or load variations; while the small WWTP has significant variations in both wastewater flows and loads.

INDUSTRIAL WASTEWATERS

Industrial wastes are the third major type of wastewaters. All industries produce domestic wastewater and some industries produce *process wastewater*. The process wastewater is the unique part of industrial wastewater. Process wastewater is produced primarily from wash water, either during production or at the end of a production cycle to clean up the equipment prior to starting another production cycle. Leaks and spills from pumps, pipes, and equipment also form part of the process wastewater stream. Leaks and spills tend to contain more concentrated contaminants than the wash waters. While similar industrial plants produce wastewaters that have the same general chemical components, each industrial plant has its own specific wastewater characteristics. Even two identical plants, designed and operated by the same company, will have different wastewater characteristics. Over the years there has been a shift in how industries handle process wastewater. It has gone from dumping wastewater on the ground or into adjacent waterways to extensive treatment and reuse. Although environmentalists favor complete elimination of industrial wastewaters, few industrial plants have been able to eliminate all wastes. Process changes have resulted in less industrial wastewaters being produced in some plants. Water conservation practices in industrial plants have usually reduced the volume of industrial wastewater while increasing the contaminant concentrations. The industrialized countries of the world have made significant progress in handling industrial wastes over the past few decades. Unfortunately, the third world countries have largely ignored industrial wastes until major environmental disasters have occurred. The future challenges with industrial wastewaters lie with the developing nations that are using rapid industrial development to increase their economy.

Where possible, industrial wastewaters have been added to the municipal sanitary sewers and diluted with domestic wastewater. Weak industrial wastewaters dilute the concentrations of various contaminants in municipal wastewater. Strong industrial wastewaters increase the concentrations of specific contaminants. Toxic industrial wastewaters adversely affect the microbial populations and subsequent treatment at the municipal WWTP. The federal EPA has developed guidelines requiring pretreatment of industrial wastes to insure that their addition to municipal sewers does not create treatment problems. The uniqueness of industrial wastewaters, in contrast to the uniformity of domestic sewage, has resulted in the development of industrial waste specialists to assist industries in solving their problems. Industrial plants located outside of municipal wastewater districts are required to develop and operate their own WWTP to produce the effluent quality established by the federal EPA. Over the years the federal EPA has held industries to higher treatment criteria than municipalities. It has been assumed that the economics of industries allows them to develop and maintain better waste treatment. Unfortunately, this assumption has only been partially correct.

Wastewater characteristics of industrial wastes are based on the same testing procedures used in analyzing domestic wastewater. Tests are normally made for BOD_5, COD, TSS, VSS, TKN, NH_3-N, NO_2-N, NO_3-N, phosphorus, pH, and alkalinity. For specific industrial wastes other tests are run to insure proper characterization of the industrial wastes. The BOD_5 analyses are the same as for domestic sewage where the waste components are largely protein and carbohydrate with a small lipid fraction. Since the bacteria in industrial wastewaters do not have sufficient time to develop a suitable microbial population for the BOD test, it is necessary to provide a suitable microbial seed. Municipal wastewater can be used for simple industrial wastewaters. For industrial wastewaters with unique chemical compounds, it is necessary to use an acclimated microbial seed to insure proper BOD_5 analyses. An acclimated microbial seed is prepared by growing an aerobic culture of bacteria that can metabolize the specific industrial waste components. The acclimated culture can either be started from domestic wastewater or from soil around the industrial plant, especially from areas where product spills have occurred. The acclimated bacteria cultures can be generated quite simply. Initially, 2.0 L of the industrial wastewater, having a COD of 1,500 mg/L or less, are put in a one-gallon bottle with the pH adjusted to 7.0 and 8.0 with sodium bicarbonate or dilute sulfuric acid, depending on the initial pH of the wastewater. The use of phosphates for pH adjustment is not recommended, as it throws the normal phosphate relationships out of balance. When the industrial wastewater contains more than 1,500 mg/L COD, the wastewater should be diluted with tap water to about 1,500 mg/L COD. The nitrogen and phosphorus concentrations in the wastewater should be sufficient for normal bacteria metabolism. A rough measure of COD/N for use in the BOD test has been found to be about 30/1. A rough measure of COD/P has been found to be 150/1. The wastewater sample must have sufficient alkalinity to neutralize all the organic acids produced during metabolism. As long as the bicarbonate alkalinity is greater than the COD of the wastes, there should be sufficient alkalinity to buffer the carbonaceous metabolism. Once the 2.0 L wastewater sample is ready, it can be seeded with 1.0 L domestic wastewater or 1.0 L of the adjusted industrial wastewater containing 10 g contaminated soil. Garden soil can be substituted for contaminated soil, if none is available. A small aeration stone connected to a suitable air supply is placed in the one-gallon bioreactor and aeration started. Sufficient air is added to keep all of the suspended solids in suspension. It is important to run all of the desired analyses on the adjusted waste sample before starting aeration. At the end of 24 hours add distilled water to make up for evaporation loss and determine the pH, alkalinity and COD of the aerated mixture. Continue aerating the sample another 24 hours and repeat the analyses. When the data stabilize, pour off 1.5 L of liquid and add 1.5 L of fresh, adjusted wastewater. Restart aeration and aerate for 24 hours. Repeat the analyses and determine if the rate of stabilization has increased. If the data are close to the minimum COD value in the initial run, repeat the feeding procedure on a daily basis. Microscopic examination of the culture at the end of the 24-hour aeration cycle should show actively motile, dispersed bacteria with some flagellated and ciliated protozoa. The bacteria are now acclimated to the industrial wastewater. The acclimated culture can be used to seed the BOD5 dilution water to provide normal BOD5 results. The acclimated culture can be maintained indefinitely with daily feedings of fresh wastewater. It may be possible to adapt bacteria to the industrial waste without growing protozoa. If microscopic examination of the acclimated culture does not show the presence of motile protozoa, the unit should be seeded again with domestic wastewater to provide a suitable protozoa population for growth in the industrial waste seed. A balanced microbial seed should be produced in another one or two days feeding. If the protozoa are unable to develop properly, dilute the industrial wastewater feed with water to reduce potential toxic effects. The BOD_5 data and the COD data can be used the same as indicated for domestic sewage to provide information on the biodegradable and the non-biodegradable fractions of the industrial wastewater. Good data are essential when evaluating the biological treatability of the industrial waste components. If the industrial wastewater components are highly volatile, the diffused aeration process for developing the desired microbial seed should not be used. When the industrial wastewater contains highly volatile organic compounds, it is necessary to use a shallow layer of wastewater in a sealed bottle or flask on a shaking table to provide the necessary mixing for good metabolism. Rapid growth of microorganisms in a 1,000 mg/L COD concentration wastewater should use between 300 and 500 mg/L oxygen, depending on the biodegradable fraction of the wastewater COD. The use of 100 to 200 ml total liquid volume in a 1,000 ml flask should provide excess oxygen in the gas space for metabolism of the volatile organics. The flask should have a tight stopper to prevent loss of the volatile organic compounds from the system. Although the volatile organic compounds will come to equilibrium with the air volume initially, as the bacteria metabolize the volatile organic compounds in the liquid phase, the volatile organic compounds in the air phase will return to the liquid phase. Repeated feedings should allow the development of an acclimated microbial seed for use in the BOD test. There is no perfect technique to use in developing seed microorganisms for industrial wastewaters. Each industrial wastewater has its own characteristics that will require the laboratory chemist to develop the procedures that work best to develop a good aerobic seed. Care should be taken in using commercial seed materials to insure that the bacteria in the commercial seed will metabolize all of the biodegradable COD in the industrial wastewater.

WASTEWATER COLLECTION

Wastewaters must be collected from their source and transmitted to the treatment plant or for discharge back into the environment. Each building contains its own wastewater collection system that discharges into a common sewer within the private property boundaries. These sewers connect to municipal sewers that lie outside of the private property boundaries in the public right-of-way. The municipal sewers are part of a larger wastewater collection system that transports the wastewaters to the treatment plant for processing. The municipal sewers are designed, as much as possible, for gravity flow to the wastewater treatment plant. Experience has shown that wastewater collection systems work best with a minimum of mechanical equipment. Like it or not, all mechanical equipment will fail, creating problems with wastewater collection until the mechanical equipment has been repaired and put back into service. Not every community has suitable terrain for gravity collection systems, necessitating the use of mechanical pumps to lift the wastewater over hills to permit gravity flow to resume. Wastewater pump stations should have duplicate pumps to allow for continuous service while one pump is being repaired. Large communities will have separate sanitary sewers and storm water sewers. Residential property owners will normally not have a storm water sewer connection; but large commercial buildings will often have a sanitary sewer and a storm water sewer. Residential property owners normally use the adjacent streets for storm water collection and street inlets to connect to the municipal storm water collection system. Until recently, municipal storm sewers discharged directly into the nearest stream or river without treatment. Concern over the potential pollution problems from storm water discharges has resulted in larger communities building storm water treatment systems prior to discharge into nearby waterways. Storm water collection systems have been gravity flow for the most part. Pump stations are required only when the discharge pipe is below the river surface elevation.

STORM WATER COLLECTION

Most storm water collection systems are constructed of reinforced concrete pipe of various sizes. Since storm water analyses indicate little biodegradable organic matter in the storm waters being collected, the major emphasis is on sizing the storm sewers to carry runoff for specific size precipitation events. The flow velocity is designed to keep all of the suspended solids moving through the collection system to the desired points of discharge. As indicated by the rate of flow shown in Figure 10-1, the storm water flow decreases to very low flow rates as the runoff ceases. Some suspended solids will settle out in the storm sewers and will remain until the next significant precipitation event produces sufficient flow to carry the suspended solids out of the storm sewer. If the storm water flow at the pipe discharge continues during dry periods, there is a groundwater leak or an illegal connection in the collection system. Chemical analyses of samples collected during the dry period will indicate if the flow is being created by groundwater or by one or more illegal connections. Bacteriological tests can be useful in determining if the illegal connection is from residences or from an industrial source. Significant concentrations of fecal coliform bacteria are an indication of illegal residential connections to the storm water collection system. Sampling storm water discharges during a storm event will show highly variable results. No two storms are identical. As previously indicated, the major contaminants are collected during the first flush of storm waters through the collection system. The bacterial contamination in storm water will come largely from animal feces washed from the urban environment. The most polluted storm water will be generated after a long dry period. Multiple storms during wet periods will show less and less concentration of contaminants with each new storm. The first storm will remove the majority of contamination from the land surface, leaving less contamination for the subsequent storms. Current concerns for storm water quality deal with potential contaminants from industrial plants and chemicals spilled on the road surfaces.

SANITARY WASTEWATER COLLECTION

We have already examined the basic problems associated with separate sanitary sewers and combined wastewater sewers. Separate sanitary wastewater sewers are being used in all new construction to minimize having to build large wastewater treatment plants that are difficult to operate with large volumes of storm water that occur from time to time. While combined wastewater sewers are being replaced in new construction, old cities are faced with combined sewers and will have combined sewers for years to come. Economics do not allow replacement of combined sewers until pipe failures require their replacement. At that time the combined sewers will be replaced with separate sewers. While separate sewers are designed to separate sanitary wastewater from storm water, it should be noted that most sanitary sewers have some infiltration from storm water runoff. Cracked pipes, leaky joints, and improper house drain connections allow extraneous water to enter the sanitary sewers. Since the storm water runoff carries few soluble pollutants, the storm water tends to dilute the contaminants in the sanitary sewers, lowering the contaminant concentrations without affecting the mass loading of contaminants. The storm water also brings additional soil microorganisms into the sanitary sewers. Storm water infiltration can be observed by sudden increases in wastewater flows after major rainfalls. Since many WWTP are located close to rivers, the main interceptor sewers may show continuous infiltration from river water if the pipes are cracked or if the joints are poorly constructed. Continuous infiltration is indicated by low contaminant concentrations with the normal mass loading of contaminants and by higher than normal wastewater flows. Leaky sewers can lose wastewater during dry periods. Lower wastewater flows during dry periods, combined with considerable infiltration during wet periods, are indicative of serious leaks in the sanitary sewers. The raw wastewater flow data can be quite useful in evaluating the wastewater collection system and in understanding variations in WWTP operations.

Sanitary sewers are designed to provide minimum velocities of flow to keep the suspended solids from settling out in the sewers as the wastewater moves from every source to the wastewater treatment plant. The minimum flow velocity for domestic wastewaters is 0.6 m/s (2 fps). Most wastewater collection systems are designed using a main trunk sewer to collect wastewater flows from a series of sewers over adjacent watershed areas. Sanitary sewers employ gravity flow as much as possible to minimize problems with pump stations. Wastewater collected in a one-mile long sewer would take 44 minutes or less to travel from one end to the other. As cities spread out over larger areas and build longer sewers, the time of flow of sewage from the farthest point increases significantly. Biological reactions will continue to occur in the sanitary sewers as the wastewaters move to the treatment plant. Bacteria metabolism begins with metabolism of the soluble organic compounds. The soluble organics are used to generate energy for the synthesis of new bacteria cells. Hydrolysis of the readily biodegradable suspended solids occurs slowly, limiting the degradation of the suspended solids. The initial bacteria metabolism in collection sewers is aerobic. As the DO is removed by the bacteria, metabolism shifts from aerobic to anaerobic with the production of low molecular weight organic compounds as end products. Aerobic conditions will continue to exist in the sanitary sewers at the liquid-air interface. Oxygen transfer from the air occurs from wastewater splashing as it enters the collection sewer and from drop manholes as the wastewater flows along the sewerage system. Most of the wastewater is devoid of DO and is anaerobic. Unfortunately, anaerobic metabolism is both slow and incomplete. The accumulation of grease and suspended solids in cracks and joints provides food for bacteria to grow over a much longer period of time. Temperature and time are important factors in the growth of different bacteria in sanitary sewers. Low temperatures and short times of flow in the sanitary sewers allows little bacteria growth as the wastewaters move from their points of origin to the treatment plant. The wastewaters arrive fresh with little change and no strong odors. Warm temperatures and long times of flow allow good bacteria growth and significant chemical changes in the wastewaters. It is not surprising that these wastewaters are septic, giving off strong odors, when they arrive at the treatment plant. Microbial growth in the collection system depends on the environment in the wastewaters. It is not surprising the vast majority of bacteria are facultative, having the ability to shift from aerobic metabolism to anaerobic metabolism and back again, as the environment changes. The bacteria growing at the bottom of the sewers tend to be anaerobic. Other microorganisms survive in the wastewater as spores or cysts, waiting for the proper environment to grow.

In areas with elevated concentrations of sulfates in the carriage water, a reasonable time of wastewater flow, and warm temperatures, growth of the sulfate-reducing bacteria can become significant with the production and release of hydrogen sulfide. The accumulation of grease deposits at the pipe joints or in bottom cracks provides a source of high-energy nutrients, favoring the sulfate-reducing bacteria, *Desulfovibrio* and others. The pH of the wastewater and the turbulence during the rapid flow allows much of the hydrogen sulfide to move from the wastewater into the air above the sewage. The moist atmosphere in the sewer produces condensation on the inside crown of the sewer pipes. Splashing wastewater adds microbes and nutrients to the condensed moisture from time to time. Dissolved oxygen from the air in the sewer pipes is absorbed into the condensed moisture, allowing aerobic metabolism to stabilize the organic matter that accumulates from the splashing sewage. The hydrogen sulfide in the air above the sewage also moves into the condensed moisture layer. Sulfur-oxidizing bacteria, *Thiobacillus*, will aerobically metabolize the hydrogen sulfide in the condensed moisture level and produce sulfuric acid as their primary end product. Over time the accumulated sulfuric acid increases and depresses the pH. The pH in the moisture layer can drop down to as low as pH 2. The dilute sulfuric acid formed at the crown of the sewer reacts with the materials making up the concrete sewer pipe, slowly destroying the strength of the concrete pipe. Eventually, external soil-pressure on the top of the sewer pipe can cause the pipe to collapse. This form of corrosion in concrete sewer pipe is known as *crown corrosion.* Reinforced concrete pipe can be protected from the sulfuric acid by coating the inside of the pipe with an asphaltic coating or using a plastic liner. The limited concentration of sulfates in the wastewater is the controlling factor in determining the overall amount of sulfide that can be formed by the sulfate reducing bacteria. Methane bacteria find the highly reduced environment at the bottom of the sewer suitable for their growth. The soluble organic acids produced by other anaerobic bacteria provide the methane bacteria with sufficient food to generate measurable quantities of methane gas in the collection sewers. Poorly ventilated wastewater collection systems can allow methane gas to accumulate over time and create a potentially explosive situation. Regular sewer cleaning to prevent the accumulation of organic nutrients in the bottom of the sewers can minimize the growth of the methane bacteria in the collection system.

Vitrified clay pipe is used for sanitary sewers in the southern part of the United States where the temperature is quite warm in the summer months and the terrain is flat, creating a long collection time. Vitrified clay pipe has a smooth salt glaze that protects the clay pipe from the sulfuric acid. In unprotected concrete sewer pipes the sulfuric acid reacts with the calcium carbonate in the pipe to form calcium sulfate. The primary problem with vitrified clay pipe is its weight and its inability to withstand bending forces. Most communities use vitrified clay pipes in the residential areas where the sewers are relatively small. Vitrified clay pipe is available in sizes up to 4 ft (1.2 m) in diameter. One of the problems with vitrified clay pipe collection sewers lies with the house connections. It is necessary to drill a hole to connect the house sewer to the collection sewer. The vitrified clay pipe requires special equipment and patience to drill a smooth hole for the house sewer. It is easy to crack the clay pipe at the house connection, allowing infiltration of storm water into the sanitary sewer. Normal procedures require sealing the house connection to the collection sewer. The seals often fail to prevent storm water infiltration. External pressures on the collection pipe from heavy trucks and heavy equipment can damage the collection sewers over time and allow storm water infiltration into the sanitary sewers. It is almost impossible to keep all storm water out of sanitary sewers. It is important to examine the raw wastewater flows at the treatment plant on a daily basis and to determine the magnitude of infiltration after each major storm. Storm water infiltration will produce a sudden rise in wastewater flow, followed by a slow return to normal flow values. If the storm water infiltration peaks become large and at frequent intervals, it is an indication that there are major leaks in the wastewater collection system that should be addressed immediately to prevent wastewater treatment problems from excessive surge flows. Major leaks will require sewer replacement to eliminate the storm water leaks. Minor leaks in old sewers are being handled by the use of plastic liners. As cities grow and expand both in population and area, it is necessary to increase the size of major sewers to handle the additional wastewater flow. The new sewers should reduce the infiltration problems.

THINGS TO REMEMBER

1. Wastewater characteristics are essential for design and operation of wastewater collection systems.

2. Separate wastewater collection systems should be used for storm water and domestic wastewater.

3. Storm water pollutants are primarily suspended solids.

4. Storm water runoff from industrial plants can carry contaminants specific to each industrial plant.

5. Storm water runoff from farms will carry soil and chemicals used on the farms.

6. Storm water runoff from rural areas will carry decomposition products from plants, trees and animals living in the rural area, as well as, pathogenic microorganisms from the animal wastes.

7. Domestic wastewater characteristics reflect the life style of the people producing the wastewater.

8. Domestic wastewater flows follow a 24 hr cycle and a 7-day cycle.

9. The chemical characteristics of domestic wastewaters are relatively uniform and are predictable.

10. The BOD_5 test is widely used as a measure of the biodegradable organic matter in streams, rivers, lakes and wastewaters.

11. The BOD_5 test is subject to significant errors and can produce misleading results if care is not taken to insure proper techniques.

12. The dichromate-sulfuric acid COD test is widely used to provide data on the total oxygen demand in wastewater.

13. The COD test measures the total oxygen demand of both the biodegradable organic matter and the non-biodegradable organic matter in wastewater.

14. Every house, building and industrial plant has its own internal wastewater collection system that is connected to an external collection system, leading to the local wastewater treatment plant.

15. Most municipalities build and operate their own wastewater collection systems to collect all of the wastewaters being produced within each municipality.

16. Wastewater collection systems are designed for gravity flow from each source to the treatment plant.

17. Wastewater pump stations are kept to a minimum in collection systems since all mechanical equipment used in the pump stations will fail eventually.

18. Vitrified clay pipes are used in warm climate areas to prevent crown corrosion that occurs in concrete sewers.

19. Wastewater collection systems should be properly ventilated and cleaned at regular intervals to prevent the formation and accumulation of methane gas and hydrogen sulfide.

REFERENCES

Balmat, J. L. (1957) Biochemical Oxidation of Various Particulate Fractions of Sewage, *Sew. & Ind. Wastes*, **29**, 757.

Bhatla, M. N. and Gaudy, A.F. (1965) Role of Protozoa in the Diphasic Exertion of BOD, *Jour. San. Engr. Div., A. S. C. E.*, **91**, SA3, 63.

Busch, A. W. (1958) BOD Progression In Soluble Substrates, *Sew. & Ind. Wastes*, **30**, 1336.

Butterfield, C.T., Purdy, W. C. and Theriault, E.J. (1931) Experimental Studies of Natural Purification in Polluted Waters IV. The Influence of the Plankton on the Biochemical Oxidation of Organic Matter, *Pub. Health Repts.*, **46**, 393.

Carson, R. (1962) *Silent Spring*, Houghton Mifflin Company, Boston (1962).

Finer, S. E. (1958) *The Life and Times of Sir Edwin Chadwick*, Methuen & Company, London.

Geldreich, E. E. (1966) *Sanitary Significance of Fecal Coliforms in the Environment*, WP-20-3, Federal Water Pollution Control

Administration, Cincinnati, OH.

Hoover, S. R., Jasewicz, L, and Porges, N. (1953) An Interpretation of the BOD Test in Terms of Endogenous Respiration of Bacteria, *Sew. & Ind. Wastes*, **25**, 1163.

Hunter, J. V. and Heukelekian, H. (1965) The Composition of Domestic Sewage Fractions, *Jour. Wat. Poll. Cont. Fed.*, **37**, 1142.

Mass. State Board of Health (1871) *Second Annual Report of the State Board of Health of Massachusetts*, State Printers, Boston, MA.

Mass. State Board of Health (1888) *Nineteenth Annual Report of the State Board of Health of Massachusetts*, State Printers, Boston, MA.

McKinney, R. E. (1978) Storm Water Quality Characteristics, *Kansas Water Quality Management Plan*, Kan. Dept. of Health & Environ., Topeka, KS.

Metcalf & Eddy (1981), Wastewater Engineering: Collection and Pumping of Wastewater, McGraw-Hill, NYC.

Mills, E. J. and Stack, V. T. (1953) Biological Oxidation of Synthetic Organic Chemicals, *Proc. 8th Ind. Waste Conf., Purdue Univ.*, 492.

Mills, E. J. and Stack, V. T. (1954) Acclimation of Microorganisms for the Oxidation of Pure Organic Compounds, *Proc. 9th Ind. Waste Conf., Purdue Univ.*, 449.

Oberton, A. C. E. and Stack, V. T. (1957) Biochemical Oxygen Demand of Organic Chemicals, *Sew. & Ind. Wastes*, **29**, 1267.

Santry, I. W. (1963) Hydrogen Sulfide in Sewers, *Jour. Wat. Poll. Cont. Fed.*, **35**, 1580.

Sawyer, C. N., Callejas, P., Moore, M., and Tom, A. Q. Y. (1950) Primary Standards for BOD Work, *Sew. & Ind. Wastes*, **22**, 26.

U S Dept. of Agriculture (1996) Agricultural Waste Characteristics, *Agricultural Waste Management Handbook,* Chapter 4.

Yao, K. M. (1976) Functional Design of Sanitary Sewers, *Jour. Wat. Poll. Cont. Fed.,* **48**, 1772.

Young, J. C. (1973) Chemical Methods for Nitrification Control, *Jour. Wat. Poll. Cont. Fed.*, **45**, 637.

Chapter 11

WASTEWATER TREATMENT

Once the wastewaters have been collected and their characteristics have been determined, the wastewater treatment processes can be evaluated and the best process can be proposed. Wastewater treatment processes can be physical, chemical, biological, or a combination of all three. Ultimately, wastewater must be treated and returned back into the environment with a minimum of damage to the environment. The treatment process should also be economical to construct and operate. Experience has shown that simplicity of design with a minimum of mechanical equipment and the least number of units provides the best opportunity for success. Ultimately, all biodegradable organic materials must undergo microbial degradation before being returned to the environment. Microbial stabilization of organic contaminants can be either aerobic or anaerobic or a combination of both. The key to success with biological treatment of all types of wastewaters lies in understanding the fundamental concepts of microbiology and the biochemistry of the different groups of microorganisms together with sound treatment plant engineering. The basic objective is to remove the contaminants from the wastewater with the least possible effort at the lowest possible cost and to return the water and the residual contaminants back into the environment with the least possible damage to the environment. Considerable progress has been made in the advanced nations of the world; but much effort remains in the developing nations. This will be the real test of environmental pollution control engineers.

STORM WATER TREATMENT

The characteristics of storm water indicate that suspended solids are the primary problem. The simplest method for storm water treatment is to prevent the suspended solids from initially accumulating in the environment. This is easier to say than to do. The wind brings much of the dust and dirt and some of the trash. Cars and trucks stir up the dust and dirt in the streets. Passersby often throw trash into the streets or onto adjacent property that does not belong to them. Domestic animals, together with non-domestic animals and birds, deposit their waste products on streets, sidewalks and yards. Homeowners use fertilizers, herbicides, and pesticides to produce nice lawns and shrubs to beautify their property. Most municipal storm sewers are located in commercial areas or densely populated residential areas. Suburban areas tend to use drainage ditches to collect the storm waters. Where storm sewers are readily available, people tend to wash the dirt off the sidewalks early in the morning when few people are around. The dirt and trash are washed into the street where they flow to the nearest storm water inlet. The amount of daily wash water is usually not sufficient to produce a discharge from storm sewer unless there is a nearby drainage way. For this reason the dust, dirt, and trash accumulate in the storm sewer until there is a significant rainfall, generating enough runoff to produce a discharge from the storm sewer. In effect, the rainfall event produces about the same amount of contaminants with or without the daily washing of the sidewalks. The best solution for storm sewer discharges appears to be some type of an end-of-pipe treatment system. The simple drainage ditches in the suburban areas will not require treatment in most instances since the grass lawns retain most of the contaminants except for very large rainfall events. Treatment of storm waters will be more physical treatment than biological treatment.

URBAN SYSTEMS

The simplest form of treatment for storm water is settling in a natural pond with skimming near the inlet to remove floating trash. A natural pond with sufficient holding time to allow for retention of several days flow could be developed as a park area for residents, as well as, for storm water treatment. The natural bacteria in the pond water should stabilize the small amount of BOD_5 in the storm water during its retention time in the pond. One of the problems with retention ponds is excessive algae growth if the storm water contains significant concentrations of nutrient elements from fertilized lawns in the drainage area. Artificial wetlands, ahead of the storm water retention pond, can remove the nutrients from the storm water and limit the growth of algae in the retention pond. Periodic harvesting will be required of the excessive plant growth in the wetlands to remove the nutrients permanently. Another form of treatment is the Swirl concentrator, employing the hydraulic flow of the storm water to create centrifugal forces that assist in separating the suspended solids. Floating solids would best be removed by screening prior to the Swirl concentrator. All of the suspended solids removed from the storm water must be collected at periodic intervals and returned back to the environment, either buried in sanitary landfills or applied to the land surface. The effluent from the Swirl concentrator is returned to the nearby receiving stream.

INDUSTRIAL SYSTEMS

Industrial plants can create serious problems with storm water runoff. Chemical leaks and spills leave residues on the ground that can be removed by storm water runoff from time to time. Industries may be required to capture the surface runoff from critical plant areas and discharge the runoff to process sewers rather than to storm sewers. Treatment of storm water containing soluble organic compounds requires the use of biotreatment systems. Currently, few industrial plants have facilities for capturing storm water containing soluble pollutants. Treating domestic wastewater and industrial process wastewater has been the primary focus for industries to date. The federal EPA has developed regulations affecting the measurement of storm water runoff and its chemical analyses from industrial plants. As information is developed, storm water regulations from industrial plants can be expected to change if serious problems are indicated.

AGRICULTURAL SYSTEMS

Agricultural runoff is difficult to treat since it occurs over such a wide area. It is not surprising that agricultural runoff has been classified as *non-point source pollution.* It is normally not possible to obtain individual flow measurements or pollutant analyses from each farm. Stream flow measurements and chemical analyses along short stretches of streams and rivers adjacent to farms can provide some measurement of agricultural runoff characteristics. Data are obtained both above and below the agricultural areas being evaluated.

Data on groundwater contamination from agricultural areas are even more limited than stream data. Chemical analyses of water from a limited number of farm wells have given the federal EPA a limited picture of groundwater contamination. To date, both the regulatory agencies and the farmers feel frustrated in their efforts to control agricultural pollution from storm water runoff. The Soil Conservation Service has made the greatest contribution in controlling agricultural runoff. The Soil Conservation Service's efforts have been directed at preventing loss of soil from farm fields. Too much topsoil was being washed into adjacent streams and rivers and lost. Over the years, the Soil Conservation Service setup programs to teach farmers to minimize soil losses. Currently, efforts are being directed to use terraces and to provide grass waterways to slow the velocity of runoff from agricultural fields. The grass waterways minimize erosion and soil loss from fields. Soluble fertilizer in the runoff provides the grass with nutrients for growth. It is important to periodically cut the grass and remove it or the nutrients will be released back into the runoff. The construction of retention ponds at the end of the grass waterways can help retain the runoff and permit its slow release into adjacent streams. Since most of the complex agricultural chemicals being used are retained on soil particles, minimizing the loss of soil particles from agricultural areas helps to significantly reduce agricultural contamination discharge.

The biggest problems facing agriculture today are wastes from confined animal operations and from animal slaughtering and meat packing operations. Confined animal operations allow more animals to be grown in less space, creating a waste disposal problem. This is the same waste disposal problem that was created when people began leaving their individual farms for life in the city. More waste is generated in a smaller space and has to be properly processed for return to the environment. Since animal manure can only be applied to crops before the growing season, the animal wastes must be collected and retained until needed. Figure 11-1 shows a confined hog finishing building with an adjacent lagoon.

Figure 11-1 CONFINED HOG BUILDING AND WASTEWATER STORAGE LAGOON

Lagoons are the simplest method of holding the animal wastes until applied to land. Improperly constructed lagoons tend to leak into the soil and create groundwater pollution, as well as creating obnoxious odors. Lagoons can fill with wastes and overflow before the wastes can be applied to land. Excess animal waste production often results in improper addition of those waste materials onto unprepared land. Sudden storms can produce runoff that carries the manure into adjacent streams and rivers, creating fish kills and serious environmental pollution problems. Failure to understand the basic problems created by animal wastes has been a major problem worldwide. Even the advanced, industrial countries with all their technology have not developed a simple solution for the animal waste problems. Confined animal wastes are directly related to the animal feed since all of the feed that is not used by the animals will appear as wastes. Feed and water are taken in by the animals and used in their metabolism. Part of the feed is oxidized for energy and part of the feed is used to create new animal mass. The remainder of the feed appears in the manure and urine. Hogs, chickens, and turkeys have been grown extensively in confined buildings over the past 30 years. Hogs have different waste characteristics than chickens and turkeys. With high-energy rations hogs convert about 0.3 of their feed into cell mass. The rest of the feed will appear as waste material. The chemical characteristics of the hog wastes will depend to a large extent upon the chemical characteristics of the feed. Feed consumption by hogs increases steadily as they grow. The hog feed contains about 10% moisture. The feed dry solids contain about 15% proteins, 3% fat, 5% fiber and 5% inorganic compounds. In addition to the feed solids, the hogs consume sufficient water to permit metabolism of the feed materials. The water intake will equal the total water in the manure and in the urine. The manure and urine combine to create a thick slurry that changes chemical characteristics over time as the bacteria metabolize the organic matter. This fact makes it very difficult to obtain satisfactory samples for chemical analyses of fresh wastes. It appears that a 91 kg (200 lbs) hog requires an average of 2.6 kg/d (5.7 lbs/d) feed. The feed contains about 0.39 kg/d (0.86 lbs/d) protein. R. Koelsch and C Shapiro of the University of Nebraska have indicated about 40% of the feed nitrogen will be converted to meat protein. This means that 60% of the feed nitrogen will be discharged in the wastes. The USDA Agricultural Waste Management Field Handbook characteristics of manure for a 91 kg (200 lbs) hog indicates that the daily combined urine and feces weight will be about 5.8 kg (12.7 lbs). About 90% of the manure slurry is water and 10% is total solids. The total solids (TS), 0.58 kg/d (1.27 lbs/d), are about 85% volatile solids (VS), 0.49 kg/d (1.08 lbs/d). The total nitrogen content of the manure, measured as TKN, is about 0.038 kg/d (0.084 lbs/d). Urine contains organic nitrogen (Org-N) in the form of urea that is rapidly broken down by bacteria to NH_3-N. The proteins in manure are from intestinal bacteria and undigested feed. The Org-N contained in the proteins will be slowly metabolized to form NH_3-N. With a chemical oxygen demand (COD) close to 1.1 times the VS, the daily COD production will approach 0.55 kg/d (1.21 lbs/d). Since much of the feed is related to carbohydrates, the biodegradable COD (BCOD) will be close to 0.33 kg/d (0.72 lbs/d) with the BOD5 being about 0.19 kg/d (0.42 lbs/d). The non-biodegradable COD (NBCOD) will be 0.22 kg/d (0.48 lbs/d). These data indicate that 60% of the volatile solids in hog manure are biodegradable. The phosphorous (P) content of the hog manure is about 0.014 kg/d (0.032 lbs/d) and the total dissolved solids (TDS) will be about 0.12 kg/d (0.26 lbs/d). From a concentration point of view the TS should be 111,000 mg/L with 94,000 mg/L VS. The COD should be 106,000 mg/L with 63,300 mg/L BCOD and 36,700 mg/L BOD5. The TKN should be 7,300 mg/L with 2,800 mg/L phosphorus. With 22,700 mg/L TDS the total suspended solids (TSS) should be about 88,300 mg/L. It can be seen that hog manure is a concentrated organic waste that acts as a semi-solid. The microbial population in the hog manure is natural, anaerobic bacteria. The readily biodegradable organics in the hog manure allow the anaerobic bacteria to continue their metabolism once the manure has been discharged from the hog.

Wastewater treatment engineers have found by experience that concentrated organic wastes are easier to treat anaerobically than aerobically. Anaerobic wastewater treatment requires a complex consortium of bacteria working together to bring about maximum stabilization of the organic contaminants in the wastewater. The complex organic suspended solids must be converted by bacteria to soluble organic acids, aldehydes, ketones, and alcohols as the first step in metabolism. The soluble organic acids, aldehydes, ketones and alcohols are then metabolized to volatile organic acids by other bacteria. The sulfate reducing bacteria metabolize some of the soluble organic compounds with sulfate to produce reduced sulfides that lower the oxidation-reduction-potential (O-R-P) to the proper level for the methane bacteria to convert the volatile acids to methane gas for discharge to the atmosphere. Metabolism of the urea from the urine results in the production of ammonium carbonate and a rise in pH unless sufficient carbon dioxide is available to convert the ammonium carbonate into ammonium bicarbonate with a lowering of the pH. Metabolism of the proteins will result in the release of ammonia ions and the formation of ammonium salts of the volatile acids, keeping the pH from dropping. Ammonium bicarbonate tends to be the major form of alkalinity produced in anaerobic systems treating hog manure. Research by McCarty and McKinney, published in 1961, found that 150 mg/L unionized NH_3 was inhibitory for the acetate utilizing methane bacteria. Since the acetate utilizing methane bacteria are essential for the production of the highest quality effluent from anaerobic systems, it is necessary to keep the unionized NH_3 concentration below the toxic level. The pH of the anaerobic environment will be a primary factor in determining the unionized NH_3 concentration. If the pH is kept between 6.5 and 7.5, there is little chance that the unionized NH_3 concentration will reach toxic levels.

Anaerobic lagoons have been designed with liquid retention times ranging from 3 months to well over a year. Too often, anaerobic lagoons have been sized to fit the available land and economics rather than being designed for the wastes to be treated. Current design criteria for anaerobic lagoons are based on field experience collected from numerous anaerobic lagoons over several decades. Anaerobic lagoons can be single cell lagoons or multi-cell lagoons. Multi-cell lagoons will have two, three, or more cells in series, depending upon site configuration. Single cell lagoons are more popular than multi-cell lagoons, primarily from an economic point of view. The characteristics of hog manure indicate that most of the pollutants are suspended solids that will settle out near the discharge from the inlet wastewater pipe, creating a sludge mound. The biodegradable organic matter in the suspended solids will stimulate the growth of the different groups of bacteria, as indicated previously. The relatively dense environment in the settled sludge mound limits the movement of bacteria, retarding rapid metabolism of the biodegradable organic compounds. Micro-environmental pockets form throughout the settled sludge mound. Some of the micro-environmental pockets will contain high concentrations of organic acids, depressing the pH well below the 6.5 level required by the methane bacteria. Some of the micro-environmental pockets will have high NH_3 concentrations with pH values above 8.5. Trace metals are not uniformly dispersed in the sludge mound, slowing the rate of metabolism of the different groups of bacteria. As the methane bacteria slowly metabolize the salts of organic acids, bicarbonate alkalinity is produced along with the methane gas. The insolubility of the methane gas allows the methane to form tiny bubbles that rise by the path of least resistance up through the settled sludge mound. As the bubbles move upward around the solids, they create localized turbulence that moves the soluble end products of metabolism and tiny bacteria upward to points where new organic compounds are available for metabolism. Excess carbon dioxide will also be formed by metabolism and discharged as a gas along with the methane gas. While the total gas production is not adequate to completely mix the settled solids, the limited mixing action helps the total bacteria population to increase. It can take anywhere from one month to a year for anaerobic lagoons to develop sufficient numbers of all types of bacteria to permit optimum metabolism of the daily load. Temperature is also an important factor in the overall rate of metabolism. Warm temperatures will speed the rate of bacterial metabolism and assist the bacteria in reaching equilibrium in the shortest time. The rising gas bubbles may be sufficient to cause some of the suspended solids to rise to the liquid air surface and form a layer of solids over the surface of the anaerobic lagoon. The surface solids layer will act as a lid over the anaerobic lagoon and prevent oxygen transfer from the air into the anaerobic lagoon. A well operating anaerobic lagoon treating hog manure can produce 90 to 95% reduction of the BOD_5 and the suspended solids from the liquid phase.

Anaerobic lagoons are started by filling the lagoon about half full with fresh water. As wastewaters are added to the anaerobic lagoon, the suspended solids settle to the bottom of the lagoon as the wastewater velocity slows on entering the lagoon. The soluble organic compounds and the soluble inorganic salts, tend to diffuse into the fresh water that was in the lagoon. The NH_3-N concentration is quickly dropped below the toxic level by simple dilution. As the bacteria population increases, the methane gas production will soon reach the expected daily production rate for the level of organic matter contained in the influent wastewaters. For a short time period, the gas production will exceed the addition of daily organic compounds, creating the impression that more methane gas is being produced than is theoretically possible. The excess gas production is generated from the accumulated organic solids that had not been metabolized during the acclimation period. The daily gas production will reach a peak and then decrease down to the normal equilibrium value. The slow rate of buildup of bacteria in anaerobic lagoons tends to lengthen the time required for the treatment system to reach equilibrium. Engineers and wastewater treatment plant operators who are used to the rapid changes created in microbial populations in activated sludge wastewater treatment plants are easily frustrated by the long time period required for equilibrium conditions in anaerobic lagoon wastewater treatment systems.

Collecting hog manure is a major problem in confined animal buildings. Confined hog buildings are normally constructed with slotted floors over a manure collection pit. The manure is dropped onto the slotted floor and eventually into the collection pit. Most confined hog operations use liquid flushing to move the manure from one end of the building to a collection pipe that discharges into the anaerobic lagoon. The effluent from the anaerobic lagoon is often used to flush the manure from the confined hog buildings. A well operating anaerobic lagoon will produce an effluent that is high in NH_3-N and a pH above 8. The turbulence created during the manure flushing will result in loss of NH_3 to the atmosphere within the house. Vent fans in the confined building will push the NH_3 gas out of the building in a short time period. Limited data indicates that most of the NH_3 lost to the atmosphere will occur in the hog buildings rather than from the anaerobic lagoons. The non-biodegradable manure solids will slowly accumulate in the settled sludge layer and will require periodic removal to keep the system operating at its maximum efficiency. The treated effluent and the excess solids from the anaerobic lagoon will have to be returned to the land. Return of these materials to agricultural land will help replenish the materials previously removed as crops. Each day, the mineral salts will increase in the liquid and try to approach equilibrium conditions. As the NH_3 concentration increases, more NH_3 is lost to the atmosphere and less TKN remains in the liquid within the lagoon. Calcium ions may increase to the point that calcium carbonate precipitates and settles into the sludge layer at the bottom of the lagoon. Even calcium hydroxyl phosphate will precipitate and settle into the sludge layer. Over time the nitrogen and the phosphorus in the treated liquid approach equilibrium values well below the concentrations in the raw hog manure. The mineral removal is the direct result of bacteria activity and their effect on the overall environment in the anaerobic lagoon.

Hog manure can also be treated by aerobic bacteria; but aerobic treatment systems are more complex to construct and operate. The value of aerobic treatment lies in a smaller treatment system and a higher quality effluent. In 1965 I designed an aerobic, oxidation ditch to treat concentrated hog manure produced in a 50 sow farrowing house. The oxidation ditch was constructed in the manure pit under the slotted floor and was one of the first oxidation ditches constructed inside a commercial hog building in the United States. The farrowing house was a standard *Honegger* building, 26 ft wide and 142 ft long. A single, mechanical, brush aerator was located across the channel at one end of the building. The aerator was a modified brush aerator, driven by a 3 HP electric motor. Since the aerator was located inside the farrowing house, there were concerns that moisture would create a wet environment and that pathogenic microorganisms would be quickly spread through the building. Normal exhaust air ventilation fans prevented moisture accumulation in the building. At no time were any of the pigs adversely affected by pathogenic microorganisms from the waste treatment operations. The manure and urine entered the oxidation ditch along the entire length of the ditch. A diverse population of bacteria quickly grew in the aerobic environment and stabilized the biodegradable materials in the manure. Because the environment was aerobic, protozoa appeared and reduced the free-swimming bacteria population. Samples of the oxidation ditch contents indicated that the microbial mixture was similar to activated sludge found in municipal wastewater treatment plants. With adequate dissolved oxygen input, nitrifying bacteria converted the excess NH_3-N to NO_3-N. Endogenous respiration reduced the biological mass quite considerably. A discharge weir controlled the effluent flow from the building to a small holding pond. Periodically, the pond contents were applied to nearby agricultural fields for ultimate disposal of the wastewaters. Research on the power requirements of the rotor aerator indicated that reducing the blade submergence into the waste liquid reduced both the power draw and the oxygen transfer. By controlling the oxygen transfer to a minimum, it was possible to oxidize the NH_3-N on the first part of the oxidation ditch and denitrify the NO_3-N on the last part of the oxidation ditch as dissolved oxygen became limiting. This technique reduced the total power requirements of the aerator and produced an effluent with less nitrogen for use as a fertilizer on crops. By increasing the power supplied to the rotor aerator, more oxygen was transferred to the water, preventing the reduction of NO_3-N. The effluent contained the maximum amount of NO_3-N for use on crops. As the hog farm increased the number of buildings and their size, it became a good site for research and for equipment development. After many years of successful operations, economic conditions for growing hogs in Kansas became unfavorable and the operations were shut down permanently.

The basic problem with aerobic treatment of confined hog wastes is the continuous power requirements of the aeration equipment. High power costs pose an extra expense that reduces the profit from raising hogs in confinement. Research continues at many agricultural experiment stations to develop new processes for treating confined animal wastes and for returning the treated residues back into the environment. To date, no one has found a simple system that can treat the wastes automatically without any mechanical equipment or any obnoxious odors. Anaerobic lagoons remain the simplest systems in use at the present time. Anaerobic treatment with methane gas capture and conversion to power for operating aerobic treatment of the anaerobic effluent offers the potential for producing a high quality effluent for return to agricultural land in large confined animal operations that can employ skilled wastewater treatment plant operators. It will be interesting to see the design of animal manure systems that are produced as a result of current research emphasis.

DOMESTIC WASTEWATER TREATMENT

Treatment of domestic wastewater utilizes physical and biological systems. Physical treatment is used first to remove the trash by screening. Screening is considered as *preliminary treatment.* The captured screenings are normally buried in sanitary landfills. If the domestic sewage is contaminated with large amounts of sand, grit chambers are added as part of the preliminary treatment system to remove the sand. Grit chamber design provides sufficient velocity to maintain the organic solids in suspension while allowing the sand to be separated by gravity or centrifugal force. The sand and grit are washed after collection to insure that the organic matter is returned to the wastewater stream for treatment. The washed sand can be returned to the land environment and used where needed.

PRIMARY TREATMENT

Primary treatment follows *preliminary treatment* and consists of physical treatment by continuous sedimentation to remove the settleable solids and the floating solids from the wastewater together with either physical or biological treatment to process the sludge solids for return to the environment. Physical treatment requires the addition of chemicals to assist in dewatering the solids and some form of mechanical dewatering to produce a relatively dry cake that can be burned in an incinerator or buried in the land environment. Biological treatment of primary sludge utilizes anaerobic digestion to stabilize the biodegradable organic solids prior to physical dewatering and return to the land environment. Primary treatment reduces the BOD_5 of municipal wastewaters about 30 to 35% and the suspended solids about 60 to 65%. Anaerobic digestion reduces the primary sludge solids about 50%.

Primary Sedimentation

Primary sedimentation tanks can be circular, square, or rectangular. Circular sedimentation tanks, as shown in Figure 11-2, are widely used in small to medium size WWTP where land is readily available. Square and rectangular sedimentation tanks are used in large plants where land is limited and common wall construction helps reduce the overall plant cost. Normally, the wastewater is retained for two hours at average design flow in the primary sedimentation tanks. The settleable solids will drop to the bottom of the tank and will be collected with slow moving, mechanical sludge scrapers. The settled sludge will thicken to between 4% and 6% total solids (TS), 40,000 to 60,000 mg/l. Since biological metabolism began prior to the wastewaters entering the collection system, metabolism will continue to occur in the primary sedimentation tanks. Metabolism will be limited by the number of active bacteria in the incoming wastewater and by the growth of bacteria during the time the primary sludge is allowed to accumulate in the sedimentation tanks. Hydrolytic reactions continue to predominate with proteins hydrolyzed to amino acids that are further hydrolyzed to NH_3-N and short chain organic acids. The starches are hydrolyzed

Figure 11-2 CIRCULAR PRIMARY SEDIMENTATION TANK

to simple sugars. The dispersed bacteria and the bacteria associated with small particles are carried out in the primary effluent in a relatively short time.

Since the incoming wastewaters are maintained in a relatively quiescent condition without aeration, bacteria metabolism will be anaerobic with the formation of soluble, short chain organic acids, aldehydes, and alcohols. Most of the bacteria growth will be by facultative bacteria. The settled sludge contains additional bacteria and provides a better opportunity for anaerobic growth with concentrated nutrients. The extent of bacteria metabolism can be estimated from the pH in the settled sludge. The excess organic acids will depress the pH to between 6 and 6.5. The settled sludge is moved by a rotating sludge scraper at the bottom of circular and square tanks and by transverse scrapers moving along the bottom of the rectangular tanks. The sludge is slowly moved to a sludge hopper located either near the center of circular and square tanks or at the influent end of rectangular tanks. Periodically, the settled sludge is pumped from the sludge hopper for further treatment, removing it from the primary sedimentation tanks. If the settled sludge is allowed to remain at the bottom of the primary sedimentation tanks for too long a period of time, anaerobic metabolism can become quite extensive. Excess organic acids will drop the pH below 6.0. Sulfate reducing bacteria can generate sufficient hydrogen sulfide to react with iron and other metallic ions in the wastewater to produce a black color and give an obnoxious odor to the sludge. Carbon dioxide can collect in the sludge as a gas and lift small clumps of solids to the liquid surface. Significant gasification in primary sedimentation tanks processing only raw wastewaters indicates that the settled sludge is not being removed quickly enough. If waste activated sludge is added to the primary sedimentation tanks, gasification can be caused by denitrification. Facultative bacteria will continue anaerobic metabolism with a conversion of solid organic compounds to soluble organic compounds. Excessive retention of primary sludge can reduce the quantity of primary sludge and increase the load on the next treatment units. Good operation requires rapid removal of the settled sludge without pumping excess water with the sludge. Positive displacement sludge pumps are more effective at removing settled, primary sludge than centrifugal sludge pumps since they remove more sludge solids and less water.

Municipal wastewaters also contain grease and other insoluble hydrocarbons that tend to float in the primary sedimentation tanks, rather than settle to the bottom. A surface skimmer is standard equipment for primary sedimentation tanks to remove the floating scum from the sedimentation tank surface. The scum is collected in a scum hopper and a scum pit before being collected and buried in a sanitary landfill. In some plants, the scum is also pumped to the anaerobic digester for further treatment.

Municipal wastewater treatment plants processing less than 44 L/s (1.0 mgd) flow often eliminate primary sedimentation tanks and anaerobic sludge treatment. The economics of using primary sedimentation and anaerobic digestion allow the raw wastewaters to be directly discharged to aerobic biological treatment units. These small treatment plants utilize either extended aeration activated sludge or aerobic digestion of the excess activated sludge in a separate tank prior to dewatering of the sludge for return to the land environment.

Primary Sludge Treatment

Primary sludge is normally pumped from the primary sedimentation tanks at periodic intervals and treated further by anaerobic digestion. Municipal wastewater anaerobic digesters consist of circular tanks having an overall retention time of 30 to 60 days, based on the concentrated sludge volume. While small WWTP may have only a single anaerobic digestion tank, medium size and large WWTP have multiple tanks. Figure 11-3 shows two of the three anaerobic digestion tanks at the WWTP in Lawrence, KS. These digesters have a brick cover over a layer of insulation around the reinforced concrete digesters to minimize the loss of heat to the atmosphere.

Figure 11-3 ANAEROBIC DIGESTION TANKS AT LAWRENCE, KS

Primary anaerobic digestion tanks are normally heated to 35°C by external heat exchangers and mixed by either gas mixers or mechanical mixers. The increased temperature and the sludge mixing permit bacteria in the anaerobic digesters to metabolize the biodegradable organics in the shortest possible time. The bacteria in anaerobic digesters are the same types of bacteria found in anaerobic lagoons. The anaerobic bacteria begin metabolism by hydrolyzing the biodegradable fraction of primary sludge to simple soluble organic compounds that can be further metabolized by other bacteria in the anaerobic environment. Facultative bacteria are responsible for the initial metabolism of the organic solids. Proteins are easily hydrolyzed to amino acids that undergo further hydrolysis to NH_3-N and short chain fatty acids. Polysaccharides are hydrolyzed to simple sugars that are metabolized to short chain fatty acids, aldehydes, ketones and alcohols. Fats are slowly hydrolyzed to glycerol and long chain fatty acids. The glycerol is quickly metabolized while the long chain fatty acids are slowly metabolized. Solubility is a major factor slowing the rate of metabolism of the long chain fatty acids.

With an excess of organic nutrients the sulfate reducing bacteria grow in proportion to the available sulfates in the fluid sludge. Since sulfates are limited in municipal wastewaters, growth of the sulfate reducing bacteria will also be limited. For practical purposes, sulfate reduction will be complete in municipal anaerobic digesters. The sulfate reducing bacteria have an important role in anaerobic digesters, creating the highly reduced environment required by the methane bacteria. Successful operation of anaerobic digesters depends upon the combined metabolism of the facultative bacteria, the sulfate reducing bacteria, and the methane bacteria. The facultative bacteria convert the biodegradable solids into soluble organic compounds. The sulfate reducing bacteria lower the O-R-P in the digester.

The methane bacteria convert the short chain fatty acids into methane gas and carbon dioxide. Metabolism of long chain fatty acids results in the production of hydrogen gas that can be used by either a second group of methane bacteria or by the acetogenic bacteria. The acetogenic bacteria use hydrogen and carbon dioxide to produce acetic acid and water. Normal methane bacteria will then metabolize the acetic acid produced as an end product by the acetogenic bacteria. The hydrogen-utilizing methane bacteria compete with the acetogenic bacteria for both hydrogen and carbon dioxide and produce methane and water as their end products. It is important to recognize that specific sequences of bacteria are required to convert the complex organic solids to methane, carbon dioxide, and water.

About 50% of the TS in the primary sludge will be destroyed in anaerobic digesters, since only 60% to 70% of the VS in primary sludge are biodegradable. Half of the primary sludge added to the anaerobic digesters is composed of inert material that remains unchanged through the digestion process. Most of the inert solids settle out in the anaerobic digestion tanks, requiring periodic removal and dewatering before being returned to the land. Examination of the data in anaerobic digesters indicates that the COD of the VS destroyed by the bacteria appears as methane gas COD. Continuous measurement of the methane gas production is a useful tool in evaluating the operational efficiency of anaerobic digesters on a daily basis. A decrease in the daily rate of methane gas production indicates a potential problem with the sludge treatment system that can become quite serious if action is not taken immediately to resolve the problem. An increase in volatile acids and a decrease in pH or an increase in toxic compounds in the feed sludge will result in a decrease in methane gas production. The acetate utilizing methane bacteria are the most sensitive group of bacteria in anaerobic digesters and will be the first group of bacteria to indicate problems. Municipal sludge digesters normally produce gas with about 65% CH_4 and 35% CO_2. Digester gas contains very little hydrogen sulfide, even though it can be detected by smell.

Examination of the metabolism of the methane bacteria indicates why the anaerobic digestion process is controlled by the growth of the methane bacteria. The acetate utilizing methane bacteria obtain very little energy from the metabolism of acetate to methane and carbon dioxide. The net effect of the low energy yield for the acetate utilizing methane bacteria is limited cell growth. Each cell must process large quantities of acetate to obtain the energy necessary to produce a new cell. The large concentration of inert solids and inadequate mixing in most anaerobic digesters make it difficult for the methane bacteria to come into contact with sufficient acetate to grow enough new cells for rapid metabolism. The addition of large quantities of primary sludge allows the rapid formation of organic acids with a localized pH decrease that adversely affects the methane bacteria. Better operations can be achieved by adding small quantities of primary sludge at frequent intervals with good digester mixing.

Methane bacteria require an environmental pH level between 6.5 and 9.0 for their maximum rate of metabolism. The pH in municipal digesters is held between 6.5 and 8.0 by bicarbonate alkalinity. Good mixing is also important for rapid reaction between the short chain organic acids and the bicarbonate alkalinity. Municipal digesters normally have bicarbonate alkalinity levels between 2,000 mg/L and 4,000 mg/L. The bicarbonate alkalinity is formed primarily by the reaction of NH_3, CO_2, and HOH. The NH_3 is released from the neutralized organic acids formed during the metabolism of proteins. The CO_2 is released from the organic acid metabolism. Ammonium bicarbonate is the natural pH buffering system found in all municipal biological wastewater treatment systems. If the anaerobic digester does not have sufficient bicarbonate alkalinity to neutralize the organic acids, it is necessary to add a chemical source of alkalinity for good operations. Fortunately, metabolism of the acetate to methane results in an increase in bicarbonate alkalinity. A well operating digester produces volatile acids that are quickly neutralized by the bicarbonate alkalinity and then metabolized to methane with the production of the previously used bicarbonate alkalinity. Problems with the methane bacteria can disrupt this operation and allow the volatile acids to accumulate. One of the problems with volatile acids lies in the fact that volatile acid salts titrate the same as alkalinity in the alkalinity test. A buildup of volatile acid salts will create the impression of having considerable alkalinity when there is very little alkalinity to neutralize additional volatile acids. As soon as there is an excess of volatile acids, the pH will plunge downward. For this reason volatile acids have been used as a major operating parameter for anaerobic digesters together with pH, bicarbonate alkalinity, and methane gas production.

Many digesters depend upon the formation of gas bubbles for natural mixing. In large, flat digesters the gas bubbles are released over a large surface area, creating limited turbulence for liquid mixing. Recent digester designs have looked at tall, cylindrical tanks with small cross sectional areas. The same volume of rising gas bubbles in the tall, cylindrical tanks produces better mixing than in the flat, shallow, circular tanks. Egg shaped digesters, developed in Germany, appear to have better natural mixing than conventional anaerobic digesters in the United States. Although the importance of mixing has long been recognized in biological systems to bring the nutrients together with the bacteria for rapid metabolism, engineers have resisted the addition of good mixers in anaerobic digesters. The original reason for this was the failure of mixing to produce additional digestion of the VS in anaerobic digesters with long retention times.

Research over the years on the effect of mixing in anaerobic digesters has produced mixed results. Laboratory studies have shown that mixing is important when the solids retention time is short, under 10 days. Professor P. F. Morgan stimulated interest in field scale gas mixing for anaerobic digesters in 1954. His research at the University of Iowa showed that complete mixing in a digester treating primary sludge resulted in greater gas production and a shorter sludge retention period. Over the years equipment manufacturers have developed various mixing systems from gas mixing to mechanical mixers. Most of the current mixing systems for anaerobic digesters do not supply the energy needed for rapid sludge mixing. The reason for the limited energy mixers lies in the field data obtained to date. Anaerobic digesters were designed to produce good digestion in the upper section of the digester. As the inert solids were released from the actively digesting solids, they settled to the bottom of the digester and were removed at periodic intervals. Complete mixing meant that two digesters would be required for anaerobic digestion of wastewater sludge, the first digester for biodegradation of the organic solids and the second digester for concentrating the digested solids. One advantage of the two-stage digestion system is the recycling of supernatant from the second digester back to the primary sedimentation tank for further treatment. Examination of operating results from municipal digesters has shown that mixing and sludge retention time are related parameters. Anaerobic digesters with long sludge retention times do not require external mixing systems with reasonable tank configurations. Only very flat and very large diameter tanks will benefit from mixing systems. Poorly mixed digesters allow the solids to accumulate in the poorly mixed areas. The available digestion volume is soon reduced to fit the mixing equipment. Loss of active volume will reduce the digester's ability to treat all the sludge being produced. Mixing becomes important as the sludge retention time is decreased in anaerobic digesters.

The ability to accumulate methane bacteria in anaerobic digesters controls the overall operations. Since anaerobic digesters are operated as single pass systems, it is not possible to increase the population of methane bacteria above the numbers supported by the organic loading rate. As the digester retention time is increased, endogenous respiration keeps reducing the active mass of methane bacteria. At 35°C the maximum methane bacteria mass will be about 1.5 times the daily increase in the entire system. As the digester retention time increases, the increased liquid volume dilutes the active microbial concentrations in a well-mixed system. Limited mixing in the active metabolism zone may allow the concentration of methane bacteria to remain high, permitting complete metabolism with poor mixing and long retention periods. Balancing mixing with the rate of metabolism of primary sludge and retention time in anaerobic digesters could be a productive area for future research. It has also been shown that trace metals affect the growth of anaerobic bacteria. If the bacteria do not have access to all the trace metals they require for metabolism, their metabolic activities will be retarded. Iron is the most important trace metal required for all bacteria. Nickel, cobalt, molybdenum, copper, zinc, and manganese in trace quantities also stimulate the metabolism of the methane bacteria. Selenium has been found to be helpful for a few species of methane bacteria. It is important to recognize that bacteria must have all the nutrients required to produce normal cell protoplasm if the bacteria are to grow at their maximum rate. In trace metal deficient environments the bacteria are limited by their ability to recycle the trace metals in the desired quantities. These heavy metal ions form metallic sulfide precipitates in anaerobic digesters, reducing the toxic effect that excess concentrations of the heavy metals produce while allowing the bacteria to obtain sufficient amounts of trace metals for metabolism.

Digested sludge is fluid and relatively inert. It can be applied onto agricultural land if the water and the inorganic salts in the water phase can be absorbed without damaging the soil, the water table, or the adjacent waterways. Soil fungi can slowly metabolize some of the residual VS in the digested sludge, creating soil humus over time. The nitrifying bacteria in the surface soil will metabolize the soluble NH_3-N to NO_3-N. The soluble phosphate will be adsorbed onto soil particles depending upon their active chemical bonds. Surface plants can use these nutrients during their normal growth periods. The soil is the ultimate receptor for the digested sludge. The sludge components originally came from the soil and must eventually return to the soil to keep the natural cycle in balance. Pathogenic bacteria survival in digested sludge is dependent upon the pathogens ability to form spores. Few vegetative cells of pathogenic bacteria are able to survive the environmental conditions created in the anaerobic digestion of primary sludge with 20 to 30 days sludge retention. Viruses that are adsorbed onto sludge particles will pass through the anaerobic digesters unless they are metabolized by bacteria needing nutrients for survival.

SECONDARY TREATMENT

Secondary treatment units follow primary treatment units and are used to biologically treat the primary effluent. Secondary treatment employs bacteria and protozoa in an aerobic environment to convert the incoming organic contaminants into microbial suspended solids that can be removed by physical treatment, secondary sedimentation. Normally, secondary treatment systems have two tanks operating in series. The first tank contains the bioreactor to metabolize the biodegradable organic compounds in the primary effluent; and the second tank is the solids separation tank. There are several treatment systems that combine all the treatment units in a single tank. Since this book deals primarily with the microbiology of the treatment systems, only the generic treatment systems will be discussed individually. The different tank configurations will be left to other books for complete descriptions of the different modifications.

Aerobic treatment systems depend upon bacteria for the stabilization of nutrients in the presence of excess dissolved oxygen (DO). Protozoa assist the bacteria by eating the dispersed bacteria, helping to produce a clarified effluent. The most efficient bacteria grow under the specific environmental conditions imposed by the wastewater characteristics and tank configurations. As indicated in the fundamentals, bacteria tend to average about 2×10^{-13} g VSS/cell. There will be 5×10^{12} bacteria/g VSS cell mass. At 25° C the bacteria require about 2.8 hrs to double their mass under optimum environmental conditions. About 2 g BCOD of balanced nutrients are metabolized to produce 1 g VSS cell mass. The bacteria will use 0.7 g oxygen for energy during the growth process. The cell mass will have a COD/VSS ratio about 1.3. Since the cell mass is not stable, the living cell mass will undergo endogenous respiration at the rate of 1.0 percent/hr without protozoa and 2.0

percent/hr with normal populations of protozoa at 20°C. Overall, 80% of the original cell VSS can be metabolized by endogenous respiration with 20% of the cell VSS remaining as dead cell mass. The first problem facing both design engineers and WWTP operators is how to translate fundamental concepts into large-scale treatment plants. Do the numbers have any validity? Experience over the past 50 years has indicated that fundamental microbiological concepts have the same validity in large scale treatment plants as in laboratory scale. Creating the same environment in full-scale treatment plants and in laboratory treatment units has been a primary problem over the years. Microorganisms create the same reactions in the same environment every time.

The second problem for design engineers and WWTP operators lies in the separation of the microbial solids produced during the metabolism of the biodegradable materials in the wastewater. Without removal of the microbial solids from the treated wastewater, only 35% to 40% of the BCOD metabolized would be removed. Bacteria flocculation is the primary mechanism involved in separating the microbes from the treated wastewater. As previously indicated, flocculation is a natural bacterial reaction that occurs when the primary nutrients are exhausted. Flocculation and gravity separation allows the microbes to separate by gravity and be removed from the treated liquid. In dispersed growth systems the separated microbial solids can be returned back to the aerobic bioreactor to increase the total population of available microbes for rapid metabolism of the biodegradable organics in the wastewater. Once the desired microbial population has been reached in the bioreactor tank, the excess microbial solids must be removed from the system for it to continue to function properly. If solid media with large surface areas are available in the bioreactor, the bacteria will attach themselves to the media surface and increase the total microbial population without having to recycle settled solids. Over time the microbial masses will accumulate on the solid media surfaces to a point where the fluid flow will shear off some of the microbial mass. The microbial chunks sheared off the surfaces settle easily in gravity sedimentation tanks. The excess secondary solids from both type systems must be collected, treated, and returned to the land environment, the same as primary sludge.

The growth of specific bacteria, fungi, algae, protozoa, rotifers, nematodes, and other higher animals in wastewater treatment systems is determined by the chemical characteristics of the wastewaters discharged to the secondary treatment system and the environment created in the secondary treatment system. Municipal wastewaters are a complex mixture of organic and inorganic compounds that are discharged at the WWTP in a continuously varying fluid flow. As previously indicated, the wastewater flow characteristics depend on the complexity of the wastewater collection system. Very large wastewater collection systems tend to produce relatively uniform wastewater flows and uniform chemical characteristics on a daily basis. As the size of the wastewater collection system decreases, the variations in wastewater flows and in chemical characteristics increase on a daily basis. Single-family houses produce the greatest variation in wastewater flows and in chemical characteristics. It is easier to design and operate wastewater treatment plants that have little variations in wastewater flow and in chemical characteristics than plants that have wide variations in flow and in chemical characteristics. The real challenges for design engineers and WWTP operators are with the small wastewater treatment plants since they require the greatest skills. In the United States, the great majority of WWTP fall in the small category. Yet, most engineers are trained to design the few large WWTP. It is not surprising that there are still lots of problems being created at small WWTP on a daily basis in the United States. Knowledge has real value only if properly applied.

There are several different types of secondary treatment systems and numerous modifications of each type. The current classification system for wastewater treatment systems defines wastewater treatment systems as either *fixed media* systems or *dispersed growth* systems. The intermittent sand filter system, developed around 1888 at the Lawrence Experiment Station in Lawrence, MA, was the first successful fixed media wastewater treatment system in the United States. The intermittent sand filter was a coarse sand filter containing five feet of sand over an underdrain system to collect the treated effluent. Municipal wastewaters were applied over the sand filter once daily. The wastewaters moved by gravity through the filter in one to two hours. In time a microbial layer grew up in the top one inch of the sand that was responsible for removing most of the organic contaminants from the wastewaters. The sand bed was allowed to remain exposed to the air until the next dosing of wastewaters the following day. The microorganisms attached to the sand grains stabilized the organic compounds aerobically. Low concentrations of bacteria were found attached to the sand grains completely through the sand filter. Nitrifying bacteria were also found throughout the sand filter. The first studies of the specific bacteria in wastewater treatment were carried out on the intermittent sand filters at the Lawrence Experiment Station. It did not take the researchers long to discover that removing suspended solids in primary sedimentation tanks ahead of the sand filters allowed longer periods of operation before the sand bed clogged at the surface.

Trickling Filters

The British found the intermittent sand filter required too much land area to be of practical value in wastewater treatment for large cities. The data on intermittent sand filters published by the MSBH in 1890 led to research in England that soon resulted in the development of the trickling filter. The trickling filter was a fixed media system employing small stones, rather than sand, as the media on which the microorganisms grew. The trickling filter quickly became the accepted system for treating municipal wastewater. The important parts of the trickling filter were the wastewater distributor, the rock media, and the underdrain system to collect the effluent. The current wastewater distributors are either a two arm or a four arm rotary distributor that discharges the wastewater over the entire diameter of a circular bed of rocks. Figure 11-4 shows a four arm rotary trickling filter in operation. The pressure of the wastewater discharge drives the rotary distributor. The best rock media is angular, granite rock having a diameter between 50 mm (2") and 100 mm (4"). Granite is very hard and holds together under the physical and the chemical stresses affecting the rock media. The depth of rock media of normal trickling filters in the United States is 1.8 m (6 ft). The rocks sit on a vitrified clay tile underdrain system that collects the effluent as it leaves the rock media and transports it to a central channel across the middle of the filter. The transport channels in the clay tile underdrain system are also designed to allow air to move though void spaces around the rock media, keeping the biological growths as aerobic as possible. In the 1950s plastic media began to replace rock media. Plastic media was lighter than rock media and had greater void spaces, allowing greater movement of air through the trickling filter. The structural characteristics of plastic media permitted trickling filters to be constructed in depths up to 6 m (20 ft). Most trickling filters currently constructed in the United States use plastic media rather

Figure 11-4 A FOUR-ARM DISTRIBUTOR ON A ROCK MEDIA TRICKLING FILTER

than rock media. The design of plastic media varies considerably. Efforts have been made to create the maximum surface area per unit volume to provide surfaces for microbial growth.

Bacteriology of Trickling Filters - It was quickly recognized that the microorganisms growing on the rock media helped to determine the efficiency of trickling filters. Some investigators labeled trickling filters as "bacteria beds". After S. Winogradsky had isolated pure cultures of nitrifying bacteria in 1890, it was readily accepted that nitrifying bacteria grew in trickling filters that produced nitrates in their effluents. Early investigators were faced with a real problem in trying to identify the bacteria in trickling filter slimes. The slimes were so dense that it was not possible to easily disperse a sample of trickling filter slime for accurate identification and counting. Microscopic examination of the trickling filter slimes showed masses of zoogleal bacteria, many rod-shaped bacteria, some filamentous bacteria, some filamentous fungi, various protozoa, and some higher animals. One of the first studies to evaluate the types of bacteria in trickling filters was carried out by M. Hotchkiss at the New Jersey Agricultural Experiment Station in 1923. Hotchkiss used various types of biochemical media with serial dilutions to measure the biochemical reactions of the important bacteria. His data evaluated the groups of bacteria at various depths in the trickling filter. The bacteria groups included: denitrifying bacteria, sulfide producers from proteins, albumin decomposers, sulfate reducers, sulfur oxidizers, nitrite formers and nitrate formers. Although the bacteria numbers were low, they gave a relative measure of the different groups of bacteria in the trickling filter slimes. It was not surprising that the largest number of denitrifying bacteria were at the top of the filter and at the 5 ft depth. Very few denitrifying bacteria were found at the 1 ft depth and at the 3 ft depth. The sulfide producers from proteins were highest at 1 ft depth and decreased through the filter. The albumin decomposers were high through out the filter depth. The sulfate reducers were highest at the surface and the sulfur oxidizers were highest at 5 ft depth. The nitrite formers increased with depth and were much greater in numbers than the nitrate formers. While this study was interesting, it fell short of producing the desired understanding of the bacterial activity in trickling filters.

S. L. Neave and A. M. Buswell made the next major study of the bacteria in trickling filters over a six month period in 1925. They followed the same procedures as Hotchkiss in selecting media for counting specific groups of bacteria. They included: peptone degraders, gelatin degraders, coagulated egg albumin degraders, sulfide producers from peptone, sulfate reducers, thiosulfate oxidizers, cellulose degraders, nitrite formers, nitrate formers, and denitrifiers. The trickling filter they used for testing was 10 ft deep. There was little difference in the bacteria numbers in the influent to the trickling filter or at the various sampling points, confirming the problems in trying to obtain quantitative numbers of the different groups of bacteria. It suffices to say that the decomposing bacteria were greatest in the upper part of the trickling filter and the oxidizing bacteria were greatest at the lower part of the trickling filter. They also made microscopic counts of higher organisms. Protozoa made up the majority of the animal population in the trickling filter slimes. The *Sarcodina* were very active, second only to the *Mastigophera* in numbers. The amoeboid protozoa were greatest near the bottom of the trickling filter while the flagellated protozoa were greatest near the surface of the trickling filter where soluble organic compounds were greatest. The free-swimming ciliated protozoa increased with depth and were the third largest group of protozoa. The stalked ciliated protozoa were highest near the top and near the bottom. The crawling ciliated protozoa were also greatest near the bottom of the trickling filter. *Suctoria* were the smallest group of protozoa with more in the upper part of the trickling filter where the small flagellated protozoa were in greatest numbers. There were more nematodes than rotifers with the bottom 2/3 of the trickling filter having the largest numbers. There were also some diatoms and *Psychoda* larvae. The more people looked at trickling filter slimes, the more they focused in on the higher animals.

In 1929 Max Levine, a bacteriologist at Iowa State University, made a study with Burke and Watkins on the stabilization of skim milk plus additional lactose to determine the changes during stabilization. Since metabolism of lactose resulted in the production of lactic acid before being oxidized to carbon dioxide and water, they measured pH at different depths through the trickling filter. Their data showed that skim milk alone showed a continuous increase in pH from 6.9 to 7.9 in 6 ft of filter media. Adding 3,000 mg/l lactose resulted in a pH change from 6.2 to 5.6 at 1 ft depth and then a pH rise to 7.7 at 6 ft depth. It appeared that the bacteria in the trickling filter metabolized the lactose to lactic acid before completing their metabolism. They isolated a number of bacteria from the treated effluent that could metabolize lactic acid. Identification of the lactic acid utilizing bacteria resulted in eight genera: *Micrococcus, Vibrio, Lactobacillus, Aerobacter, Escherichia, Pseudomonas, Flavobacterium, Alcaligenes, and Achromobacter.* All of these bacteria were common soil bacteria. The nitrifying bacteria, also common soil bacteria, fascinated microbiologists. These autotrophic bacteria worked together to oxidize NH_3-N to NO_2-N and then to NO_3-N. Since organic wastes contained organic nitrogen, the heterotrophic bacteria had to metabolize the Org-N to NH_3-N first.

Levine, Burke and Linton measured the changes in nitrogen in a 6 ft deep trickling filter fed skim milk. The protein in skim milk is casein. As the casein was metabolized, NH_3-N was released. The NH_3-N was converted back into Org-N in the form of bacteria cell mass. About half of the Org-N was removed in the top one-foot of the trickling filter. Half of the Org-N removed appeared as NH_3-N. As the diluted skim milk moved downward in the trickling filter, nitrification began by the second foot. NH_3-N peaked at 15 mg/l at 2 ft depth. By the 6 ft depth only 1.6 mg/l NH_3-N remained. NO_2-N peaked between 3 ft and 4 ft depth at 1.2 mg/l. NO_2-N was quickly oxidized to NO_3-N by the *Nitro-* bacteria. Overall, 54 mg/l total nitrogen was added in the applied wastewaters and 22 mg/l total nitrogen appeared at the bottom of the 6 ft deep filter. The overall microbial reactions were observed, but little was known about what specific biochemical reactions occurred through the trickling filter.

Although everyone who observed trickling filter slimes under the microscope saw the zoogleal masses of bacteria, no one was able to isolate the zoogleal bacteria in pure cultures until C. T. Butterfield did in 1935. Working at the U. S. Public Health Service laboratory in Cincinnati, Ohio, Butterfield isolated *Zooglea ramigera* in pure culture from activated sludge. In 1941 C.T. Butterfield and Elsie Wattie reported on the isolation of zoogleal forming bacteria from trickling filters. The zoogleal forming bacteria isolated from trickling filters were similar to the zoogleal forming bacteria isolated from activated sludge. In 1942 Elsie Wattie reported on the biochemical characteristics of the zooglea-forming bacteria that had been isolated in pure culture from both activated sludge and trickling filters. Although she found that the biochemical characteristics were different enough that she could divide the cultures into nine groups, Elsie Wattie stopped short of complete identification of the bacteria groups, stating only that all the zooglea-forming bacteria were related.

R. H. Holtje presented a paper summarizing the biology of trickling filters in 1943. The primary bacteria were considered as being zoogleal bacteria with nitrifying bacteria occurring in low rate trickling filters. The filamentous bacteria were easier to identify than the zoogleal bacteria and included *Beggiatoa* and *Sphaerotilus*. *Beggiatoa* were easily identified since they were filamentous sulfur oxidizing bacteria that deposited sulfur granules inside the cells. The fungi were mostly *Fusarium* and *Leptomitus*. There were green algae, *Stigeoclonium,* and blue-green algae, *Oscillatoria*. The algae occurred near the top of the trickling filter where light was readily available. Amoeboid protozoa, flagellated protozoa, free-swimming ciliated protozoa, crawling ciliated protozoa and stalked ciliated protozoa were found in different parts of the trickling filter slimes. Rotifers were found in lightly loaded trickling filters. Nematodes and other worms were extensively found. *Psychoda* fly larvae soon became flies that were part of trickling filter biota everywhere. The basic problem with biologists was they were good at identification of each and every species, but they lacked an understanding of the significance of all the various organisms in either the design of trickling filters or their operations. Trickling filters simply worked. Design and operation of trickling filters were entirely based on past experiences. Most of the operational data consisted of input and output data. Engineers developed a series of empirical equations, based on the operating results, to help understand how to design trickling filters for different wastewater characteristics.

Biological Concepts for Trickling Filters - It is not surprising that trickling filters stimulate the growth of a complex mixture of microorganisms. A 2-arm rotary distributor provides a quick wastewater loading over the complete diameter of the circular trickling filter, followed by a rest period until the distributor arm passes over the same area again. The feed-rest concept was a carryover from the intermittent sand filter. Many engineers believed that the bacteria responsible for metabolism of the contaminants in the wastewater needed additional time to complete metabolism. The wastewaters are quickly pulled by gravity through the trickling filter along the path of least resistance. The effluent is discharged from the bottom of the trickling filter in just a few seconds after being added at the surface. The 1.8 m filter depth does not slow the wastewater flow significantly. Once the media has been wetted, surface tension will hold a thin liquid layer on the rock surface. The subsequent wastewater additions simply slide over the attached liquid layer as it provides the least resistance to flow. It is not surprising that the bacteria begin their growth and attachment in the cracks and crevices on the rock media surfaces and on the edges where rocks come into contact with each other. As the bacteria growth extends out from the cracks and crevices along the rock media surface, the attached layer of water expands and extends beyond the bacteria surfaces. The added wastewater flow now moves as a wave over the water layer attached to the bacteria surfaces. There is an exchange of liquid between the moving water wave and the attached water layer over the bacteria. The traveling water wave containing the waste contaminants mixes with the attached surface water layer. A small amount of the contaminants are transferred into the fixed water layer that is left attached to the surface of the bacteria. The traveling water wave picks up waste products from the attached liquid layer. This exchange process results in a decreasing concentration of contaminants in the traveling wastewater as it flows through the trickling filter. The effluent quality from the trickling filter is determined by the amount of contaminants remaining in the liquid displaced from the bottom of the rock media. If the attached water layer over the bacteria can be maintained at a low organic concentration by rapid bacteria metabolism, the treated effluent will contain few of the influent organic compounds.

Because the rock surfaces prevent the bacteria from penetrating into the rocks, the bacteria growth is horizontal and vertical, depending upon the least resistance for cell growth. Once the entire surface of the rock media is covered with bacteria, new growth is primarily outward. Although the surface of the bacteria growth appears smooth, close examination of the bacteria growth indicates that the outer surface is not uniformly smooth. The microbial layer has a rough surface with pockets of dense growth and areas with thin growth, depending on the hydraulic flow characteristics through the trickling filter. Shear forces are created as the wastewater waves flow through the void spaces in the trickling filter. Lightly held bacteria are easily sheared off the attached bacteria mass. The bacteria on the surface layer take up the soluble organic compounds together with dissolved oxygen and create new cells on a semi-continuous basis. The bacteria growth becomes vertical with the new bacteria forming a layer over the previous layer of bacteria. Since bacteria average about 1.0 μ thickness, it would take 1000 layers of living bacteria to produce a layer 1.0 mm thick. Adsorption of suspended solids allows the microbial layer to increase in thickness with many fewer bacteria than 1000 layers.

As the bacteria layers accumulate over time, only the outer few layers are able to remain aerobic. The dissolved oxygen is removed by the metabolism of the bacteria on the outer surface, allowing little DO to diffuse into the next layer of bacteria. There are even less nutrients to reach the second layer of bacteria. The bacteria below the aerobic surface layer shift their metabolism from aerobic to anaerobic. The adsorbed organic particles are engulfed by bacteria and slowly metabolized under anaerobic conditions to soluble organic compounds. The anaerobic end products slowly diffuse towards the surface layer, providing additional organic nutrients to the surface layer of bacteria. Over time, the sulfate reducing bacteria can be found on the rock surface as conditions favor their growth. The hydrogen sulfide produced as a metabolic end product of the sulfate reducing bacteria reacts with iron and other metallic elements in the rock surface to produce a black precipitate of metallic sulfides. The soluble sulfides slowly diffuse upward between the bacteria to the surface layer where sulfur-oxidizing bacteria can be found under aerobic conditions. Here again, the specific environment created in the trickling filter determines the growth of specific groups of bacteria. The most efficient bacteria for the specific environment will always predominate. The bacteria growing on the trickling filter media are primarily common soil bacteria with few exciting characteristics. The bacteria tend to metabolize the soluble organic compounds in the wastewaters to carbon dioxide, water, ammonia, and new cells. For this reason few studies have been made to identify the different bacteria growing in trickling filters. Growth of nitrifying bacteria in low rate trickling filters has stimulated more interest in bacteria identification than the heterotrophic bacteria have.

The aerobic growth of bacteria on the media surfaces provides a good environment for the growth of many different types of animals. Protozoa, rotifers, nematodes, higher worms, insect larvae, crustaceans, and snails all find the trickling filter environment suitable for growth. Essentially, the animals feed on the bacteria, reducing the bacteria mass in proportion to their growth. Each organism requires a specific environment in which to grow. As the environment changes, the animal populations change. Since the animals are easier to observe and count under the microscope at relatively low power magnification, more data are available on the animal populations in trickling filters than on the bacteria. The higher animals are much easier to wash out of the trickling filter and are largely removed in the secondary sedimentation tank following the trickling filter. Insect larvae that are allowed to reach the adult stage can create nuisance conditions around trickling filters. Snail shells can also become a nuisance when treating secondary sludge. The snails are easily washed from the trickling filters and settle with the sludge in the final sedimentation tanks. If the sludge is pumped to an anaerobic digester for treatment prior to final disposal on the land, the anaerobic digester will slowly fill with snail shells. Since the animal growth is dependent on the available bacteria, the animal mass is a small fraction of the total bacteria growth. N. F. Gray presented a representative list of animals found in trickling filters in his 1989 book on wastewater treatment. Since the animals are aerobic microorganisms, they grow best in the attached liquid layer over the surface layer of actively growing bacteria. Each animal must consume large numbers of bacteria to grow and maintain normal metabolic activity. If the animals eat too many bacteria, the animal population will begin to decrease, allowing the bacteria population to return to normal. The bacteria population controls the overall operation of the trickling filter wastewater treatment systems.

With the top layer of rock media in a trickling filter exposed to the atmosphere, algae will be able to grow on the upper layer of rocks. The algae obtain their energy from sunlight and their carbon from the carbon dioxide released by bacteria metabolism. With excess nitrogen and phosphorus in the incoming wastewaters, algae growth can become quite significant. Filamentous algae growth can be sufficient to fill the voids between small rock media, clogging the filter and preventing normal fluid flow through the trickling filter. If the combined growth of algae and bacteria clog the trickling filter sufficiently, the incoming wastewater will actually form ponds on the surface of the trickling filter media. Surface ponding prevents the normal passage of air through the trickling filter and upsets the normal aerobic metabolism in the trickling filter. Obnoxious odors are often generated as a result of ponding. Treatment efficiency decreases rapidly. It is necessary to kill the algae by the use of acid or chlorine and to flush the dead cells out of the trickling filter to restore good bacteria growth and normal treatment efficiency.

Plastic media began to replace rock media when the demand for good rock media exceeded the supply. Poor quality rock media broke down more rapidly and allowed the microbial growths to clog the trickling filter media easier than the higher quality, higher priced rock media. Overall economics soon favored plastic media over rock media. The plastic media allowed the use of tall filters and higher loading rates. Microbial growths formed thinner layers on plastic media than on rock media. The higher hydraulic loading rates produced greater shear stresses over the plastic media surfaces. Many of the plastic media trickling filters employed a high rate of fluid recirculation to produce an effluent with a low BOD_5. The high rate of fluid recirculation created a dispersed growth phase to go along with the attached growth on the plastic media. As a net result, the plastic media trickling filter was definitely different than the rock media trickling filter. The thin fluid film over the plastic media transferred oxygen from the air into the flowing liquid at a higher rate than in rock media trickling filters. Plastic media trickling filters allowed greater rates of microbial metabolism than the rock media trickling filters when loaded to the maximum rates.

Trickling filters are not used very much in the United States for municipal wastewater treatment at the present time. The inability of trickling filters to produce a high quality effluent in a small area at a low cost resulted in their replacement by dispersed growth biological systems. In many areas of the world trickling filters still have a place in municipal and industrial wastewater treatment.

Activated Sludge

Activated sludge is an aerobic, dispersed growth system in contrast to the fixed media trickling filter. The ability to produce a high quality effluent, 90 to 99+ percent BOD5 reduction and 90 to 95 percent TSS reduction, has made activated sludge the most popular wastewater treatment system currently in use in the United States. Activated sludge treatment plants use circular, square, or rectangular aeration tanks. Oxygen is supplied to the microbes by diffused air, mechanical surface aerators, submerged turbines, or impingement jets. Pure oxygen has also been used in place of air for systems with high oxygen demand rates. The keys to activated sludge are excess microorganisms, excess dissolved oxygen (DO), sufficient time, and adequate mixing in the aeration tank to promote rapid metabolism of the biodegradable organic compounds. Excess microorganisms can be maintained in the aeration tank as a result of the ability of bacteria to form floc after the nutrients have been metabolized and to settle quickly by gravity under quiescent conditions before being collected and pumped back to the aeration tank as *return activated sludge* (RAS). The continuous stabilization of organic matter in wastewater results in the production of more microorganisms than are needed to maintain the activated sludge at its desired concentration in the aeration tank. The extra microbial production has been termed *excess sludge* or *excess activated sludge*. The excess sludge must be removed from the activated sludge system on a continuous or a semi-continuous basis to maintain the desired microbial population.

A small amount of excess sludge will be lost in the final effluent, about 10 to 30 mg/L. Most of the excess sludge must be wasted from the system as *waste activated sludge* (WAS). The WAS must be treated before being returned to the land environment, as it contains living microorganisms. Maintaining excess DO in the aeration tank is the most difficult part of the activated sludge process. Oxygen transfer is a function of the aeration tank configuration, the aeration equipment, the mixing characteristics produced by the aeration equipment, the chemical characteristics of the incoming wastewater, and the hydraulic retention time (HRTa) in the aeration tank. Experience in the United States has shown that municipal wastewater can be easily treated with a 6 hr HRTa after primary sedimentation. It is not surprising that the mixing characteristics in the aeration tanks are determined by the aeration tank configuration and the aeration equipment used in the aeration tank. All of these important characteristics will be presented in detail to provide a better understanding of the activated sludge process.

Activated Sludge Discovery - Like the trickling filter process, the British discovered the activated sludge wastewater treatment process. Edward Ardern and William T. Lockett published the results of their research on the basic concepts of activated sludge in 1914. The key that Ardern and Lockett discovered was retention of flocculated solids in the aeration tank. They called the flocculated solids "activated sludge", since the solids were able to quickly stabilize both the carbonaceous organics and the nitrogenous compounds in the wastewater. Initially, Ardern and Lockett carried out their studies on a fill-and-draw basis with aeration of municipal wastewater. When the microbes had grown and oxidized the organic compounds in the wastewater, they stopped the aeration and allowed the microbial solids to settle. Very few microbial solids settled after the first feeding cycle. They poured off the supernatant after settling and added fresh wastewater to the system. Aeration was restarted and the treatment cycle began again. The feed-settle cycle was repeated after analyses indicated oxidation of the organic compounds in the wastewater. The settled solids accumulated at a faster rate after each feeding. Soon, the settled supernatant was almost clear. The microbes had finally accumulated to a suitable level to stabilize the organic compounds and to oxidize the NH_3-N in a short aeration time. The problem of microbial accumulation in a dispersed aeration system had now been solved. Analysis of the activated sludge, using gelatin plates for counting the active bacteria, showed 30 million bacteria/ml of mixed liquor. Microscopic examination of the activated sludge particles indicated an active population of protozoa. The chemical analyses of their activated sludge showed 65% VSS, 4.6% N, 2.6% P_2O_5 and 5.8% lipids. Incorporation of 35% NVSS in the activated sludge greatly assisted its rapid settling when the air was turned off. The concept of biological treatment of municipal wastewaters without large tanks of rock media was very exciting to engineers faced with designing wastewater treatment plants for large cities. Yet, much research had to be done before activated sludge could be used in full-scale wastewater treatment. World War I slowed British efforts to develop activated sludge and stimulated American interest in this new wastewater treatment process. Professor E. Bartow of the University of Illinois visited the Manchester WWTP and observed the laboratory activated sludge units during the summer of 1914. On his return to the University of Illinois, Professor Bartow immediately began research on activated sludge. It did not take long before American engineers realized that activated sludge had a place in municipal wastewater treatment.

Basic Process – In the United States the conventional activated sludge process consists of a rectangular aeration tank and a secondary sedimentation tank. The aeration tank provides a suitable environment for a mixture of bacteria and other microorganisms to aerobically metabolize the biodegradable contaminants in the incoming wastewater in a relatively short period of time, 6 to 8 hrs for settled municipal wastewater. The ability of activated sludge to flocculate under quiescent conditions and to separate from the treated wastewaters is an essential characteristic of the activated sludge process. The final sedimentation tank is designed to provide a suitable quiescent environment for the activated sludge to settle and concentrate on the tank bottom while allowing the clarified effluent to be discharged at the top of the tank. The settled activated sludge is continuously collected and returned to the aeration tank to maintain an excess of active bacteria in the aeration tank. Since there is a continuous production of new microorganism, a point is reached where the concentration of microorganisms in the aeration tank reaches an optimum level. The excess activated sludge that is produced in the aeration tank must be removed from the system on a continuous basis to keep the system in equilibrium. A small amount of activated sludge will normally be carried out in the final effluent, 5 to 30 mg/L TSS. The rest of the excess sludge is removed by wasting concentrated sludge from the final sedimentation tank. The basic activated sludge process is shown in the schematic diagram in Figure 11-5. While there have been numerous variations in activated sludge systems, the basic process for activated sludge systems has not changed.

The activated sludge in the aeration tank is called *mixed liquor suspended solids* (MLSS). The settled sludge returned to the aeration tank is called *return activated sludge* (RAS) and the sludge removed for wasting is called *waste activated sludge* (WAS). The suspended solids carried out in the final effluent are called *effluent suspended solids* (Eff SS). Activated sludge plants treating settled domestic wastewaters in the United States will normally contain 2,000 mg/L MLSS, ranging between 1,500 mg/L and 2,500 mg/L. The raw wastewater aeration time (HRTa) ranges from 6 to 8 hrs with a 2hrs sedimentation time. The RAS concentration will be about 10,000 mg/L for a normal activated sludge with a return sludge flow rate (Qr) equal to 25% of the incoming wastewater flow rate (Qi). The WAS is often

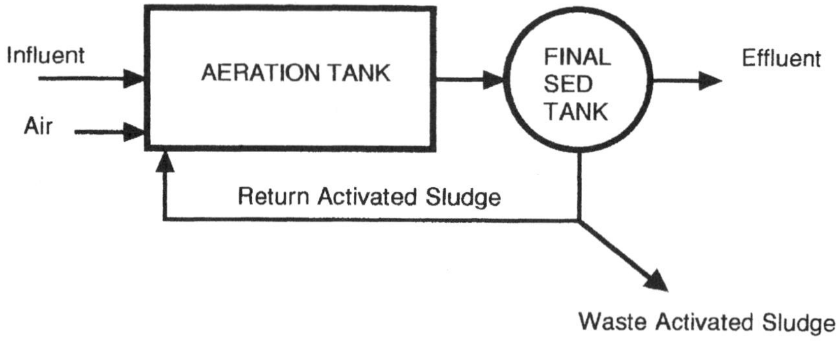

Figure 11-5 SCHEMATIC DIAGRAM OF ACTIVATED SLUDGE PROCESS

removed from the RAS flow for simplicity of operations.

Microbiology of Activated Sludge – The first American studies on the microbiology and biochemistry of activated sludge were made by the Illinois State Water Survey, which was associated with the University of Illinois. A. M. Buswell and H. L. Long reported the findings of their studies on activated sludge microbiology in 1923. They used daily microscopic examinations of the mixed liquor to obtain gross information on activated sludge organisms. Most of their data dealt with the protozoa, since they were easily counted under the microscope. They found that the initial population of small flagellated protozoa and small ciliated protozoa quickly gave way to larger free-swimming ciliated protozoa and stalked ciliated protozoa. As the system became more stable, the crawling ciliated protozoa appeared. Only a few suctoria and rotifers were observed. Nematodes suddenly appeared and then slowly decreased. The overall bacteria were simply classified as zoogleal masses with filamentous bacteria of various types. They felt that both bacteria and protozoa were important in activated sludge.

Bacteria - Over the next decade a number of general studies were carried out to determine if activated sludge was basically a biological process or a physical-chemical process with the assistance of microorganisms. Essentially, there was a difference of opinion between the biologists and the chemists. The engineers had no interest in the basic concepts. Their interest was purely in the application of the process to wastewater treatment. The center of the intensity of the arguments between chemists and biologists was the USPHS laboratory located at Cincinnati, Ohio. E. J. Theriault represented the chemical point of view. C.T. Butterfield represented the biological point of view. In 1935 Theriault published a paper stating that activated sludge was a biozeolite with bacteria playing a minor role, only in the regeneration of the zeolite. The chemical analyses of Cincinnati activated sludge had a chemical formula very similar to zeolites. Unfortunately, the chemical formula of the Cincinnati activated sludge was affected by the discharge of alum sludge from the water treatment plant into the sanitary sewer and was not normal for municipal activated sludge. The same year, Butterfield isolated a bacterium in pure culture, *Zooglea ramigera*, which formed activated sludge and stabilized soluble organic compounds the same as a mixed culture activated sludge. The isolation of a pure culture of bacteria that resulted in floc similar to activated sludge established the biological theory of activated sludge once and for all. Only C.T. Butterfield, his assistant, Elsie Wattie, and H. Heukelekian at Rutgers were able to isolate and study *Z. ramigera* in the laboratory. The teasing technique that Butterfield had developed was a very difficult method for isolating floc-forming bacteria that few bacteriologists could duplicate. The photomicrograph, Figure 11-6, shows the two different growth forms of *Zooglea ramigera*. The lower mass is the typical finger-like growths with the bacteria concentrated at the end of the finger growths. The upper mass is the dense, bulbous growth. The photomicrograph is from unstained specimens of *Zooglea ramigera*, grown in a soluble substrate and observed at about 100X total power magnification with a phase contrast microscope.

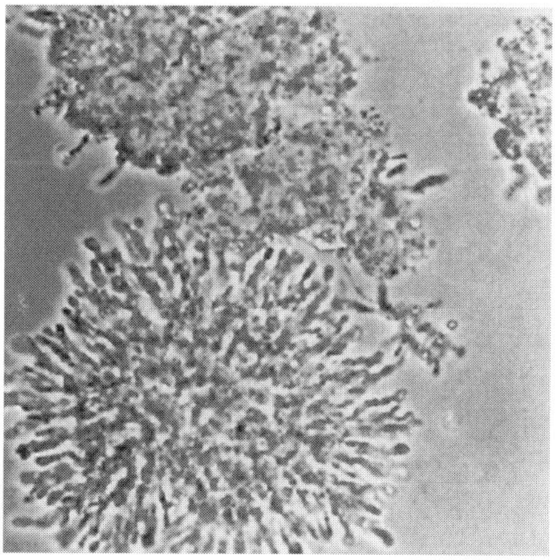

Figure 11-6 *Zooglea ramigera*, UNSTAINED, PHASE CONTRAST MICROSCOPE

In 1944 L. A. Allen in England examined the bacteria in activated sludge by homogenizing the sludge and plating the bacteria on nutrient agar. He found that activated sludge contained 2.2 x 10^9 bacteria/ml. Identification of the major groups of bacteria included *Achromobacter, Chromobacterium* and *Pseudomonas.* Allen was unable to isolate *Z. ramigera* from his samples of activated sludge. Increased application of activated sludge in the 1950s and early 1960s stimulated further studies on the bacteria in activated sludge. E. F. Dias and J. V. Bhat of the Indian Institute of Science in Bangalore, India, in 1964 examined the bacteria in activated sludge. They isolated over 300 bacteria from seven different activated sludge samples. The majority of bacteria were either *Zooglea* or *Comamonas*, gram negative bacteria. Their biochemical tests indicated many different species within *Zooglea* and *Comamonas*. They found that polyhydroxybutyric acid was a major storage product in over half of the bacteria isolated.

In the Netherlands in 1964 H. W. van Gils examined the bacteria in the activated sludge from the Zeist WWTP. He washed the activated sludge to remove extraneous bacteria and then homogenized the sludge to disperse the flocculated bacteria. Washing the activated sludge removed about 30 percent of the bacteria in the initial activated sludge, indicating many bacteria were loosely attached to the sludge particles. The washed activated sludge contained about 10^{11} bacteria/g activated sludge. He found the bacteria were 25% - 35% *Alcaligenes*, 23% - 30% *Flavobacterium*, 6% - 16% *Achromobacter* and only 2% - 5% *Pseudomonas*. There was no *Zooglea* or polyhydroxybutyric acid in the bacteria. The Dutch research contradicted the Indian research on predominant bacteria in activated sludge. A short time later, A. D. Adamse examined activated sludge from a dairy waste oxidation ditch. He found three major groups of bacteria: *Arthrobacter*, *Achromobacter*, and *Pseudomonas*. *Arthrobacter* were the predominant group of bacteria in the dairy waste activated sludge.

In the United States Lighthart and Oglesby examined the activated sludge at the Renton WWTP in Seattle, Washington, and found that the primary bacteria were *Flavobacterium*, *Achromobacter* and *Pseudomonas*. In a later evaluation of the data, using numerical analysis and cluster grouping, Lighthart and Loew found nine major groups, confirming the diversity of the bacteria in activated sludge. The data to date have all shown that the predominant bacteria in activated sludge are common soil bacteria with specific organic compounds in the wastewaters having a major impact in determining the specific groups of bacteria isolated.

One of the problems in isolating and identifying the bacteria in activated sludge is finding a suitable growth medium. For general groups of bacteria, tryptone-glucose-agar (TGA) is the most universal medium for isolating individual bacteria cultures. For isolating specific groups of bacteria from unusual industrial waste activated sludge, agar can be added to a sample of the industrial wastes to provide a solid medium. Synthetic media can be prepared using a single organic compound and inorganic nutrients in sterile dilution water with agar for solidification. Since the bacteria in activated sludge are aerobic, surface growth on agar media will produce the best results. The spot plate technique has proven quite effective in isolating the predominant bacteria from all types of activated sludge. Because of the large population of bacteria in activated sludge and their flocculated state, it is necessary to disperse the bacteria by shaking the activated sludge sample with glass beads for several minutes. Serial dilutions in sterile dilution water are made to produce a dilution with about 500 bacteria/ml. A 0.2 ml graduated pipette allows the application of 0.02 ml spots to be applied to the surface of agar plates that have been air dried for 48 hrs under sterile conditions. A total of 10 spots can be applied to the surface of the media in a standard petri dish. Growth of individual cells on the agar surface allows easy counting of individual colonies, as well as, isolation for identification. Spot plate bacteria counts have proven to be as accurate as those from normal pour plates. A good lab technician should be more accurate with spot plate counts than with pour plate counts.

The initial isolation and identification techniques were based on microscopic examinations and biochemical tests. Currently, some researchers are using rRNA to create oligonucleotide probes for direct determination of specific populations in activated sludge. Snaidr et al in Germany reported on their research in 1997. Their phylogenic analyses resulted in quite different types of bacteria than reported by previous investigators. It is apparent that rRNA analysis is going to create a different set of bacteria identification than that created by biochemical testing. The rRNA analysis of bacteria will eventually replace biochemical identification. Yet, it should be remembered that the rRNA identification is a specialized laboratory technique that yields little information on the biochemical characteristics of the bacteria. While rRNA identification of bacteria will become popular, biochemical reactions will control bacteria use in environmental microbiology. During the transition to rRNA identification of bacteria, there will be considerable confusion in the area of bacteria identification in the environmental area.

Activated sludge depends upon its ability to form large floc particles that rapidly settle when moved from the rapidly mixed aeration tanks to the quiescent sedimentation tanks. Floc formation has been the subject of much research and debate. When C. T. Butterfield first isolated *Zooglea ramigera* from activated sludge and demonstrated in 1935 that *Z. ramigera* formed floc similar to activated sludge, it was assumed that *Z. ramigera* was the floc-forming bacteria in activated sludge. Isolation of *Z. ramigera* in pure culture was a difficult task. It was necessary to wash all the bacteria off of the floc and then to use a micropipette to remove the bacteria from the floc. The difficulty in isolating *Z. ramigera* in pure culture was in stark contrast to the ease with which activated sludge could be

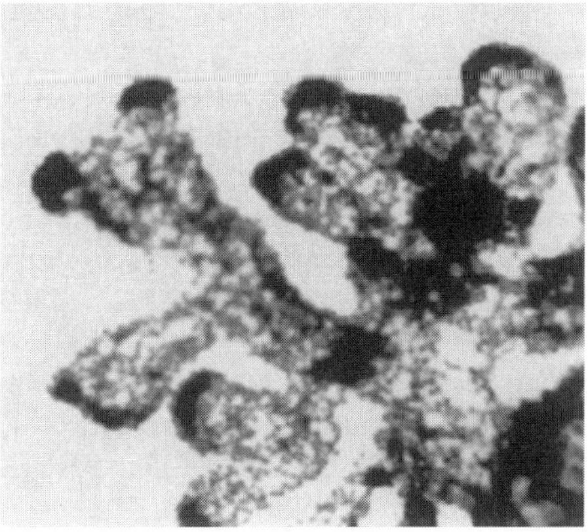

Figure 11-7 *Zooglea ramigera*, STAINED WITH ALCIAN BLUE

formed in any organic wastewater. The photomicrograph, Figure 11-7, shows a stained specimen of *Z. ramigera* at 970X magnification. Individual bacteria cells can be seen in the polysaccharide slime. Note the dense masses of cells that aggregate at the end of the finger-like projections. Centrifugal forces in the aeration tank cause the cells to concentrate at the end of the gelatinous polysaccharide that surrounds the mass of *Zooglea* cells. The polysaccharide material around the cells makes it difficult to separate and isolate individual cells. It did not seem logical that a single group of bacteria would grow and form floc in a wide variety of municipal and industrial wastes. Research at MIT, begun 15 years after Butterfield reported on *Z. ramigera,* was based on these concerns. Dr. Butterfield doubted that we would find any floc-forming bacteria other than *Z. ramigera.*

The 1950 research at M.I.T. used a different approach than the one used by Butterfield, Wattie, and Heukelekian. Since activated sludge formed so quickly under aeration, it was felt that small aeration systems should yield the floc-forming bacteria very easily. A series of glass culture tubes were used as aeration tanks with 6 mm glass capillary tubing used as diffused aerator. The entire system was constructed on a 2x6 wooden board, 18 inches long, suitable for sterilization in the laboratory steam sterilizer. Each culture tube became its own sterile activated sludge system capable of aeration and settling. A 0.1 dilution of nutrient broth was used as the nutrient media for the aeration tubes. The dilute nutrient broth was completely soluble, making it easy to determine floc formation. Activated sludge from Leominster, MA, was maintained in the laboratory by regular feeding with fresh Boston sewage. Samples of activated sludge were treated by various physical techniques, indicated in the literature to release the bacteria from the floc. Serial dilutions of the dispersed floc samples were made in sterile tap water. One ml samples from the different dilutions were aseptically placed into the culture tubes of sterile media and aerated. After 24 to 48 hours aeration the culture tubes showing floc at the greatest dilutions were used for inoculating nutrient agar streak plates. After 24 to 48 hours incubation at room temperature individual colonies were selected from the streak plates and streaked onto fresh nutrient agar plates. This procedure was repeated until pure cultures were obtained on the nutrient agar plates, based on microscopic examination. Stock cultures were prepared from the pure cultures for testing and identification.

Once bacteria were isolated in pure culture, each pure culture of bacteria was inoculated into fresh batches of sterile 0.1 dilution of nutrient broth and aerated to determine floc formation. This technique resulted in the isolation of four bacteria capable of forming floc similar to activated sludge: *Bacillus cereus, Nocardia actinomorpha, Paracolobactrum aerogenoides,* and *Escherichia intermedium*. All of these bacteria were soil bacteria. This technique did not yield *Z. ramigera*. Later, it was possible to isolate *Z. ramigera* in pure culture using the washing and dilution technique that had been used by other investigators. Because coliform bacteria showed floc formation, a pure culture of *Aerobacter aerogenes* was tested and found capable of floc formation. It was apparent that the aeration technique used to normally form activated sludge could be used to quickly isolate floc-forming bacteria other than *Z. ramigera.* Additional studies on different activated sludges yielded 10 more floc-forming bacteria. Evaluation of the biochemical characteristics of the *Zooglea* bacteria, previously isolated by Elsie Wattie in 1942, indicated that her floc-forming bacteria included *Achromobacter, Alcaligenes, Bacillus, Flavobacterium,* and *Pseudomonas*, in addition to *Zooglea.* All of these bacteria are common soil and water bacteria. It became apparent that many different bacteria had the ability to form floc similar to activated sludge under the proper environmental conditions.

Further studies on floc formation showed that bacteria did not flocculate when excess nutrients were available for metabolism. Floc was produced when the bacteria had metabolized the organic nutrients and had shifted from rapid growth to endogenous metabolism. It was also noted that some bacteria remained dispersed even though most bacteria clumped and formed floc. Even mixtures of bacteria did not produce a perfectly clarified effluent until all the bacteria died off. The addition of ciliated protozoa to pure cultures of bacteria produced floc at a faster rate than with the bacteria alone and resulted in a clarified supernatant. The protozoa ate the dispersed bacteria that were creating the turbid effluent. The inert, residual materials from the bacteria metabolized by the protozoa were contained in the protozoa waste products and accumulated in the activated sludge floc. It appeared that bacteria and protozoa worked together to produce normal activated sludge in wastewater treatment plants..

Although these research studies demonstrated that many different bacteria were capable of forming activated sludge, the magic that belonged to *Zooglea ramigera* persisted. *Z. ramigera* was an interesting organism because of its growth characteristics. Although *Z. ramigera* was supposed to rapidly metabolize organic matter, its growth characteristics limited its ability to obtain nutrients in mixed cultures. Studies on the slime material showed it was largely polysaccharide material with few reactive chemical groups to attract nutrients when compared with the motile soil bacteria. All nutrients had to pass through the slime before reaching the individual bacteria. Growth and persistence of any bacteria in wastewater treatment systems depends on their competition for nutrients. The scanning electron photomicrograph, Figure 11-8, shows the finger-like growth of *Zooglea* bacteria at 5,000X power magnification. Unfortunately, the pretreatment with ethyl alcohol dehydrated the polysaccharide slime material, changing the appearance of the *Zooglea* floc. Individual cells are apparent at the end of the finger-like floc and along the floc stem. The polysaccharide slime is not as readily apparent as in the other photomicrographs with stained preparations of *Zooglea* floc. The difference shows that care must be taken in the interpretation of all microscopic observations. Depth of field in optical microscopy and various pretreatments used in electron microscopy can adversely affect what we think we see. Further studies are definitely needed in all aspects of microscopy, as it affects the microorganisms found in activated sludge.

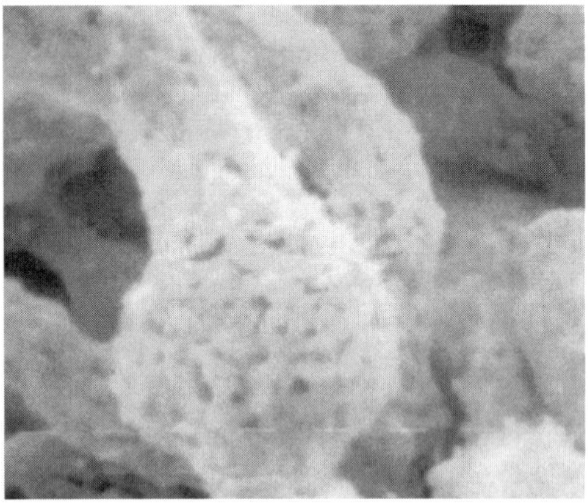

Figure 11-8 ELECTRON SCAN OF ZOOGLEA FLOCS (U. of Kansas)

Recently, a German study used three oligonucleotide probes from the 16S rRNA of three different strains of *Z. ramigera*. They tested 11 different WWTP to determine the magnitude of *Z. ramigera* in the systems. Most activated sludge systems showed positive results for at least one strain of *Z. ramigera*. The maximum content of *Z. ramigera* was 10 percent of the total bacteria population in one plant. These results added further evidence that *Z. ramigera* are not the primary bacteria in activated sludge systems. Based on my research on floc-formation, it appears that *Z. ramigera* is not the major floc-forming bacteria in activated sludge systems. Yet, Z. ramigera is still an interesting bacteria for study. During a study on the metabolism of alkylbenzene sulfonate (ABS) detergents by activated sludge at MIT, it was observed that the growth of *Z. ramigera* was greater in the activated sludge fed biodegradable ABS detergents than in the control activated sludge not fed biodegradable ABS detergents. Later observations indicated that a simple aromatic substrate, such as sodium benzoate plus nutrients, stimulated *Z. ramigera* growth. It appeared that the metabolism of the benzene ring structure increased the growth of *Z. ramigera.*

From an engineering point of view it is not important to know which bacteria form floc, only to know the environmental conditions necessary for floc formation. It takes time to create a sufficient mass of floc for good settling. Where possible, activated sludge from a well operating activated sludge plant should be brought in by tank truck and used as the microbial seed to start a new activated sludge plant. The smallest aeration tank should be used initially for feeding and accumulating the activated sludge. The aeration tank is filled with wastewater and activated sludge. It is aerated for several hours until the soluble COD has been reduced to a minimum. The aeration is stopped and the activated sludge is allowed to settle in the aeration tank. The settled supernatant is removed by a siphon or pumped out. The tank is filled with fresh wastewaters and aeration started again. This process is repeated until the aeration tank contains 400 to 600 mg/l MLSS. Once the MLSS builds up to between 400 and 600 mg/l, the wastewaters can be fed continuously with mixed liquor from the aeration tank being displaced into the secondary sedimentation tanks. The settled sludge should be pumped back to the aeration tank at the lowest possible rate to prevent loss of excess suspended solids in the effluent. Once the MLSS accumulate to the desired level in the aeration tanks, the excess activated sludge should be wasted from the system on a semi-continuous basis in small to medium sized plants and on a continuous basis in large plants. This procedure for starting activated sludge plants will work for municipal wastewaters and non-toxic industrial wastewaters. Industrial wastewaters that are normally toxic will require special startup procedures.

The overall environment in the aeration tank determines which microorganisms grow and to what extent they grow. An optimum environment for the production of a good activated sludge with a diverse population of bacteria includes a pH between 7 and 8, a temperature between 20°C and 30°C, a dissolved oxygen above 1.0 mg/l at all times in all parts of the aeration tank, sufficient agitation to keep the suspended solids in uniform suspension, a readily biodegradable substrate with adequate carbon, nitrogen, phosphorus, and trace nutrients, and adequate time for complete metabolism. The bacteria best able to metabolize the substrate and produce new cell mass will automatically predominate. Normal wastewater load variations will allow the bacteria population to adjust to the changing organic loads. The predominant bacteria will be motile, rod shaped, bacteria that can metabolize the maximum amount of organic contaminants to the greatest extent. This does not mean that spherical bacteria, spiral-shaped bacteria, or filamentous bacteria will not grow in the activated sludge. All types of aerobic bacteria will grow to the extent that they can obtain nutrients. The motile, rod shaped, bacteria are simply more efficient at metabolism than the other shaped bacteria. Growth of new bacteria is simply a response to the environment. If the environment is changed, the bacteria predomination will change. It is important to recognize that the only way to change the bacteria predomination in a given activated sludge system is to change the environment in the aeration tank.

Activated sludge, treating municipal wastewaters that are predominantly residential, will have bacteria acclimated to proteins, carbohydrates, and lipids. The relative population of acclimated bacteria will depend upon the rate of addition of the three groups of organic compounds. The treatment of unusual organic compounds in industrial wastewaters can pose a challenge in accumulating the desired acclimated bacteria for efficient treatment. In addition to providing a good environment in the aeration tank, it is important to maintain the specific bacteria necessary for metabolism of the industrial contaminants. Fortunately, the bacteria found in activated sludge are capable of metabolizing most organic compounds found in nature. A few complex organic compounds cannot be metabolized in a reasonable time period and are considered as being *non-biodegradable*. Lignin and bacteria polysaccharides are two complex organic compounds that cannot be metabolized in activated sludge systems in a reasonable period of time. The complex, non-biodegradable, organic compounds tend to be insoluble materials that accumulate in the activated sludge as non-biodegradable, volatile, suspended solids. Some of the bacteria polysaccharides will be soluble, yielding a soluble, non-biodegradable, organic fraction in the treated effluent.

The organic loading rates and the total mass of microbial removal rate from the system in the treated effluent and the wasted solids will have a definite effect on bacteria predomination. For a given set of environmental conditions, there will be a definite amount of activated sludge accumulating each wastewater displacement period. Increasing the rate of activated sludge removal from the system will lower the MLSS and shorten the solids retention time (SRT) in the system. The bacteria predomination will shift towards the most efficient bacteria for the substrate being treated. If too much activated sludge is removed from the system, the bacteria will shift to a dispersed growth phase. Decreasing the rate of removal of activated sludge will increase the MLSS and increase the SRT in the system. The bacteria predomination will favor the least efficient bacteria capable of metabolizing the organic nutrients. Often, filamentous bacteria will predominate at very long SRT periods. If insufficient activated sludge is wasted from the system, the MLSS will rise to a point where they occupy too much of the final sedimentation tank volume and will be washed out in the effluent. The system will seek its own optimum equilibrium point automatically, depending on the waste loading rate, oxygen transfer, and the sludge removal rate.

The growth of filamentous bacteria is normal in activated sludge and will not be a problem unless the filamentous bacteria population increases sufficiently to adversely affect the settling rate of the activated sludge. Excessive growth of filamentous bacteria will cause activated sludge to *bulk*. Bulking sludge is light, fluffy, and slow to flocculate and settle in the final sedimentation tanks. While the filamentous bacteria stabilize the organics the same as the non-filamentous bacteria, the filamentous bacteria occupy more volume in the sedimentation tanks and are more easily washed out in the effluent by surge hydraulic flows than normal activated sludge. Normally, the non-filamentous bacteria predominate over the filamentous bacteria in activated sludge. Figure 11-9 shows a stained photomicrograph of dense bacteria floc and heavy filamentous growths. The specimen was stained to allow easy observation of the different bacteria. It shows how the less dense filamentous bacteria keep the floc particles from forming larger

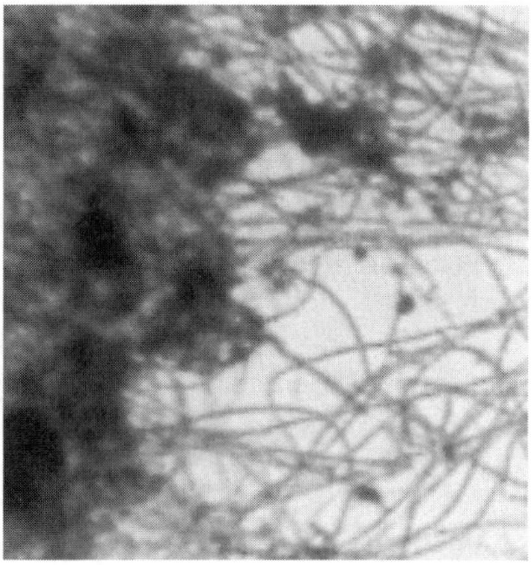

Figure 11-9 FILAMENTOUS BACTERIA IN FLOC

masses.

Filamentous bacteria are able to predominate over the normal bacteria when oxygen is limiting, in high carbohydrate wastewaters, under nitrogen limitation, at low pH, and in excessively long SRTa systems. About 95% of the filamentous bulking problems in activated sludge are the result of an inadequate oxygen supply during stabilization of the organic compounds in wastewaters. Once filamentous bacteria have been established, they are hard to remove. The primary cause for the filamentous bacteria predomination must be removed first and then the filamentous bacteria must be wasted from the system. Since the mass of filamentous bacteria is large to start with, the filamentous bacteria will continue to grow in accordance with their ability to obtain nutrients. Patience is required during the period of time required to reduce the filamentous bacteria population back to normal levels. It has been found that chlorination of the RAS will kill both the filamentous bacteria and the normal bacteria on the surface of activated sludge floc, but not the bacteria within the floc. The normal bacteria remaining will out-grow the surviving filamentous bacteria and good biological conditions will be returned in time, if there is adequate oxygen for proper metabolism and good mixing. Without adequate DO the filamentous bacteria will quickly predominate again.

Recently, it has been proposed to use an anaerobic selector to prevent growth of filamentous bacteria. With the anaerobic selector, the incoming wastewaters and the return activated sludge (RAS) are mixed without aeration to allow the facultative bacteria to begin to grow under anaerobic conditions. If the filamentous bacteria are strict aerobes, they will not grow and will continue to die off under the anaerobic conditions. After the short retention time in the selector, about one hour, the mixed liquor enters the aeration section where the facultative bacteria shift their metabolism from anaerobic to aerobic and rapidly metabolize the remaining organic compounds under aerobic conditions, limiting nutrients for growth of the filamentous bacteria. Anaerobic selectors have shown mixed results in controlling filamentous bacteria in the field.

One concept of activated sludge floc formation advanced in California by D. Jenkins and others utilizes filamentous bacteria as a *backbone* for the other bacteria to be attached. This concept was developed from observation of a large number of activated sludge flocs in the field. Filamentous bacteria are a normal part of the bacteria population in field scale plants and will be observed in the floc under the microscope. Because of the large size of the filamentous bacteria, the non-filamentous bacteria become attached to the filaments as well as to each other. It gives the appearance of the filaments forming a backbone for the bacteria to attach to. Previous research on floc forming bacteria has established that filamentous bacteria are not required to form activated sludge floc. Research has also established that all bacteria growing in the aeration tank can contribute to the activated sludge floc. No one group of bacteria has been found to be solely responsible for floc formation in activated sludge.

Over the years efforts have been made to correlate the growth of *Nocardia* with foaming on the surface of aeration tanks. As previously indicated, *Nocardia* is an actinomycetes that tends to be filamentous, but easily fragments into small segments. A number of articles have been published concerning the observation that *Nocardia* are found in very large numbers when foam occurs in aeration tanks and conclude that *Nocardia* causes the foam. One should be careful when establishing cause-and-effect relationships with microorganisms from field observations alone. Foaming in aeration tanks is caused by the accumulation of slowly, biodegradable, hydrophobic materials that cannot be discharged to the final sedimentation tanks. Hydrophobic materials in wastewaters are attracted to the lipids on the bacteria surfaces. The slowly biodegradable hydrophobic materials accumulate with time and produce a scum on the surface of the aeration tank unless the tank has been designed for continuous removal of surface scum. Surface baffles in aeration tanks and submerged drawoffs tend to allow scum to accumulate for very long periods of time. Air bubbles become trapped in the scum, making it difficult for the scum to be mixed with the activated sludge moving around the aeration tank. The air bubbles also create an aerobic environment of poorly degradable, hydrophobic materials, stimulating the growth of *Nocardia*. The net effect is that *Nocardia* grow as a result of the foam rather than being the cause of the foam. It is not surprising that activated sludge plants with major foaming problems show good growths of *Nocardia*. *Nocardia* is an organism that finds the foam a perfect environment for growth. It simply takes advantage of the environmental conditions. Continuous removal of scum from the surface of aeration tanks prevents scum accumulation unless aerobic digestion scum is recycled back into the treatment plant. This material must be removed from the treatment plant with the WAS on a continuous basis or it will buildup on the surface of the aeration tanks and even on the surface of final sedimentation tanks. Once the scum has been removed, few activated sludge plants will have *Nocardia* in their activated sludge. *Nocardia* do not compete very well with the normal bacteria in activated sludge.

The presence of excess ammonia nitrogen in wastewaters is necessary to insure complete metabolism of the carbonaceous organic compounds and growth of the bacteria in the aeration tank. The bacteria create VSS with 11% to 12% organic nitrogen. Endogenous respiration releases some of the cell nitrogen back into the environment as ammonia nitrogen. Overall, the nitrogen fraction of activated sludge can drop from a maximum of 11% to 12% to a minimum of 5% to 6% nitrogen with aerobic digestion. The presence of amino sugars in the dead cell mass prevents the cell nitrogen from dropping to zero during digestion. Excess ammonia nitrogen and dissolved oxygen in the aeration tank will stimulate the growth of nitrifying bacteria. Unfortunately, the nitrifying bacteria do not obtain much energy from their metabolism and cannot produce much cell mass.

Nitrifying bacteria are autotrophic bacteria, using carbon dioxide as their primary source of cell carbon. There are two major groups of nitrifying bacteria. The first group of bacteria oxidizes ammonia to nitrous acid and has been labeled as *Nitroso-* bacteria. The second group of bacteria oxidizes the nitrous acid to nitric acid and has been labeled as *Nitro-* bacteria. There are many different species of nitrifying bacteria in the soil. The specific environment determines which nitrifying bacteria grow and the extent of their growth. Activated sludge systems provide a good environment for both carbonaceous metabolism and nitrification. Ammonia nitrogen in municipal wastewaters tends to exist as ammonium bicarbonate. It is formed from the hydrolysis of urea and proteins together with carbon dioxide and water. Ammonium bicarbonate is a major part of the buffer system in municipal wastewaters. When the bacteria oxidize ammonia, they obtain energy to make cell mass, the same as the heterotrophic bacteria. The interesting part of the oxidation of ammonia to nitrous acid is the limited energy yield for the nitrifying bacteria. The nitrifying bacteria only obtain about 42% of the total energy available from the oxidation of ammonia. The other 58% of the energy is tied up in the nitrous acid. Oxidation of the nitrous acid to nitric acid is also a low energy yielding reaction. Even less energy is available for synthesis by the *Nitro-* group of bacteria than is available for the *Nitroso-* group of bacteria. The lower energy yield for the *Nitro-* group of bacteria results in rapid metabolism of nitrites in activated sludge systems, giving the appearance that ammonia nitrogen is oxidized directly to nitrate nitrogen. The nitrifying bacteria undergo endogenous respiration and release ammonia nitrogen that can be reused if adequate DO is available. With long SRT periods in the aeration tanks most of the excess ammonia nitrogen will be oxidized to nitrate nitrogen and only a small residue will remain in the dead cell mass accumulating in the activated sludge. The lack of complete nitrogen balances in activated sludge plants has limited the development of good mass balances for nitrification.

The growth of nitrifying bacteria is much slower than the heterotrophic bacteria in activated sludge systems. Although the only real competition between the nitrifying bacteria and the heterotrophic bacteria is for DO, the greater growth of heterotrophic bacteria limits the ability of the nitrifying bacteria to obtain nutrients. Dense floc keeps the nitrifying bacteria from the necessary nutrients for rapid growth. Experience has shown that complete nitrification requires a SRT in aeration of 3 to 5 days at 20°C. At warmer temperatures in the aeration tank the SRT can be shortened. At cold temperatures in the aeration tank the aeration SRT must be increased for complete nitrification. A rough rule of thumb is a doubling of the rate of metabolism for each 10°C temperature increase between 5°C and 35°C.

One of the simplest techniques for evaluating nitrification in biological wastewater treatment systems is to measure the change in soluble alkalinity between the influent to the aeration tank and the effluent from the aeration tank. Theoretically, the soluble alkalinity decreases 7.1 mg/l as $CaCO_3$ for each mg/l ammonia nitrogen oxidized to nitrites. The oxidation of nitrites to nitrates has no requirement for additional alkalinity. Soluble alkalinity is required since suspended solids in both the aeration influent and aeration effluent will react with the titrant the same as the alkalinity. Centrifuging the aeration influent and effluent samples for a short time and using the centrate for determining the soluble alkalinity is adequate for removing the suspended solids that could adversely affect the alkalinity measurements. The small amount of turbidity remaining in the centrate is not sufficient to significantly affect the results. The change in soluble alkalinity works quite well in most cases. If the activated sludge has a high pH, above 8.5, some of the ammonia nitrogen can be removed by air striping. Some of the alkalinity can also be precipitated as calcium carbonate and removed with the suspended solids during centrifuging. In the early days of wastewater treatment the production of nitrates was considered as the indicator of complete stability of the wastewaters being treated. Later, it was found that the carbonaceous organic matter could be stabilized without complete nitrification. It appeared that nitrification occurred as the second part of a two-part system. The carbonaceous organic matter was stabilized first and then the ammonia nitrogen was stabilized. Complete nitrification required considerable additional aeration. By stopping the stabilization process at the end of the carbonaceous phase, it was possible to save considerable costs for aeration. Economics dictated the limited nitrification in activated sludge plants until recently, when the EPA decided that NH_3-N was toxic to fish in the receiving waters. As NPDES permits were renewed, NH_3-N criteria were added to the permits for many treatment plants, stimulating renewed interest in nitrification. When activated sludge systems used the conventional, long, narrow aeration tanks, it was necessary to add a second aeration tank and a second final sedimentation tank for nitrification. The oxygen transfer relationships in the long narrow aeration tanks were not sufficient for both carbonaceous metabolism and nitrogenous metabolism. If the activated sludge system used complete mixing activated sludge with adequate DO, both carbonaceous metabolism and nitrification could occur in a single aeration tank. Demonstration of these relationships in the field ended further discussion of "one tank" versus "two tanks" for nitrification. Both systems can provide the desired effluent quality.

Protozoa – Although protozoa were observed regularly in activated sludge, their role in activated sludge was not fully recognized for a number of years. Initial studies simply identified the various protozoa that were observed. Over time it was noted that there was a clear succession of protozoa during the development of activated sludge treating municipal wastewater. The initial growth of amoeboid protozoa and small flagellated protozoa gave way to small free-swimming ciliated protozoa and then, to larger free-swimming ciliated protozoa. Finally, stalked ciliated protozoa and crawling ciliated protozoa predominated as the activated sludge reached normal operating conditions.

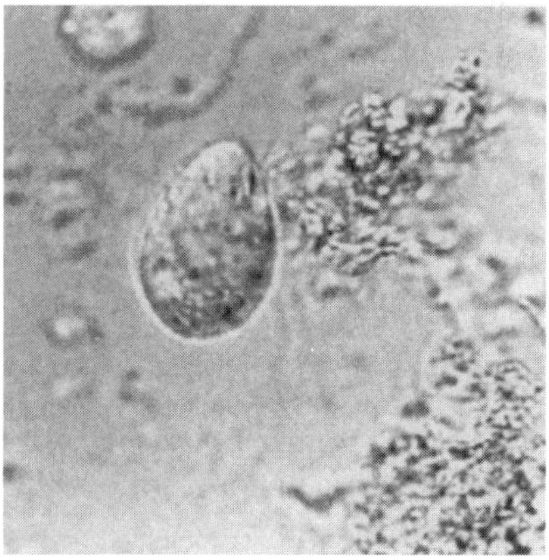

Figure 11-10 A FREE-SWIMMING CILIATED PROTOZOA ON FLOC

The photomicrograph, Figure 11-10, shows a free-swimming ciliated protozoa eating bacteria on the surface of a small floc particle. The specimen was unstained and observed at 500X magnification. Because the predomination of stalked ciliated protozoa correlated with a low effluent BOD_5, efforts were directed towards using the number of stalked ciliated protozoa to estimate the activated sludge effluent BOD_5. Research by Cramer in 1931 indicated that the numbers of stalked ciliated protozoa in activated sludge could not be used to estimate the effluent BOD_5.

Research at MIT in the 1950s confirmed previous studies that the normal growth of protozoa in activated sludge systems follows definite relationships. During the initial growth of bacteria the flagellated protozoa are able to find sufficient numbers of bacteria to grow and predominate. The flagellated protozoa are not competitive with the small free-swimming protozoa and soon drop in numbers as the small free-swimming protozoa increase their population. As the number of dispersed bacteria decrease, the large free-swimming ciliated protozoa are able to eat some of the bacteria in small floc particles. Since the large free-swimming ciliated protozoa require large numbers of bacteria for growth, the dispersed bacteria population quickly drops to very small numbers, followed by the numbers of large free-swimming ciliated protozoa. The stalked ciliated protozoa attach themselves to floc particles and use very little energy in removing some of the remaining dispersed bacteria. For these reasons, the stalked ciliated protozoa predominate at low BOD_5 levels. The crawling ciliated protozoa have a series of *cirri* on the underside of their bodies. The cirri helped these protozoa to move over the floc particle surface and eat lightly attached bacteria that the stalked ciliated protozoa cannot reach.

The growth of protozoa in activated sludge systems is aerobic, requiring positive DO levels in the aeration tanks. The protozoa obtain their energy for cell synthesis by eating the biodegradable nutrients contained in bacteria. Limited data indicates the energy-synthesis relationships for protozoa are similar to that of bacteria with about 1/3 of the biodegradable portion of the bacteria being oxidized with 2/3 of the biodegradable portion of the bacteria being converted into protozoa cell mass. The non-biodegradable portion of the bacteria cell mass is discharged as waste material by the protozoa and accumulates as part of the activated sludge floc particles. Since protozoa metabolism cannot be easily separated from bacteria metabolism, protozoa metabolism is normally considered as part of endogenous respiration. Normal bacteria endogenous respiration utilizes DO and reduces the active cell mass. Protozoa metabolism also utilizes DO and reduces the bacteria active cell mass the same as endogenous respiration of the bacteria. It appears that the endogenous rate of bacteria is about 1.0 percent/hr at 20°C based on active cell mass. Protozoa metabolism adds another 1.0 percent/hr metabolism rate to the endogenous respiration rate, giving an overall endogenous respiration rate in activated sludge systems of 2.0 percent/hr at 20°C. The overall rates of endogenous respiration are a function of temperature in the activated sludge systems, increasing as the temperature increases and decreasing as the temperature decreases. A rough correlation of metabolic rates indicates that the rates change by a factor of two for each 10°C temperature change between 5°C and 35°C.

It is not necessary to learn all the different species of protozoa in activated sludge any more than it is necessary to learn the specific species of bacteria. The growth of the various species of bacteria and protozoa is determined by the environmental conditions in the aeration tanks. It is only necessary to know the basic groups of protozoa in the activated sludge to understand the biochemistry of the system. The photomicrograph, Figure 11-11, illustrates a typical municipal activated sludge with free-swimming ciliated protozoa and *Vorticella*, an individual, stalked, ciliated protozoa, The free-swimming ciliated protozoa eat the free, dispersed bacteria and bacteria on the floc surface. *Vorticella* has a stalk which contracts like a metal spring. The length of the stalk appears to be related to the available nitrogen in the mixed liquor. Long stalks, as shown in Figure 11-11, are indicative of a high nitrogen concentration. Since floc particles are simply clumps of bacteria, living and dead, and inert suspended solids, the activated sludge floc has no specific size, form, or shape.

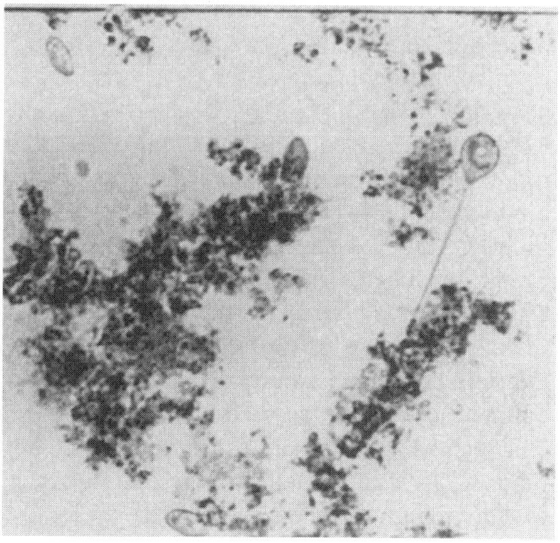

Figure 11-11 PROTOZOA ON TYPICAL ACTIVATED SLUDGE FLOC

While research has shown that specific numbers of protozoa could not be correlated to the effluent quality from activated sludge systems, the relative numbers of protozoa in each group can be used as an indicator of effluent quality. The lack of protozoa in the MLSS is a clear sign of toxicity. Toxicity can be created by toxic chemicals in the incoming wastewater or by a lack of DO. With normal municipal wastewater the lack of DO in the aeration tanks is the primary reason for the lack of protozoa. While protozoa cannot grow in an absence of DO in activated sludge systems, the protozoa can survive 12 to 24 hrs without DO. Protozoa slow their metabolism in the absence of DO for long periods of time and create cysts that allow their nucleuses to survive for growth when favorable conditions return. The activated sludge environment allows all of the different groups of protozoa to survive at their appropriate level. When the organic level is high and the residual DO is low, the flagellated protozoa will predominate over the other groups. As the available nutrients decrease, the predomination of protozoa shifts accordingly. The activated sludge system is a dynamic system with all of the microorganisms shifting as the environment changes. Routine microscopic examination of the protozoa in the MLSS on a daily basis is good practice for activated sludge operators. By recording the relative numbers of each group of protozoa and the structure of the bacteria floc, the operator can see how the change in microorganisms affects his activated sludge system. Correlation of the relative groups of protozoa and the structure of floc with effluent quality will provide the operator with information on how his activated sludge system is operating with this very simple test.

Rotifers - Rotifers also find some activated sludge systems suitable for growth. In the early days of activated sludge, the effluent quality was determined by the extent of nitrification. Complete nitrification was considered essential for a stable effluent. Complete nitrification meant less than 1.0 mg/L biodegradable organic matter in the treated effluent, a long aeration period and a long SRTa.

Microscopic examination of MLSS from activated sludge systems showing complete nitrification indicated growth of large numbers of rotifers. The complex metabolism and the large size of the rotifers allowed them to eat the bacteria on the surface of floc particles and even some small floc particles. Rotifers require 3 to 4 mg/L DO for good growth. For this reason the presence of rotifers in MLSS samples indicates an activated sludge effluent with few soluble, biodegradable, organic compounds. Figure 11-12, shows a group of rotifers attached by their tails to a dense activated sludge floc particle. The flexible body of rotifers allows the rotifers to bend around and feed on bacteria on the floc surface. The cilia on the head of the rotifers in Figure 11-12 create the impression of two rotating wheels. When the rotifers relax their hold on the floc particle, the cilia supply the motility for the rotifers. When it was found that activated sludge systems could be designed and operated without complete nitrification and produce a satisfactory effluent, rotifers did not find the new activated sludge environment adequate for their growth and were no longer seen in MLSS samples. It was not until the *Extended Aeration* modification was introduced at the end of the 1940s that rotifers were observed again in activated sludge samples. Extended aeration systems had 24 to 48 hrs aeration with a long SRTa, creating conditions for complete nitrification. The extended aeration systems allowed the rotifers to return and be observed in MLSS samples. With the federal EPA requiring nitrification in activated sludge systems in the 1990s, rotifers are observed in many MLSS samples on a regular basis.

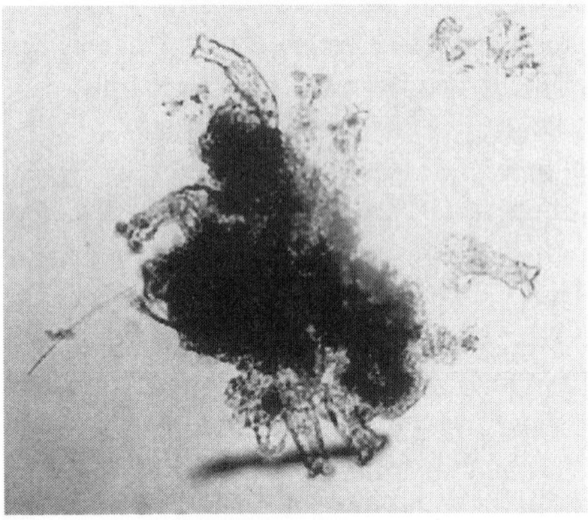

Figure 11-12 MANY ROTIFERS ATTACHED TO ACTIVATED SLUDGE FLOC

Nematodes and Other Worms - Other higher animals found in activated sludge include nematodes and round worms. Nematodes are complex animals that consume large numbers of bacteria to survive. Although nematodes are seen in various activated sludge systems, nematodes do not normally occur in activated sludge systems. Observations have indicated that nematodes occur in environments subject to oscillating levels of DO from zero to 2 mg/L over several days. Nematodes are very motile worms, breaking up floc particles with their rapid thrashing motion. A high population of nematodes can create a highly turbid effluent in an otherwise normal activated sludge system. Occasionally, bristle worms will be seen in activated sludge. Bristle worms are relatively large and overpowering when observed at 100X magnification, the normal power for observing animals in activated sludge. The bristle worms observed in activated sludge samples have orange spots along their flexible bodies. The environment in activated sludge systems does not favor the growth of bristle worms except in unusual circumstances that have yet to be identified. For the most part, worms are not normally found in activated sludge and indicate the environment is not normal.

Biochemistry of Activated Sludge - It is important to recognize that the biochemistry of activated sludge is the same as the biochemistry of pure cultures of bacteria and pure cultures of protozoa and higher animals. The basic difference is that activated sludge is a mixture of bacteria, protozoa, and higher animals with the most efficient microorganisms predominating in the activated sludge environment. Much of the biochemical research on pure cultures has application in activated sludge systems; but one must be careful in translating the data from the field of microbiology to the field of biological wastewater treatment. It is equally important for researchers dealing with activated sludge and other biological wastewater treatment research to recognize the environmental differences between laboratory studies and field scale plants. The microorganisms will produce the same biochemical reactions in laboratory studies as in field scale plants if the environments in both systems are identical. When activated sludge studies are made in the laboratory to simulate field scale plant operations, the laboratory environment must duplicate the field plant environment to produce the same quantitative biochemical reactions. Too often the laboratory environment has better oxygen transfer, better mixing, and a better controlled pH than the field scale plant. The wastewater treatment literature has illustrated a number of pilot plant studies that performed better than real treatment plants. Laboratory scale pilot plants will always produce better results than field scale plants when the laboratory units are pushed to their maximum capability.

Early Studies - The biochemistry of activated sludge has evolved from batch fed and continuously fed laboratory activated sludge pilot units. The early research was with batch fed activated sludge systems. C.C. Ruchhoft and his staff at the USPHS, H. Heukelekian and his students at Rutgers University, and C.N. Sawyer and his students at MIT were the primary researchers on batch fed activated sludge systems. My students at MIT used both batch fed systems and continuously fed systems. By the time I moved to the University of Kansas in 1960, emphasis was on continuously fed systems. The development of a slow speed peristaltic feed pump was important in making the shift from batch fed activated sludge units to continuously fed activated sludge systems. Important information was obtained from both type systems.

Between 1935 and 1947 the USPHS researchers published 17 papers on their studies of sewage purification. Many of those papers dealt with activated sludge metabolism. Their study on glucose metabolism by activated sludge showed that glucose was removed from solution at a higher rate than it was oxidized. At 1.5 hrs aeration 80% of the glucose removed was assimilated into protoplasm. Their 1947 study examined the metabolism of 36 different pure organic substrates. They found the carbohydrates were oxidized 13%. The alcohols were oxidized 30% with the amino acids oxidized 42% and the organic acids were oxidized 50%. A later study by N. Porges *et al* on skim milk metabolism showed that carbohydrates could be stored as glycogen inside the cells. Continuously fed biotreatment systems with a dispersed growth showed no storage of carbohydrates, only complete metabolism to cell mass. The creation of internal polysaccharide storage products, as well as, the creation of external polysaccharide slime products has been noted in batch fed activated sludge systems. It appears that with limited oxygen for normal metabolism, the excess carbohydrate nutrients are converted into different storage products that may or may not be metabolized with adequate DO. If the carbohydrates are stored as glycogen inside the bacteria cell, the bacteria will be able to metabolize the glycogen when adequate DO is available. If the carbohydrates are processed into extracellular slime on the outside of the bacteria, the bacteria will be unable to metabolize the carbohydrates.

For many years Rutgers University was a major producer of biological wastewater treatment research. W. Rudolfs, H. Heukelekian, and their students carried out research in all areas of wastewater treatment starting in the 1920s. In 1951 Heukelekian *et al* proposed one of the first equations for activated sludge synthesis from municipal wastewater, Equation 11-1.

$$\text{VSS Synthesis (g/d)} = 0.5(\text{g/d BOD}_5 \text{ Removed}) - 0.055(\text{g MLVSS}) \qquad (11\text{-}1)$$

This equation recognized that activated sludge growth was related to the biodegradable organic compounds metabolized and to endogenous respiration by the microorganisms in the aeration tank. It is not surprising that design engineers began using VSS as a measure of active microbial mass in activated sludge WWTP designs. In the course of their studies Heukelekian's students looked at the basic concepts activated sludge and the chemical characteristics of municipal wastewater. The research at Rutgers University contributed significantly to understanding the basic concepts of activated sludge.

C. N. Sawyer and his students at MIT concentrated on the nitrogen and phosphorus requirements of activated sludge and trickling filters. They determined that the N requirement for activated sludge was 6 lbs/100 lbs BOD_5 removed and the P requirement was 1 lb/100 BOD_5 removed. Since municipal wastewaters have more N and P than is needed for cell synthesis in activated sludge systems, these values

Figure 11-13 A WARBURG RESPIROMETER

have importance with nutrient deficient industrial wastewater.

As a graduate student at MIT in 1950, I began to use the Warburg respirometer to measure oxygen uptake rates in my research on floc-producing bacteria from activated sludge. Dawson and Jenkins had indicated in 1949 that the Warburg apparatus was a good tool for studying activated sludge oxidation. When I returned to the MIT faculty in 1953, the Warburg respirometer, Figure 11-13, became a major part of my research efforts with my graduate students. Richard Englebrecht looked at the energy relationships for the metabolism of different pure organic compounds in low MLSS activated sludge systems. One part of his study dealt with a Warburg respirometer study of oxygen uptake and its correlation with changes in the MLSS and substrate removal in a glucose fed activated sludge system. He too found that glucose was stored without complete synthesis of cell mass. The metabolism of carbohydrates in batch fed activated sludge systems produces an entirely different set of reactions than in continuously fed activated sludge units. Fortunately, many major groups of organic compounds in wastewaters are metabolized the same in batch fed and in continuously fed systems. During the 1950s graduate students at MIT examined the metabolism of various pure organic compounds by acclimated activated sludge with the Warburg respirometer. John Jeris looked at the metabolism of the short chain alcohols by activated sludge. William A. Cawley examined the single carbon compounds: methanol, formaldehyde, and formic acid. A. C. E. Oberton studied the various factors affecting the metabolism of isopropanol for his undergraduate thesis. Robert L. Wilcox studied the metabolism of phenol; while H. D. Tomlinson acclimated a series of activated sludges to different aromatic compounds: benzyl alcohol, mandelic acid, anthranilic acid, benzene, toluene, phenol, p-cresol, m-cresol, and o-cresol. Warburg studies showed the rates of oxygen uptake of the different organic compounds and their metabolic relationships to one another. John E. O'Brien took time to examine the general relationships between activated sludge and the Warburg apparatus. In one series O'Brien took activated sludge and washed it to remove all excess soluble nitrogen. Next he fed glucose and varying amounts of NH_3-N, NO_2-N, and NO_3-N in different flasks. The flasks with NH_3-N used the most oxygen for metabolism, followed by NO_2-N and then by NO_3-N. The reduction of nitrites and nitrates to produce ammonia for synthesis of new cells used part of the glucose and reduced the amount of DO required for normal oxidation. As continuously fed activated sludge units displaced the batch fed systems in the laboratory, the Warburg apparatus lost its popularity with many researchers. It was not until I moved to Kansas University that I had the opportunity to direct more Warburg studies.

In 1966 Carl E. Burkhead made a Warburg study using a complete set of flasks for each pure substrate. He measured the oxygen uptake rate and then removed a flask and quickly sampled for total COD, TSS, VSS, and soluble COD. This provided a complete set of data for the primary types of soluble organic compounds found in wastewater. Later, Frank B. Nelson examined how batch fed and continuously fed activated sludge responded at varying MLSS concentrations and substrate concentrations. Allan E. Williams used the Warburg to study acetate metabolism by batch fed activated sludge systems operated at varying SRT periods. The Warburg respirometer was also useful in following algae metabolism. David H. Stoltenberg and Eugene W. Nelson examined the algae under light and dark environments. Much has been learned about microbial metabolism in activated sludge and other wastewater treatment systems using the Warburg respirometer. As previously indicated, the Warburg respirometer simulates batch fed activated sludge systems under completely aerobic environments or partially limiting oxygen environments during rapid metabolism. The ability of the Warburg respirometer to transfer oxygen is related to the amplitude of the shaking, the frequency of shaking, and the sample size. It has been found that the Warburg respirometer can transfer oxygen from the gas phase to the liquid between 200 and 400 mg/L/hr, almost 10 times the oxygen transfer rate observed in conventional activated sludge plants. For this reason, the data derived from the Warburg respirometer show what would happen in a batch fed system provided adequate oxygen was present. It is not surprising that increased organic loading rates and high MLSS concentrations exceeded the oxygen transfer rate in some experiments. Even slower rates of shaking produced oxygen-limiting conditions with different metabolism in some substrates. Nitrification affected many activated sludge studies, making it difficult to determine the carbonaceous oxygen demand without corrections. The shift to continuously fed activated sludge units eliminated the need for Warburg studies. Small laboratory units provided more basic data on metabolism with less effort to collect data and allowed data to be collected over a longer period of time for confirmation of the results obtained.

The small, laboratory, continuously fed activated sludge units were all completely mixed systems. Solids separation compartments were built into the aeration units to allow the flocculated activated sludge to drop back into the aeration compartments. Figure 11-14 illustrates a schematic diagram of a simple, continuously fed

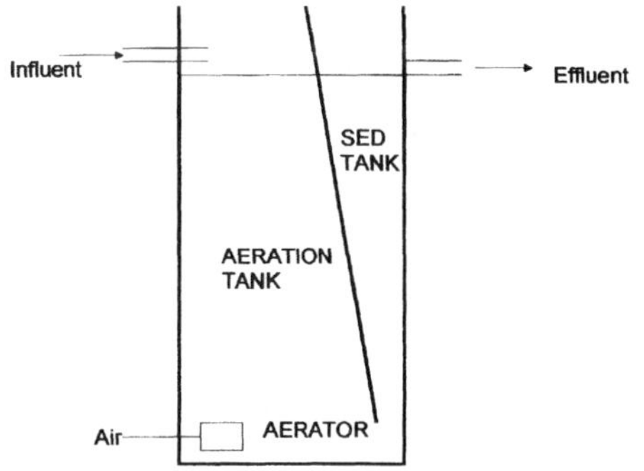

Figure 11-14 DIAGRAM OF A CONTINUOUSLY FED, LAB ACTIVATED SLUDGE UNIT

laboratory activated sludge unit. These small units were constructed of Lucite plastic with liquid volumes from 1.5L to 3.0L. The sedimentation sections ranged from 20% to 33% of the total liquid volume. Small, porous air stones were placed near the bottom of the aeration section and attached to the laboratory air supply or to a small mechanical air compressor. The wastewaters being treated were added continuously by a peristaltic pump through the influent tubing, located just above the liquid in the aeration section. The treated effluent was displaced through the effluent tubing which controlled the fluid depth in the unit. The treated effluent was collected in a suitable effluent container.

The air flow was adjusted to provide sufficient mixing to keep all of the activated sludge dispersed throughout the aeration tank. The diffused air created a circular, fluid motion that rose along the wall next to the aerator, moved across the liquid surface to the inclined baffle, and then moved down the inclined baffle to the bottom of the aeration tank. When the fluid reached the bottom of the aeration tank, most of the fluid moved across the bottom of the aeration tank to the aerator, completely the circular flow pattern in the aeration tank. Part of the fluid hitting the bottom of the aeration tank moved to the wall opposite the aerator, creating an eddy current as the fluid energy was dissipated. The wastewaters being treated moved through the aeration tank and then into the bottom of the sedimentation (sed.) tank. The wastewaters moved up the outside wall to the effluent line for discharge from the unit. Once activated sludge formed in the unit, the treated wastewaters and the activated sludge moved into the sed. tank on a continuous basis. The quiescent conditions in the sed. tank allowed the activated sludge to flocculate and separate from the treated water. The sludge formed a blanket in the lower section of the sed. tank. The eddy currents at the entrance to the sed. tank helped to keep the sludge blanket in motion. As fresh sludge was added to the sludge blanket, the dense sludge was returned back to the aeration tank. Equilibrium was established by wasting a fixed volume of mixed liquor each day. The characteristics of the activated sludge determined how well these small laboratory units operated. Excessive growth of filamentous bacteria created the same sedimentation problems as in full scale activated sludge plants.

Problems with sludge accumulation in the sed. tank at night caused some researchers to remove the sedimentation baffle and add an external sed. tank outside the aeration tank. The external sed. tank required a separate pump to continuously remove the settled sludge and pump it back to the aeration tank. While the separate aeration tank and the separate sed. tank simulated the activated sludge process, the separate sed. tank did not perform as well as field scale sed. tanks. The problems with laboratory units occur because activated sludge floc cannot be readily scaled down in size and the hydraulic currents in the sed. tank cannot be simulated. Yet, the results from the continuously fed activated sludge units gave results that were more comparable to field scale treatment plants than the batch fed units.

General Metabolism - When activated sludge systems are fed wastes that have both soluble organic compounds and suspended organic compounds that are biodegradable, the bacteria metabolize the soluble organic compounds first. The suspended organic compounds must be converted to soluble organic compounds before the bacteria can metabolize them. Bacteria must come into direct contact with the suspended organic compounds before the suspended organic compounds can be converted to soluble organic compounds by hydrolysis.

The salts of short chain organic acids are the easiest organic compounds for the bacteria in activated sludge to metabolize. Acetic acid salts are one of the major intermediate compounds formed during the metabolism or complex organic compounds. Acetate is easily oxidized to carbon dioxide and water. The reactive group in acetate is the *carboxyl group*. Bacteria begin metabolism at the terminal carboxyl group in organic acids containing 12 carbon atoms or less. Metabolism results in the shearing of carbon-carbon bonds with the formation of acetate units by a process known as *beta oxidation*. For organic acids containing over 12 carbon atoms, bacteria begin metabolism at the terminal carbon atom opposite the primary carboxyl group. If the terminal carbon group is a methyl group, it will be oxidized to a carboxyl group before the organic acid is metabolized into a series of acetate groups. If the organic acid contains an odd number of carbon atoms, one mol of propionic acid will be formed before being metabolized to acetic acid and formic acid. Together, acetic acid and formic acid form the basic building blocks for all the chemicals produced in bacteria protoplasm. The oxidation of a terminal methyl group follows normal chemical oxidation from the methyl group to the hydroxyl group to the aldehydic carbonyl group to the carboxyl group. Oxygen does not react directly with the organic compound; but rather, oxygen reacts with the hydrogen removed from the carbon atoms to form water. Loss of two hydrogen atoms from two adjacent carbon atoms creates an unsaturated double bond that reacts with water to form a hydroxyl group. The organic compound has become oxidized by the addition of water. Further removal of hydrogen atoms and the addition of water allow the organic compound to become oxidized without the direct addition of oxygen to the compound itself. Dissolved oxygen reacts with the hydrogen atoms to form water. The oxidation reaction yields energy to the bacteria for the synthesis of new cell mass. Bacteria metabolize straight chain organic compounds easier than branched chained compounds. Multi-branched organic compounds are most difficult for bacteria to metabolize.

Benzene is a six carbon aromatic ring compound with three double bonds. The resonance structure of benzene makes it difficult for bacteria to metabolize. Yet, bacteria can add water and remove two hydrogen atoms to produce phenol, a toxic organic compound that is easier to metabolize than benzene. Phenol is oxidized to catechol, a dihydroxyl aromatic compound. The benzene ring is then split to yield a 6 carbon acid that is easily metabolized. Adding a carboxyl group to benzene creates benzoic acid that can be easily metabolized. Adding a second ring structure to benzene decreases the solubility of the new compound and slows its metabolism by bacteria. Much of the information on activated sludge metabolism came in the 1950s and the 1960s. Extensive research on alkyl benzene sulfonate (ABS) detergents at this time contributed considerable information on metabolism of aliphatic compounds, as well as, on aromatic compounds. The alkyl side chain used in the ABS detergents was developed from low cost petroleum residues. Unfortunately, the alkyl radical was not a pure compound, but consisted of a mixture of isomers that averaged about 12 carbons. Half of the commercial ABS detergent isomers used in the 1950s were biodegradable. The other half of the detergent isomers were essentially non-biodegradable in activated sludge systems. The non-biodegradable ABS created serious foaming problems in municipal activated sludge plants and stimulated research on developing ABS detergents that were completely biodegradable.

Research on pure chemical compounds, related to ABS detergents, showed that bacteria began metabolism of benzene sulfonate at the sulfonate group rather than at the benzene group. Adding short chain alkyl radicals, having 1-6 carbon atoms, to benzene sulfonate allowed the bacteria metabolism to start at the sulfonate group. The sulfonate group was removed and oxidized to sulfate before the benzene ring was metabolized. The short chain alkyl radical was metabolized last. As the number of carbon atoms in the alkyl side chains increased above six carbons, the metabolic characteristics of these ABS compounds changed. Bacteria metabolism did not start at the sulfonate group. Bacteria metabolism began at the terminal carbon atom on the alkyl side chain attached to the benzene ring. The terminal carbon atom on the alkyl side chain was a methyl radical that was attracted to the lipids located in the bacteria cell wall structure. Enzymes located inside the bacteria begin metabolizing the methyl radical, eventually oxidizing it to a carboxyl group. Bacteria metabolism continues at the new carboxyl group and proceeds by β-oxidation. The ABS molecule is pulled into the bacteria as metabolism continues until the molecule is completely metabolized. Branched carbon groups along the alkyl chain make bacteria metabolism difficult. A quaternary carbon with three methyl groups, located at the end of the alkyl side chain, created a block to metabolism of this ABS compound in laboratory activated sludge systems. Eventually, synthetic detergent manufacturers shifted ABS production to a linear dodecyl benzene sulfonate that could be completely metabolized by the bacteria in activated sludge systems without creating detergent foam. The shift from a compound that could not be completely metabolized to one that could be completely metabolized served notice on industrial plants that biodegradability was an important parameter in the development of new chemical products.

Proteins in wastewaters are composed of α-amino acids in various combinations. Most of the amino acids are short chain organic acids that have been combined with ammonium nitrogen to form amino acids. A few amino acids have more than one amino group. Some amino acids have reduced sulfur as part of the molecule. There are mono- and dicarboxylic amino acids as well as branched alkyl groups and benzene rings. The ability of bacteria to metabolize proteins as well as their ability to synthesize proteins in their cell protoplasm means that bacteria have the necessary enzymes to metabolize all of the chemical groups found in the various amino acids. This concept is very important for the adaptation of activated sludge to many of the synthetic compounds found in municipal and industrial wastewaters. The amino group is easily removed from amino acids by hydrolysis with water to form ammonia and the α-hydroxyl acid. Amines are also natural organic compounds that are easily metabolized by bacteria in activated sludge. Ammonia is quickly released and the residual molecule is metabolized as an alcohol. The ammonia reacts with water and carbon dioxide to form ammonium bicarbonate, the primary buffer system in biological wastewater treatment systems.

Alcohols are common chemicals that can be metabolized by bacteria. Alcohols are neutral compounds that form organic acids when oxidized. If the organic acids are not neutralized with existing alkalinity, the pH in the environment will decrease. Carbohydrates are the most common, complex alcoholic organic compounds in wastewater. Starches are easily hydrolyzed to simple carbohydrates. Glucose is the basic structure used in most natural starches and can be quickly metabolized to organic acids, the same as the other alcohols. Balancing the formation of organic acids with the existing alkalinity in wastewaters is very important for normal metabolism and has often been overlooked as the cause of many problems. Cellulose is also a common polysaccharide in wastewater. While the human digestive system lacks sufficient bacteria to metabolize cellulose, soil contamination contains sufficient numbers of cellulosic bacteria for slow metabolism of cellulose during the warm temperatures. It is interesting that cellulose also degrades to form glucose. The glucose from cellulose is metabolized the same as the glucose from starch. The different linkages for the glucose in starch and cellulose affect the rates of metabolism of these two common organic compounds in wastewater. As temperatures drop below 5°C, the increased fluid viscosity prevents significant metabolism of cellulose fibers in activated sludge systems. In effect, the cellulose fibers become part of the inert organic suspended solids until the environment creates suitable conditions for metabolism. The short chain alcohols are easily metabolized with the alcoholic carbon group oxidized to the aldehydic carbon group and eventually to the carboxyl carbon group before being metabolized by β-oxidation to acetic acid.

Energy-Synthesis Reactions - Basic biochemical reactions produced by the bacteria in activated sludge follow a common reaction pattern with carbohydrates, proteins, lipids and similar organic compounds being oxidized to organic acids before being oxidized to carbon dioxide and water plus ammonia from proteins. The purpose of bacteria metabolism is to yield energy for the synthesis of new cells. The energy reactions are closely coupled with the synthesis reactions. New cell mass is generated as quickly as energy is produced. It is not surprising that the primary intermediates in the energy reactions, acetate and formate, are the primary intermediates in the synthesis reactions. The synthesis reactions simply take the energy produced by the energy reactions and create all of the compounds required to produce the bacteria cell. The primary constituents for the bacteria are carbon, hydrogen, oxygen and nitrogen with sulfur and phosphorus as secondary constituents and iron as the major trace metal with molybdenum, copper, zinc, manganese, cobalt, and nickel as minor trace metals together with potassium, magnesium, and calcium as essential elements.

The oxygen utilization rate in activated sludge systems is related to the oxygen required for cell synthesis by the carbonaceous bacteria, the endogenous respiration by the carbonaceous bacteria, cell synthesis by nitrifying bacteria, endogenous respiration by the nitrifying bacteria, cell synthesis by protozoa and higher animals, and endogenous respiration by the protozoa and higher animals. Experience has shown that the cell synthesis and endogenous respiration reactions of the protozoa and higher animals can be expressed mathematically by doubling the endogenous respiration rate of the bacteria. The general relationships for the carbonaceous bacteria indicate about 1/3 of the BCOD metabolized in activated sludge plants is oxidized to provide the energy to convert the other 2/3 of the BCOD metabolized to cell mass as VSS. With a VSS/COD ratio of 1.3/1 the VSS production will be about 0.5 the BCOD metabolized on a mass basis. At an endogenous respiration rate of 2%/hr at 20°C, the active cell mass will decrease about 32% each 24 hrs in the aeration tank under completely aerobic conditions. The death of bacteria results in the accumulation of dead cell VSS, about 6.4% of the active cell mass each 24 hrs. It is important to recognize that activated sludge is a dynamic biological system that undergoes continuous growth and death. Since the rate of growth exceeds the rate of cell destruction, some cell mass must be removed from the system on a daily basis to keep the system at equilibrium.

Limited Dissolved Oxygen – Activated sludge systems operate best in an excess of DO in the aeration tank. Unfortunately, not all activated sludge plants operate with sufficient oxygen to maintain a positive DO level in all parts of the aeration tank 24 hours a day. When the DO level drops to 0 mg/L in part of the mixed liquor, bacteria metabolism is adversely affected in that part of the aeration tank. The oxygen deficient areas in the aeration tank occur when the rate of oxygen demand by the microorganisms exceeds the rate of oxygen transfer. Overloading the aeration tank with BCOD causes the microorganisms to increase their rate of metabolism and their demand for oxygen. From a microbial point of view the DO is the primary electron acceptor to keep metabolism moving at its maximum rate. When the DO level drops to zero, the bacteria must find a new electron acceptor. Many bacteria in activated sludge can use nitrates or nitrites as their electron acceptors and can continue normal metabolism. Sulfates and phosphates in wastewaters are not readily available as electron acceptors. Only specialized bacteria can use sulfates and phosphates as electron acceptors. Since excess DO is needed to produce nitrites and nitrates, municipal activated sludge systems do not have very much oxidized nitrogen for the bacteria to use for metabolism. The bacteria turn next to the organic compounds in the wastewater being treated in the aeration tank for their electron acceptors. Carbohydrates provide the largest amount of electron acceptors for bacteria metabolism. Proteins provide a small amount of electron acceptors. Fats are insoluble in water and provide no significant amount of electron acceptors. The carbohydrates and some proteins are metabolized to short chain fatty acids, amino acids, aldehydes, ketones and alcohols to maximize the energy for bacteria synthesis of new cells. In addition to producing the critical cell components, some of the bacteria store excess organic nutrients within the cell for later metabolism. Glycogen and poly-β-hydroxylbutyric acid polymer are two of the most common bacteria storage products. As the BCOD is reduced in the aeration tank by metabolism, a point is reached where the rate of oxygen demand no longer exceeds the rate of oxygen transfer and excess DO exists in the remainder of the aeration tank.

The hourly variations in wastewater flow in municipal WWTP produces the greatest oxygen demand when the BCOD loading rate is a maximum and the highest DO in the aeration tank when the BCOD loading rate is a minimum. The diurnal flow rate for municipal wastewater creates the variations in the BCOD loading rate every day. Low rates of oxygen transfer in aeration tanks often means low fluid mixing rates. The net effect of oxygen limited operations during the day has been the stimulation of filamentous bacteria over the normal bacteria in activated sludge. The filamentous bacteria are a normal part of activated sludge floc. Under oxygen limiting conditions the individual filaments project out into the liquid and obtain the available oxygen and organic nutrients for additional growth. As the filaments increase, the bacteria floc masses are entrained within the filaments, allowing only the bacteria on the outside of the floc to come into contact with the limited oxygen. Slowly but surely, the filamentous bacteria increase until the settling characteristics of the activated sludge in the final sedimentation tank are adversely affected. Excessive accumulation of filamentous bacteria in the activated sludge floc eventually results in increased effluent TSS losses and violation of effluent permits. Poor sludge settling characteristics occur slowly over time and require a long time period for their reduction. Oxygen limitations in activated sludge plants are related to the aeration tank configuration and the aeration equipment. It is very important to understand how aeration tank designs and aeration equipment designs affect the microbiology and biochemistry of activated sludge operations.

Design and Operation – Activated sludge design practices in the United States have evolved more from practical experience than from an understanding of the basic fundamentals of microbiology and biochemistry. By 1950 the consensus of design engineers indicated that activated sludge should only be applied to large wastewater treatment plants, was difficult to operate efficiently, and was capable of 90% BOD_5 and SS reductions. It was believed that further improvement in effluent quality would require tertiary treatment units. Research at MIT, coupled with field evaluations, demonstrated that the efficiency of activated sludge systems could be improved by proper application of basic science with sound engineering. It was shown that activated sludge plants worked just as well in small plants as in large plants; were not difficult to operate; and did not require tertiary treatment to improve BOD_5 and SS removals of 97 to 99%.

Aeration Tank Configurations – The first aeration tanks for continuous wastewater flow were long narrow tanks to simulate batch feeding as close as possible. Engineers were greatly influenced by the feed-starve-feed-starve concept used in the design of intermittent sand filter plants. The rule of thumb for tank design indicated the length of aeration tanks (L) should be greater than 5 times the width (W) to prevent short-circuiting of incoming wastewater through the aeration tanks. Large aeration tanks had L:W ratios of 10 to 20 and larger. The depth of aeration tanks (D) ranged from 0.5 to 1 times W.

The incoming wastewater and the return activated sludge were added at one end of the long narrow aeration tank with the treated mixed liquor removed at the far end of the aeration tank. Many activated sludge plants had long aeration tanks, operating in parallel. The common wall construction between the parallel tanks reduced the plant cost. Limited space sometimes required that the aeration tank be bent in the middle, creating a two pass aeration tank. The first half of the aeration tank allowed the wastes to flow down the first pass to the middle of the aeration tank, turn and return down the second pass to the discharge end of the aeration tank. This design allowed the influent and effluent pipes to be placed in a common tunnel for easy access. Some aeration tanks have been constructed with three, four, and more passes. A ten pass aeration tank has also been used for activated sludge.

It was assumed that the hydraulic flow pattern in the long narrow aeration tanks was *plug flow*, simulating the batch feed activated sludge process. Research by H. Thomas and J. McKee at Harvard University in 1944 demonstrated that the hydraulic flow pattern in long narrow tanks was not plug flow. The hydraulic mixing pattern was a combination of transverse mixing and longitudinal mixing, creating axial dispersion in the aeration tank. Unfortunately, Thomas and McKee did not explain their results in simple enough terms for design engineers to understand and use in their tank designs. In 1967 research by K. Murphy and P. Timpany in Canada confirmed the earlier research of Thomas and McKee that the axial flow model was the best model for activated sludge aeration tank design. A research study by B. Boyko on mixing in a rectangular aeration tank in Canada, using a dye tracer to follow the flow characteristics, was published in 1968. Boyko concluded that mixing in rectangular aeration tanks was essentially complete mixing rather than plug flow. None of these studies had any significant impact on aeration tank design as the authors failed to explain how to use the new information for aeration tank design.

The addition of the incoming wastewater and the RAS at the head end of the long narrow aeration tank creates a high oxygen demand rate and a serious oxygen deficit in this part of the tank. If the oxygen deficiency is great enough, filamentous bacteria will grow sufficiently to create bulking sludge. The resulting activated sludge will show poor settling characteristics and create operational problems with sludge collection and MLSS. It appears that 95% of the filamentous bacteria problems in activated sludge plants in the United States may be attributed to an oxygen deficiency in the aeration tanks. While the aeration tank configuration plays a definite role in the development of filamentous bacteria, part of the blame lies with the aeration equipment.

Aeration Equipment – British engineers used both diffused aeration and mechanical aeration in their aeration tanks. American engineers found diffused aeration worked best in their aeration tanks. The first diffused aerators were porous plates. Initial placement of the porous air diffusers across the width of the aeration tank required too much air to keep the activated sludge in suspension. Placement of

Figure 11-15 TYPICAL DIFFUSED AERATION TANK

the air diffusers along one of the longitudinal walls produced excellent mixing and a satisfactory effluent quality. The air bubbles rising next to the wall created an air lift pump effect that caused the water to move up to the water surface and then move across the tank to the opposite water. The water moved down the opposite wall from the aerators to the tank bottom before moving across the tank bottom to the aerator. With the wastewater moving down the aeration tank from the influent to the effluent, a *spiral flow* pattern was created in the aeration tank with the cross flow being greater than the longitudinal flow. Less air was required for spiral flow aeration than for the cross-tank aeration used originally. Figure 11-15 shows the surface of a typical diffused aeration tank. Air is being supplied by diffusers located on the left side of the tank. The rising air generates a fluid flow across the aeration tank. The pipe on the right side of the aeration tank is used to spray water on the right side of the aeration tank if excessive amounts of foam are generated. The porous plate air diffusers became clogged over time and required replacement. It was found that the porous plate air diffusers clogged from the inside as a result of dirty air. Extensive air filtration equipment was required to remove the tiny particulates that clogged the porous plates.

Maintenance of air diffusers became a major issue for WWTP operators. The equipment manufacturers responded to the maintenance problem by producing air diffusers that could be removed from the aeration tank without draining the entire aeration tank. The ability to lift a group of air diffusers and to replace the defective diffusers without shutting the tank down was a major step forward for WWTP operators. The next step came with the replacement of the heavy diffusers tubes with light weight air diffusers. Equipment manufacturers came up with a variety of light weight air diffusers that did not clog. The oxygen transfer efficiency of the non-clog diffusers was less than the efficiency of the porous plate diffusers; but the effluent characteristics were essentially the same. Increased power costs in the early 1960s resulted in a shift from diffused aeration to mechanical surface aeration. Operational economics became a major issue for aeration equipment.

Mechanical surface aeration produced greater oxygen transfer at less power cost than the diffused aeration systems. Mechanical surface aeration systems consisted of an electric motor, a gear reducer, and an aeration blade and were mounted on a concrete platform. The mechanical surface aerators operated by pumping the mixed liquor into the air, creating a definite mixing pattern around the aerator. Competition for sales resulted in a mechanical surface aerator mounted on floats to eliminate the concrete platforms. Next, the gear reducer was eliminated by some manufacturers in an effort to be more competitive in sales. The direct drive units used a slow speed electric motor and a small aeration blade. Other equipment manufacturers reduced the quality of the gear reducers. As the quality of mechanical aerators decreased, it was just a matter of time before equipment failures began occurring with the surface aerators, putting them in disfavor with design engineers.

The development of a light weight dome air diffuser in England stimulated American engineers to return to fine bubble diffused aeration with the dome diffusers. Instead of placing the dome diffusers along one side of the aeration tank, design engineers expanded the size of the aeration tanks and placed the fine bubble, dome air diffusers over the entire bottom of the aeration tank. Other manufacturers developed different fine bubble aerators using elastomer diffusers over a light weight metal frame that was attached to air pipes running across the bottom of the large aeration tanks. Fine bubble, diffused aeration is the current aeration system for many new activated sludge plants.

Jet aeration units have been manufactured for many years, but never gained much popularity until they were combined with a new treatment plant design as the *Sequencing Batch Reactor* (SBR). The SBR utilizes a single tank for aeration and for solids separation and will be discussed in greater detail in a separate section covering all of the different activated sludge modifications. The jet aerator combines the MLSS and diffused air together in a jet for rapid oxygen transfer into the pumped fluid.

Aeration systems have two functions in activated sludge, oxygen transfer and mixing. These two functions are both separate and related. The rate of oxygen transfer into water can be expressed by Equation 11-2. The oxygen transfer equation is very straightforward. The rate of oxygen transfer from the air to the water provides DO for the microorganisms to use during metabolism. The alpha factor is related to the soluble chemicals in water that affect oxygen transfer. Clean water has an α of 1.0. Municipal wastewater can have an α value around 0.1 to 0.2. A good effluent from an activated sludge plant will have an α value close to 1.0. The oxygen transfer factor, K_La, is the reciprocal of the time required to saturate the water with oxygen and is a measure of the mixing intensity in the aeration tank. The β value is related to the water quality affecting the saturation value of oxygen in the aeration tank. The β value of clean water and a well treated activated sludge

$$dO/dt = \alpha(K_La)(\beta Cs - C) \qquad (11\text{-}2)$$

where: dO/dt = rate of oxygen transfer into the water, mg/L/hr
α = alpha factor, dimensionless number related to the chemical quality of the water in the aeration test tank that affects oxygen transfer
K_La = oxygen transfer factor for oxygen into water, 1/hr
β = beta factor, dimensionless number related to the chemical quality of the water in the aeration test tank that affects oxygen saturation
Cs = concentration of O_2 at saturation in the aeration test tank, mg/L
C = measured O_2 concentration in the aeration test tank, mg/L

effluent is 1.0. Most β values will be between 0.9 and 1.0. The saturation value for oxygen in water is affected by temperature and atmospheric pressure. Standard conditions for DO saturation in water is measured at 20°C and 1.0 atm pressure. As the temperatures drops below 20°C, Cs increases. As the temperature rises above 20°C, Cs decreases. Atmospheric pressure at sea level is 1.0 atm. Tests run at locations above sea level will show less than 1.0 atm pressure and a lower value of Cs. Diffused aeration tests are made with air compressed to overcome the resistance of the air piping and the head of water over the air diffusers. The actual air pressure affecting Cs in diffused aeration systems is about 1/3 of the head of water over the air diffusers. Mechanical surface aerators pump the water at the surface and are only affected by atmospheric pressure at the water surface. Maximum oxygen transfer occurs when C, the DO concentration in the water, is zero. Engineers often design aeration systems for a minimum value of C between 1.0 and 2.0 mg/L DO. Examination of the oxygen transfer equation indicates that the K_La value is the critical value for evaluating aeration equipment in aeration tanks.

Although a few researchers had looked at oxygen transfer in aeration tanks, design engineers and equipment manufacturers had shown limited interest in the oxygen transfer rates of aeration equipment until the popularity of activated sludge systems began to increase in the 1950s and 1960s. Many technical papers were published on various oxygen transfer studies during the 1950s and 1960s. While these research studies contributed to a better understanding of oxygen transfer in aeration tanks, the research had little impact on the design of aeration equipment in activated sludge plants. R. A. Conway and G. W. Kumke published a paper in 1966 evaluating the different methods available for testing oxygen transfer in aeration tanks. They found that the sulfite method in clean water was best for determining whether or not the aeration equipment met the design specifications for oxygen transfer and the biological method was best for operators, once the plant had been placed in operation. In 1964 H. Benjes and I developed a specification for testing aeration equipment in activated sludge plants under plant operations to determine if the equipment could meet the future oxygen demand. This specification was one of the first specifications that delineated how the aeration equipment was to be tested and evaluated and indicated what the equipment manufacturer had to do if the equipment failed the test. The aeration tests were run after the activated sludge had been operational for several months. In 1967 the results of the aeration testing at the Grand Island, Nebraska, WWTP were published to show that the biological method could be used as the primary method for evaluating the aeration equipment.

Previously, aeration equipment manufacturers had made claims for the oxygen transfer efficiencies of their aeration equipment; but design engineers did not test the equipment to determine if the equipment met the manufacturer's claims. The lack of a simple method for testing aeration equipment kept design engineers from testing aeration equipment. Biological testing was a relatively easy method that design engineers could use for testing aeration equipment. Unfortunately, the aeration equipment manufacturers were not comfortable with the biological testing method. There were too many unknown factors that could affect the results of biological testing, as far as the manufacturers were concerned. Reluctantly, the equipment manufacturers accepted the sulfite test in clean water as the only satisfactory method for testing the oxygen transfer characteristics of aeration equipment. Most major aeration equipment manufacturers constructed aeration tanks for testing their own equipment and recommended that design engineers use the manufacturer's data or use the manufacturer's aeration tanks to meet specification testing of the aeration equipment. Some design engineers chose to use the actual plant aeration tanks as their test tanks for the clean water testing prior to the plant startup.

Aeration equipment testing during the 1970s and 1980s yielded so much new information on oxygen transfer by the different aeration systems that it all but eliminated the need for routine testing of aeration equipment. During the 1980s the American Society of Civil Engineers and the federal EPA joined together in an effort to establish a standard procedure for testing aeration equipment in clean water. The ASCE Standard for clean water aeration testing with sodium sulfite deoxygenation was officially adopted in 1990. Although the design profession has accepted the clean water test to determine if the aeration equipment meets design specifications, biological testing in actual plant operations is the only method to determine the real oxygen transfer characteristics of the aeration equipment being used in the wastewater treatment plant. Treatment plant operators should make DO surveys throughout their aeration tanks while determining the oxygen uptake rates at numerous points in the aeration tanks on an annual basis. The oxygen data should be quite useful in predicting when the aeration tanks will reach their design load, requiring expansion of the WWTP.

The complexity of evaluating oxygen transfer in long, narrow aeration tanks has limited meaningful data collection on conventional activated sludge systems. Everyone assumed the oxygen transfer was adequate as long as the effluent quality was satisfactory. In the 1970s I was able to examine oxygen transfer in a number of conventional, long, narrow aeration tanks having various aeration tank configurations. The aeration tank configurations ranged from a single pass tank to a ten pass tank with different diffused aeration equipment. All of the aeration tanks showed the same general data pattern for DO and oxygen uptake rates. The addition of primary effluent and RAS at the influent end of the aeration tanks stimulated the bacteria to create a high initial oxygen demand rate and to remove the DO from the mixed liquor. The water just above the air diffusers had excess DO as oxygen was transferred from the air bubbles being formed to the water around the bubbles. Microbial metabolism quickly removed the DO from the water before it returned to the aerators for more air. The rapid rise of the air bubbles to the surface allowed a little additional oxygen transfer to the water. Moving along the aeration tank to each new sampling point resulted in an ever decreasing oxygen demand rate until the nutrients were all metabolized and only the endogenous respiration of the microorganisms created the oxygen demand rate. The DO at the bottom of the aeration tank remained near 0 mg/L until the oxygen transfer rate exceeded the oxygen demand rate. As soon as the oxygen transfer rate exceeded the oxygen demand rate, the DO in the aeration tank began to rise, indicating that the aeration tank had shifted from an oxygen limiting condition to an aerobic environment. Under the oxygen limiting conditions there was little oxygen available for endogenous respiration or for nitrification. The oxygen transferred to the water, under oxygen limiting conditions, was used primarily for cell synthesis. Once aerobic conditions were established, cell synthesis was completed, endogenous respiration proceeded, and nitrification occurred. The BCOD in the primary effluent was all metabolized, producing a high quality effluent. The limited time for endogenous respiration meant that more excess sludge was being produced than would have been produced under completely aerobic conditions. Nitrification was limited to the short aerobic time and simply confirmed that aerobic conditions were established by the time the MLSS left the aeration tank. The portion of the aeration tank that has excess DO actually moves back and forth along the aeration tank as the BCOD loading rate changes. Increasing the BCOD loading rate pushes the excess DO closer to the effluent end of the aeration tank. Too much BCOD loading will result in loss of DO throughout the aeration tank. Decreasing the BCOD load allows the excess DO in the aeration tank to move closer to the influent end of the aeration tank.

A Design For Microorganisms – The MIT research on the microbiology and biochemistry of activated sludge in the 1950s resulted in a suggested change for aeration tank design. The conventional activated sludge system produced a shock organic loading at the head end of the aeration tank and an oxygen limiting environment. The resulting environment did not favor the most efficient growth of bacteria and protozoa. By changing the aeration tank to a completely mixed aeration tank, it was possible to maintain the aeration tank in an aerobic environment at all times and to distribute the organic load uniformly to all the active bacteria in the aeration tank. The uniformity of the organic load allowed the bacteria population to develop the most efficient population for the specific organic compounds being treated and allowed the optimum protozoa population to predominate. The complete mixing environment permitted both the carbonaceous bacteria and the nitrifying bacteria to grow at the same time. The concepts of complete mixing activated sludge (CMAS) were developed in the laboratory to best fit bacteria metabolism; but it required demonstration in field units before it could be accepted as a valid process.

The design criteria for municipal activated sludge plants in the 1950s were regulated by each State Health Department. To obtain State Health Department approval for a new design process, it was necessary to construct and operate a full size treatment plant for several years to prove the validity of the new concept. Since it was not practical construct a full size treatment plant for test purposes, the initial efforts to apply CMAS design concepts were directed towards industrial wastewater treatment plants. In the 1950s State Health Departments allowed industries to design and operate their own WWTP, only requiring that the industries meet specific effluent requirements and submit reports on the effluent quality at regular intervals. The CMAS design concepts were demonstrated by a number of engineers with different industrial wastes without their realizing the significance of their aeration tank designs. My first CMAS plant treated a starch desizing wastewater from a small cotton bleachery. This design was unique in its ability to treat a high pH, nutrient deficient, carbohydrate waste. The CO_2 produced during metabolism was used to reduce the pH of the incoming wastes waters from 11 to 9 in the mixed liquor. Reuse of nitrogen and phosphorus released during endogenous respiration allowed minimum addition of these two critical nutrients. Excess DO and a controlled organic load allowed normal bacteria to predominate in the activated sludge rather than filamentous bacteria, normally found in carbohydrate waste treatment plants. Additional CMAS plants were soon constructed to treat other types of textile wastes, ice cream manufacturing wastes, mixed phenolic wastes, antibiotic manufacturing wastes, and domestic wastewaters. The success of the CMAS plants resulted in its acceptance by some industrial design engineers. By the 1960s CMAS was sufficiently established that it was accepted for both industrial wastewater and municipal wastewater treatment.
Examination of the published literature indicated that the first CMAS plant was built and operated at Bury, England in 1919. The Bury CMAS plant was not constructed to demonstrate the validity of the CMAS process. The Bury CMAS

plant was built to demonstrate the value of a new mechanical surface aerator, the "Simplex" aerator. The success of the Simplex aerator at Bury demonstrated the merits of CMAS as a municipal wastewater treatment system. Unfortunately, the focus of attention was on the Simplex aerator rather than on the aeration tank configuration. After the testing was completed, new Simplex aerators were placed at regular intervals in long narrow aeration tanks since long narrow aeration tanks were considered essential for activated sludge. Most of the small activated sludge plants, constructed between 1919 and 1950, were CMAS plants even though no one recognized them as such.

The 1960s saw a number of researchers looking at the mixing concepts within activated sludge aeration tanks. Theoretical studies indicated that plug flow activated sludge systems were more efficient than completely mixed activated sludge systems. Unfortunately, the theoretical studies failed to consider the effect of oxygen limitations in the plug flow aeration tanks. Field results showed that CMAS systems produced carbonaceous effluents of equal or higher quality than conventional activated sludge systems. The equal carbonaceous effluent quality data caused some investigators to state that conventional activated sludge plants were actually CMAS plants. Field data confirmed that conventional activated sludge plants were not completely mixed, even though the carbonaceous effluent quality from some conventional plants were the same as the effluent quality of CMAS plants.

Research published in Europe in 1973 indicated that CMAS systems stimulated the growth of filamentous bacteria, creating poor settling activated sludge; and that conventional, long narrow aeration tanks did not stimulate the growth of filamentous bacteria. Experience with long narrow aeration tanks in the United States was just the opposite. Filamentous bacteria growth has been a real problem with conventional activated sludge plants. Later, European research indicated that the addition of an anaerobic *selector* tank ahead of the CMAS system would prevent filamentous bacteria growth. The shape of the aeration tanks is a critical factor in the design of aeration tanks only as shape affects the environmental conditions created within the aeration tank. The environmental conditions in the aeration tank will determine which bacteria will grow and the rate at which each group of bacteria will metabolize the pollutants in the wastewaters. Field results with anaerobic selectors have been mixed, ranging from no effect to a reduction in filamentous bacteria. The anaerobic selector cannot eliminate filamentous bacteria completely. Fortunately, the addition of the anaerobic selector has not created any adverse operational conditions in activated sludge systems. On the negative side, anaerobic selectors increase both the capital cost and the operating cost of the activated sludge plant without producing a significant change in bacteria predomination.

Aeration Tank Mixing – Aeration tank mixing is a very important parameter in the design and operation of activated sludge systems. Good mixing is essential to continuously break up floc particles and to generate new floc. Mixing is responsible for bringing the wastewater contaminants, the key nutrients, and DO into contact with the bacteria and other microorganisms of importance in activated sludge. Too many design engineers and equipment manufacturers look at the hp/aeration tank volume as the measure of mixing. Design efforts are often directed at minimizing the power applied to the aeration tank volume. It is important to realize that mixing is more than a power/volume function. The key to mixing is how that power is applied to the water. The specific mixing device will determine the effectiveness of the power applied by the aeration equipment.

The simplest technique to study mixing in activated sludge aeration tanks is to measure the DO and the oxygen uptake rate throughout the aeration tank. A completely mixed activated sludge system will have an aeration tank with a variable DO and the same oxygen uptake rate throughout the aeration tank under a constant BCOD loading rate. The DO will be highest at the aerator discharge and will decrease as the mixed liquor moves from the aerators out into the aeration tank and returns back to the aerators. Differences in oxygen uptake rates indicate that the system is not completely mixed. Conventional activated sludge systems will show 0 mg/L or very low DO values as the mixed liquor returns to the aerator, when oxygen transfer is limiting. The DO values will rise quickly in the aeration tank as the aeration tank becomes completely aerobic. The oxygen uptake rate will be the highest at the influent end of the aeration tank and will decrease along the aeration tank until the organic contaminants have been metabolized. After the organic contaminants have been metabolized, the oxygen uptake rate will level off at a relative constant rate, being generated by endogenous respiration and nitrification. When evaluating aeration equipment in activated sludge systems, it should be recognized that the measured oxygen uptake rate data are *not* the actual oxygen uptake rates occurring in the aeration tank until there is positive DO at all points across the aeration tank where the oxygen uptake rate sample is collected. The oxygen uptake rate test is always made with an excess of DO in the sample of mixed liquor being tested and reflects the aerobic metabolism of the available substrate at the point where the sample was taken. Once there is excess DO throughout the aeration tank section being evaluated, the measured DO and the measured oxygen uptake rate data reflect the actual oxygen transfer in the aeration tank. Understanding the basic oxygen transfer relationships and DO levels in aeration tanks is important for design engineers concerned with sizing the aeration equipment in activated sludge systems.

Design engineers and equipment manufacturers have had problems in recent years designating aeration tanks as completely mixed. Large aeration tanks have been constructed with multiple mechanical aeration units or with large numbers of fine bubble air diffusers. These large tanks are not completely mixed aeration tanks as the test procedure given in the previous paragraph would clearly show. It is important to recognize that it is not possible to have instantaneous mixing throughout an aeration tank. There will always be a lag time between the addition of wastewater and its dispersion throughout the aeration tank. A well mixed aeration tank will distribute the incoming wastewaters throughout the aeration tank without creating a DO deficiency anywhere in the aeration tank or a significant difference in MLSS concentrations. CMAS systems were designed for small to medium size aeration tanks with the incoming wastewater discharged under the aeration system for rapid mixing. Aeration tanks can be circular, square or rectangular. Rectangular aeration tanks should use multiple wastewater feed points along one of the longitudinal walls with mixed liquor discharge from the aeration tank along the opposite longitudinal wall. Transverse baffles can be placed across the rectangular aeration tank to create a group of square aeration tanks operating in parallel.

Quantitative Relationships – Ultimately, it is necessary to examine the quantitative relationships that exist in activated sludge aeration tanks. The quantitative relationships exist both as concentration units (mg/L) and as mass units (kg or lbs). The MLSS in the aeration tank are measured as mg/L. The mass of MLSS under aeration are determined by the product of the MLSS concentration and the aeration tank volume. The MLSS characteristics are determined largely by the characteristics of the incoming wastewaters, the aeration tank environment, and the MLSS retention time in the aeration tank.

Municipal wastewater in the United States averages about 200 mg/L BOD_5, 200 mg/L TSS, 40 mg/L TKN, and 6 mg/L P without industrial wastewaters or excessive infiltration. The TSS are about 80% VSS and 20% NVSS. Close to 62% of the VSS in municipal wastewater are biodegradable. These relationships indicate that half of the TSS is biodegradable and half is inert. This means that 100 mg/L TSS can be metabolized by the bacteria in the biological treatment tanks and 100 mg/L TSS will be unchanged. The biodegradable COD is estimated from the BOD_5 data. The BCOD will be about 1.7 times the BOD_5 in normal municipal wastewater, yielding 340 mg/L BCOD. Municipal WWTP utilize primary sedimentation tanks to remove about 60% TSS, leaving 40% TSS to go to the aeration tank along with the soluble compounds. The primary BCOD to the aeration tank will average 200 mg/L.

With an aeration tank having a volume equal to 6 hours wastewater retention at average flow, it is possible to achieve 98 to 99% BCOD metabolism under aerobic conditions. This means about 197 mg/L BCOD will be metabolized in the aeration tank. With normal diurnal flow variations greater BCOD metabolism will occur at low flows and less BCOD metabolism will occur at high flows. A 24 hr flow composite sample will average the variations in flow and metabolism. The theoretical maximum cell mass synthesis (Max Ma) in the aeration tank will be about 2/3 of the BCOD metabolized. With a COD/VSS ratio of 1.3, the Max Ma produced will be 158 mg/L COD and 121 mg/L VSS. Cell synthesis uses 1/3 of the metabolized BCOD as DO from the aeration tank. An average of 79 mg/L oxygen will be required for the synthesis of new cells. With 6 hrs aeration the synthesis oxygen demand will only average about 13 mg/L/hr. Experience has shown that activated sludge plants treating municipal wastewaters after primary sedimentation have a solids retention time in the aeration tank (SRTa) about 20 times the wastewater aeration time. With a 6 hr aeration time (t) the SRTa will be 120 hrs, 5 days.

Since the microorganisms in the aeration tank are living creatures, they undergo endogenous respiration at about 2% of the active microbial mass/hr (Ke) at 20°C in the aeration tank with a balanced population of bacteria and protozoa. Thus, synthesis increases the mass of active microorganisms while endogenous respiration reduces the active microbial mass. Equation 11-2 can be used to calculate the active mass of microbes remaining in the aeration tank (Ma).

$$\text{Ma} = (\text{Max Ma})(\text{SRTa}/t)/(1 + (\text{Ke})(\text{SRTa})) \qquad (11\text{-}2)$$

$$\text{Ma} = (121)(120/6)/(1 + (0.02)(120)) = 712 \text{ mg/L VSS}$$

Endogenous respiration results in the formation of dead cell residue (Me), about 20% of the change between the Max Ma and Ma, Equation 11-3.

$$\text{Me} = (0.2)(\text{Max Ma} - \text{Ma}) \qquad (11\text{-}3)$$

$$\text{Me} = (0.2)(2420 - 712) = 342 \text{ mg/L VSS}$$

The actual increase in VSS over 5 days, related to metabolism, will be Ma + Me, 1,054 mg/L. Since bacteria metabolism produces about 10% NVSS, the MLSS related to metabolism will be 1,160 mg/L, as shown in Equation 11-4.

$$\text{Synthesis MLSS} = (1.1)(\text{Ma} + \text{Me}) \qquad (11\text{-}4)$$

$$\text{Synthesis MLSS} = (1.1)(712 + 342) = 1{,}159 \text{ mg/L}$$

The remainder of the MLSS is related to the accumulation of the non-biodegradable VSS and the NVSS in the primary effluent. The NBVSS in the primary sedimentation tank effluent are 0.38 VSS, 24 mg/L. The NVSS are 20% of the TSS in the primary sedimentation tank effluent, 16 mg/L. These two inert fractions of MLSS accumulate as shown in Equation 11-5.

$$\text{Inert SS Accumulation} = ((0.38)\,(\text{Inf. VSS}) + \text{NVSS})\,(\text{SRTa/t}) \quad (11\text{-}5)$$
$$\text{Inert SS Accumulation} = ((0.38)\,(64) + 16)\,(120/6) = 800\ \text{mg/L}$$

The MLSS inert SS will accumulate in the aeration tank to 800 mg/L, 480 mg/L NBVSS and 320 mg/L NVSS. Overall, the MLSS in the aeration tank will average 1,960 mg/L for these waste characteristics. The MLNVSS will average 425 mg/L with 1,535 mg/L MLVSS, 22% NVSS and 78% VSS. The oxygen uptake for endogenous respiration can be calculated by the change in the COD of the microbial VSS, as shown in Equation 11-6.

$$\text{Microbial}\ \Delta\text{COD} = (1.3)(\text{Max Ma} - (\text{Ma} + \text{Me})) \qquad (11\text{-}6)$$

With 120 hrs retention in the aeration tank, the average endogenous oxygen uptake is about 15 mg/L/hr. The total carbonaceous oxygen uptake rate in the aeration tank at equilibrium would be the sum of the synthesis oxygen demand plus the endogenous oxygen demand, averaging 28 mg/L/hr. At maximum wastewater flow rates, twice the average wastewater flow, the carbonaceous oxygen demand rate would be close to twice the synthesis oxygen demand plus the endogenous oxygen demand, 41 mg/L/hr. At minimum wastewater flow rates, 50% of the average flow rate, the carbonaceous oxygen demand rate would be half of the average synthesis oxygen demand plus the endogenous oxygen demand, about 22 mg/L/hr. These quantitative relationships are for CMAS systems that have adequate DO throughout the entire aeration tank 24 hrs a day, 7 days a week at equilibrium conditions.

Temperature changes the rate of microbial metabolism, making it necessary to adjust for temperature. Between 5°C and 35°C the rate of metabolism changes by a factor of two for each 10°C temperature change. The previous relationships were determined for 20°C. At 10°C the soluble effluent BCOD will increase from 3 mg/L to 6 mg/L and will drop to 1.5 mg/L at 30°C. Essentially, metabolism of the wastewater contaminants is almost complete over this temperature range. This means that the amount of new cell mass will be the same, within the experimental error of the TSS and VSS measurements. The major difference will be in the endogenous respiration. The real impact of endogenous respiration is generated by the long retention time in the aeration tank. If the SRTa is kept constant at 120 hrs, the MLSS at 10°C will increase to 2,320 mg/L with an overall dO/dt of 24 mg/L/hr. At 30°C the MLSS will drop to 1,710 mg/L with an overall dO/dt of 30 mg/L/hr. More sludge will have to be removed from the activated sludge system at 10°C than at 30°C to maintain the same SRTa. The active microbial mass in the MLSS changes from 48% of the MLSS at 10°C to 24% at 30°C. It is important to realize that the accumulation of inert SS in municipal treatment plants reduces the active microbial fraction. Some WWTP operators increase the SRTa in hopes of increasing the active mass of bacteria for stabilization the wastewater contaminants. Doubling the SRTa from 120 hrs to 240 hrs would increase the active microbial VSS from 712 mg/L to 834 mg/L at 20°C, a 17% increase in active mass at equilibrium. At the same time, the dead cell mass and the inert SS would increase, pushing the MLSS from 1,960 mg/L to 3,420 mg/L. The active mass fraction would drop from 36% of the MLSS to 24%. The higher fraction of inert SS requires greater mixing to insure the contact between the organic contaminants and the bacteria for adequate metabolism.

Conventional activated sludge aeration tanks tend to be oxygen deficient part of the time, depending upon the oxygen transfer rate of the aeration equipment and the BCOD loading rate. There is no simple method to evaluate the operational characteristics of conventional activated sludge aeration tanks. The controlling factor affecting the microorganisms is the percentage time the aeration tank is aerobic. If the aeration tank is aerobic when the mixed liquor reaches the effluent end of the aeration tank, the soluble BCOD discharged from the aeration tank should be the same as the comparable CMAS system; but the excess sludge removed from the mixed liquor will be greater than from a CMAS system with the same aeration time, SRTa, and temperature. In effect, the microbial synthesis reaction will be complete. Limited endogenous respiration will reduce only a portion of the active microbial mass produced by normal cell synthesis.

For illustrative purposes, we will assume our conventional activated sludge plant averages only 15 mg/L/hr oxygen transfer, 13 mg/L/hr oxygen used for cell synthesis and 2 mg/L/hr oxygen used for endogenous respiration. Using the same wastewater characteristics as in the CMAS example with 6 hrs aeration tank retention, 20°C, and a SRT of 5 days in the aeration tank, the BCOD metabolism would be the same. Only 183 mg/L VSS would be oxidized during endogenous respiration, leaving 2,240 mg/L (Ma + Me) in the aeration tank. Adding the NVSS for cell synthesis brings the cell MLSS to 2,460 mg/L. The inert SS in the primary effluent would raise the MLSS by 800 mg/L, producing 3,260 mg/L MLSS. The active microbial VSS would make up 67% of the MLSS, much greater than the 36% in the CMAS system. The trade off between the two systems is about 66% more sludge removed from the conventional activated sludge system than from the CMAS system and 46% less oxygen used in the conventional activated sludge system. Because of the differences in α values between the two types of aeration tanks, the conventional activated sludge system would require greater aeration than the CMAS system and transfer less oxygen.

The simplest method for evaluating activated sludge in conventional activated sludge plants is with its endogenous respiration rate. A sample of activated sludge obtained from the discharge end of the aeration tank should be in the endogenous respiration phase if the aeration tank shows DO completely around the cross-section of the tank. The oxygen uptake rate of the activated sludge in the endogenous phase is directly proportional to the active mass of microorganisms in the sample. With an active cell mass COD of 1.3 times the active mass VSS, 80% oxidation of active mass, and an endogenous respiration coefficient, Ke, of 0.02 at 20°C, the modified Ke_0 will be 0.021. The value of Ma in the MLSS can be calculated as shown in Equation 11-7.

$$Ma = (dO/dt)/Ke_0 \tag{11-7}$$

With a $(dO/dt)_e$ of 20 mg/L/hr, the value of Ma would be 950 mg/L VSS at 20°C. A temperature correction factor can be made to Ke_0 as shown in Equation 11-8 for temperatures between 5°C and 35°C.

$$Ke_0 \text{ at any temperature} = (Ke_0 \text{ at } 20^\circ C)(1.072)^{(T-20)} \tag{11-8}$$

These equations appear to give a reasonable approximation of the active carbonaceous microbial mass in field plants. A more accurate value of the active microbial mass can be made by a direct determination of the endogenous respiration rate over time. A sample of sludge can be aerated continuously for three days with determination of the (dO/dt), total COD, TSS, and VSS at daily intervals. The (dO/dt) values represent the instantaneous oxygen uptake by the existing active mass at the time when the analysis was made. The total COD data indicate the changes in total oxygen used. The TSS and VSS data indicate the changes in suspended solids during the test period. Microscopic examination of the MLSS would indicate the continued presence and activity of the protozoa. A drop in protozoa population would result in a drop in the values of Ke and Ke_0 with time. Normally, the numbers of dispersed bacteria will decrease during endogenous respiration, providing less food for the protozoa. The net result will be a decreasing rate of apparent endogenous respiration with time. Considering the importance of understanding the overall biochemical changes in the activated sludge, very few detailed studies have been reported in the literature with sufficient data to be of real value. Nitrification can also create errors in the direct determination of the endogenous respiration oxygen uptake rate. Nitrogen analyses can be used to confirm the impact of nitrification and to allow the application of suitable correction for the oxygen used for nitrification.

An evaluation or estimation of the percentage of time the aeration tank is completely aerobic can be used to modify Equation 11-2 to calculate a reasonable value of the active mass of microbes in the activated sludge MLSS. Generally speaking, most conventional aeration tanks will show between 10% and 30% aerobic conditions. The value of Ma can be calculated from Equation 11-2a.

$$\text{Ma} = (\text{Max Ma})(\text{SRTa}/t)/(1 + (\%\ \text{Aerobic})(\text{Ke})(\text{SRTa})) \qquad (11\text{-}2a)$$

Assuming 10% aerobic and using the data in the previous analyses,

$$\text{Ma} = (121)(120/6)/(1 + (0.10)(0.02)(120)) = 1{,}950 \text{ mg/L}.$$

Calculating Me from Equation 11-3,

$$\text{Me} = (0.2)(2{,}420 - 1{,}950) = 94 \text{ mg/L}.$$

The synthesis MLSS can be calculated from Equation 11-4.

$$\text{Synthesis MLSS} = (1.1)(1{,}950 + 94) = 2{,}250 \text{ mg/L}$$

Adding the 800 mg/L inert MLSS from the primary influent SS accumulation gives a MLSS concentration of 3,050 mg/L in contrast to the 1,960 mg/L under completely aerobic conditions. The active microbial mass will be 64% of the MLSS. The microbial Δ COD can be calculated from Equation 11-6.

$$\text{Microbial } \Delta\text{COD} = (1.3)(2{,}420 - 2{,}040) = 494 \text{ mg/L}$$

The actual endogenous respiration rate will average about 4 mg/L/hr, giving a total oxygen uptake rate average around 17 mg/L/hr. Direct measurement of the MLSS endogenous respiration rate would yield almost 41 mg/L/hr. These calculations show how easy it is to produce results that appear contradictory. Care must be taken in evaluating any activated sludge system having partial aerobic conditions.

Nitrification - The aerobic environment in activated sludge systems will stimulate the growth of nitrifying bacteria in the presence of excess NH_3-N. As previously indicated, nitrifying bacteria are autotrophic bacteria capable of obtaining their energy for cell synthesis from the oxidation of NH_3-N. Although there are two major groups of nitrifying bacteria, their overall metabolic activity in activated sludge systems is considered as a single, metabolic reaction. Considerable research has been carried out over the past 40 years on both pure culture studies of nitrifying bacteria and on mixed culture studies of nitrifying bacteria in fixed media and dispersed growth systems. All of these studies have demonstrated that nitrifying bacteria are aerobic bacteria that oxidize about 96% of the NH_3-N to obtain energy to convert the remaining 4% of the NH_3-N to new cell mass. Since the end product of the energy reaction is nitric acid (HNO_3), sufficient alkalinity must be available to neutralize the HNO_3 to maintain a suitable pH in the environment for continued growth of the nitrifying bacteria and the carbonaceous bacteria. The nitrifying bacteria also undergo endogenous respiration like the carbonaceous bacteria. The ability of nitrifying bacteria to form very dense clumps of cells limits the penetration of DO into the inner layers of bacteria and reduces the overall endogenous respiration rate. The dense clumps of nitrifying bacteria also limit the ability of protozoa to eat the nitrifying bacteria. The overall effect of the clumping of nitrifying bacteria is a significantly reduced rate of endogenous respiration. Ke values for nitrifying bacteria in activated sludge systems will be 0.005/hr or less. The mixture of carbonaceous bacteria and nitrifying bacteria in activated sludge floc, adversely affects the nitrifying bacteria ability to come into contact with DO and NH_3-N for rapid metabolism. The normal SRTa for carbonaceous bacteria in CMAS systems treating municipal wastewater allows sufficient accumulation of nitrifying bacteria to obtain 98% to 99% nitrification. As previously indicated, nitrification in conventional activated sludge systems is dependent upon the available time for aerobic conditions in the aeration tank.

Theoretically, 35 mg/L of the 40 mg/L TKN in the raw wastewaters could be available for nitrification with all of the liquid recirculation streams from sludge processing facilities included. Loss of NH_3 to the atmosphere from the treatment tanks would probably reduce the available NH_3-N for nitrification to 30 mg/L. With normal synthesis and endogenous respiration of the nitrifying bacteria in a CMAS system with a 5 day SRTa about 3% of the nitrogen would be tied up as nitrifying bacteria with 97% of the nitrogen oxidized to NO_3-N. The nitrifying bacteria would increase the VSS by 165 mg/L and the NVSS by 16 mg/L, raising the MLSS in the aeration tank to about 2,120 mg/L. The oxygen uptake rate for nitrification would increase by 22 mg/L/hr for a total oxygen demand rate, averaging 50 mg/L/hr. The high rate of oxygen demand for nitrification without any improvement in the carbonaceous effluent quality limited interest in nitrification until the federal EPA showed that NH_3 in the treated effluent could be toxic to some fish and began to require nitrification in some NPDES permit renewals to reduce effluent NH_3-N.

Denitrification – Many WWTP operators found out the hard way that activated sludge contained a considerable population of facultative bacteria that could use nitrates for their electron acceptors when undergoing endogenous respiration in the final sedimentation tanks. The facultative, denitrifying bacteria reduced NO_3-N to N_2 gas in the final sedimentation tanks. The N_2 gas collected under the settled sludge blanket, causing large pieces of the thickened sludge solids to rise to the surface of the settling tank. The N_2 gas bubbles were distributed throughout the sludge clumps, preventing the sludge from settling back to the bottom of the sedimentation tank. Since final sedimentation tanks were not designed to collect sludge from the tank surface, denitrification created real problems for the plant operators. It was not possible to maintain the desired MLSS under aeration and excess sludge was discharged in the final effluent. As this denitrification problem became more common, it was called the *rising sludge* problem.

Eventually, it was recognized that denitrification could be used to reduce the excess nitrates in an unaerated mixing tank prior to the final sedimentation tanks. Unfortunately, endogenous respiration by the MLSS required too long a retention time to complete the reduction of nitrates. It was found that a small amount of biodegradable organic matter could be added to the MLSS to stimulate rapid metabolism with the corresponding denitrification reaction reducing the nitrates to nitrogen gas. The most economical source of biodegradable organic matter for denitrification turned out to be methanol. Large plants used a separate system for denitrification that had a separate sedimentation tank following the mixing tank to keep the methanol acclimated bacteria at their maximum concentration.

Where complete denitrification was not necessary, WWTP operators found that they could put the anaerobic mixing tank ahead of the aeration tank and recycle sludge at a faster rate through the final sedimentation tank to prevent rising sludge and bring the RAS back into the mixing tank with nitrates and primary effluent. Denitrification in the mixing tank resulted in reducing the amount of oxygen required to complete the stabilization of the carbonaceous organic matter in the aeration tank. The anaerobic mixing tank was called an *anoxic tank* to minimize confusion with the anaerobic digestion tanks. In an effort to increase the denitrification reaction and reduce the carbonaceous oxygen load on the aeration tank, some WWTP operators recycled MLSS from the effluent end of the aeration tank back to the anoxic tank.

Denitrification in the presence of organic matter results in the synthesis of new bacteria cell mass the same as with DO. Only 83% of the energy contained in the nitrates is available for the bacteria to use for their metabolism. While the metabolism of nitrates is the same as DO, the synthesis of new cell mass with nitrates is 83% of the synthesis obtained by bacteria with DO. The 17% of the energy not available to the bacteria is used to produce an equivalent amount of alkalinity. Nitrification destroys alkalinity and denitrification generates alkalinity. Theoretically, nitrification destroys 2 mols of alkalinity for each mol of NH_3-N oxidized to NO_3-N; and denitrification generates 1 mol of alkalinity for each mol of NO_3-N reduced to N_2 gas. One of the simplest methods to determine the extent of either nitrification or denitrification is to measure the soluble alkalinity before and after the aeration tank for nitrification and before and after the anoxic tank for denitrification. Since denitrification also occurs in the final sedimentation tanks and the RAS pipes, the extent of denitrification can be measured by determining the soluble alkalinity between the activated sludge flowing out of the aeration tank and the RAS flow entering the aeration tank. Soluble alkalinity is best measured on the centrifuged supernatant of samples containing significant amounts of suspended solids. The suspended solids react with the sulfuric acid used in the alkalinity titration and must be removed to obtain valid alkalinity data for evaluating the extent of nitrification and/or denitrification.

Biological Phosphorus Removal – Biological phosphorus removal, called *Bio-P*, is one of the most complex biological wastewater treatment processes and one of the most difficult to operate. During the 1950s and 1960s the construction of new biological wastewater treatment plants resulted in the removal of large quantities of organic pollutants from wastewater, but failed to solve the water pollution problems in receiving streams and lakes. Research indicated that excess phosphorus in the treated effluents helped to create massive plant growth in the receiving water. Initially, it was believed that chemical precipitation was the only method that could be used to remove the excess phosphorus from the treated effluents.

Research in 1955 by A. E. Greenberg *et al* at the University of California-Berkeley demonstrated that bacteria in activated sludge could store more phosphorus inside of the cell than required for normal cell synthesis. This information did not create much interest at the time when this research data was published. It was not until 1965 that Levin and Shapiro demonstrated that activated sludge from the District of Columbia WWTP was able to remove more phosphorus than the normal amount required for bacteria synthesis. They termed the extra phosphorus removal, *luxury uptake* of phosphorus. Their research showed that high DO concentrations were required in the aeration tank for good phosphorus removal and that phosphorus leaked out of the sludge under anaerobic conditions. Thus, it was possible to remove the phosphorus and to release it under controlled environments. Some experts felt that the additional phosphates were being chemically precipitated with the activated sludge. By 1970 the concept of biological removal of excess phosphorus in activated sludge systems was slowly being accepted. Efforts to capitalize on biological phosphorus removal resulted in the development of the *Phostrip* process. The Phostrip process was demonstrated at Seneca Falls, NY, in 1973. It would take another 8 years before additional Phostrip plants were constructed and placed into operation.

The need for greater wastewater treatment in South Africa led to research by J. L. Barnard on nitrification-denitrification and his discovery that excess phosphorus was being removed in his activated sludge pilot plant. His research led to the

Bardenpho process in 1976 for removing carbon, nitrogen and phosphorus in a single activated sludge system. The key to nitrification and biological phosphorus removal was excess DO in the aeration tanks. Further studies indicated that the excess phosphorus removed by the bacteria was released back into solution when the environment shifted from being aerobic to being anaerobic. The complexity of the initial Bardenpho process quickly led to various modifications to simplify the process. Research at the University of Cape Town resulted in the UCT process, a modification of the Bardenpho concepts that were believed to make it easier on the WWTP operators. By 1984 there were about 30 WWTP under design for nutrient removal in South Africa.

Examination of the bacteria from Bio-P activated sludges by G. W. Fuhs and M. Chen indicated the Bio-P bacteria belonged to the genus *Acinetobacter*. These results stimulated additional research into pure culture studies of different species of *Acintobacter*. As pointed out in 1994 by M. Wagner *et al*, *Acinetobacter* are non-motile, Gram negative, and strict aerobes. They developed a 16S rRNA targeted oligonucleotide probe to study the microbial populations in a Bio-P WWTP in Germany. Their research indicated that *Acinetobacter* was only of secondary importance in the Bio-P process. They also believed that the filamentous bacteria described as type 1863 by Eikelboom was affiliated with *Acinetobacter*. In 1999, P. L. Bond *et al* used fluorescence in situ hybridization (FISH) staining and 16S and 23S rRNA probes to examine the bacteria in activated sludge. They found their Bio-P sludge showed 35% *Acinetobacter*, leading to the conclusion that *Acinetobacter* were important in the Bio-P process. G. R. Crocetti *et al* reported on research utilizing the same techniques to evaluate the bacteria in laboratory SBR activated sludge in 2000. Their results indicated that the majority of the Bio-P bacteria were β *Proteobacteria* with the *Acinetobacter* in the minority. The latest microbiological study by Y. Barak and J. van Rijn showed that *Paracoccus denitrificans* was able to remove excess phosphorus without the metabolism of poly-hydroxybutyrate. They also found that a *Pseudomonas* denitrifying species could form polyphosphates without being able to form poly-β-hydroxybutyrate. Since denitrifying *Pseudomonas* bacteria are quite common in activated sludge, they may play a more important role in phosphate removal than has been recognized. It appears from the research to date that several groups of bacteria are capable of producing polyphosphates and removing excess phosphates from the wastewaters. The Bio-P bacteria definitely represent a major area for bacteriological research in the activated sludge area.

Biological phosphorus removal requires three different environments for all of the bacterial reactions to occur. There must be an anaerobic environment with sufficient amounts of acetate to produce the poly-β-hydroxybutyrate storage product within the Bio-P bacteria. The polyphosphate storage product within the phosphate removing bacteria supplies the energy to form the poly-β-hydroxybutyrate storage product with the release of phosphate into the water around the Bio-P bacteria. There must be an aerobic environment with excess DO sufficient for complete metabolism of the remaining organic matter and for the production of sufficient polyphosphate polymer to account for all the phosphate released in the anaerobic environment plus the new phosphorus contained in the incoming wastewater. The metabolism of the poly-β-hydroxybutyrate polymer yields the energy for synthesis of new phosphorus removing bacteria and the production of the intracellular polyphosphate polymer. The excess DO in the aerobic environment will also produce complete nitrification for the excess nitrogen in the wastewater being treated. With increased effluent criteria requiring nitrogen removal and phosphorus removal, it has been necessary to have an anoxic environment to reduce the nitrates to N_2 gas. It should be recognized that the high polyphosphate containing activated sludge must be wasted from the system for phosphorus removal to occur. The activated sludge must be kept aerobic or the phosphorus will start to leak out into the liquid. This means that the activated sludge must be dewatered and returned to the land environment to keep the phosphorus from being returned back to the treatment plant with the recycled water removed from the sludge. Much research still needs to be done to understand all of the various biological reactions occurring in these three different environments, as well as, in sludge processing. While the excess nitrogen has been changed to a form that can be readily absorbed by the environment without a problem, the phosphorus has only been removed from the liquid phase. The phosphorus itself has not been changed and remains as biologically active as ever.

The concept of Bio-P and nitrogen removal appears to be a significant step forward in the use of applied bacteriology to solve an environmental pollution problem. Although nutrient removal WWTP are being designed and placed into operation, the trial-and-error approach to design and operation is still the basis for these new plants. Nutrient removal WWTP are more complicated than conventional activated sludge WWTP. With a minimum of 3 cells, two recycle streams, and specific organic waste requirements, the number of variables affecting the biological processes increases significantly, making the nutrient removal process much more difficult to design and operate than the other activated sludge processes. Understanding the actual wastewater characteristics and the biochemistry in the multiple biological tanks is essential for success with the biological nutrient removal process. Further studies of full-scale biological nutrient removal WWTP operations are needed to develop proper design and operational strategies for these plants. The variability of municipal wastewater characteristics in the United States has made the use of a single set of design criteria of limited value when designing biological nutrient removal activated sludge plants.

Final Sedimentation Tanks - The mixed liquor from the aeration tank is discharged into final sedimentation tanks for separation of the suspended solids from the treated liquid. The final sedimentation tanks are designed to maximize solids separation, collection, and removal. The incoming flow from the aeration tank consists of the primary effluent flow and the RAS flow together with the MLSS. It is necessary to separate the MLSS from the total flow and to concentrate the suspended solids in the RAS flow for return to the aeration tank to maintain an excess bacteria population in the aeration tank. The clarified effluent is discharged from the sedimentation tank and returned back into the environment. A small WAS flow, containing concentrated suspended solids, is wasted from the final sedimentation tank at periodic intervals to maintain the desired MLSS concentration in the aeration tank. Often, the WAS flow is taken from the RAS flow for ease of operations. The entire mass of fluid flow will enter the final sedimentation tank at a single point. The velocity of the fluid flow at the discharge into the sedimentation tank must be quickly reduced to minimize short-circuiting of the fluid flow through the sedimentation tank. The initial change in velocity is designed to create eddy currents to enhance both liquid dispersion in the sedimentation tank and flocculation of the dispersed MLSS. The flow is directed downward to the bottom of the tank where the settled sludge has concentrated. The sludge blanket is compacted by the mass of the water and the flow direction of the fluid is forced to change from downward to outward. It appears that the downward moving floc particles impinge on the settled sludge blanket and become part of the sludge blanket. As the fluid moves across the settled sludge blanket, the floc particles are left behind and only the tiny particles are carried outward over the sludge blanket to the outer wall. The moving fluid hitting the wall has a small downward flow that reaches the bottom and turns back towards the inlet and a larger upward flow to a weir at the edge of the tank. The tiny particles carried by the flow generally are carried upward and out over the weir, producing the effluent suspended solids. The energy contained in the incoming mixed liquor is largely absorbed by baffles, the settled sludge, the outer wall, and the mass of water in the sedimentation tank. The remaining energy is used to move the liquid through the sedimentation tank to the effluent weir and out of the final sedimentation tank. Examination of fluid flow patterns in final sedimentation tanks indicates that suspended solids separation does not follow the conventional concepts of gravity sedimentation. The suspended solids are removed by entrainment in the settled sludge blanket on the bottom of the sedimentation tank. Entrainment is the only possible mechanism for concentrating the MLSS from 1,500 to 3,000 mg/L to 8,000 to 10,000 mg/L in only 3 to 5 m (10 to 16 ft) depth.

Final sedimentation tanks can be circular, square, or rectangular. Large WWTP tend to use rectangular sedimentation tanks to take advantage of common wall construction with multiple tanks and shorter sludge collection pipes. Rectangular sedimentation tanks use the smallest area. Circular sedimentation tanks became popular because of the simplicity of the wastewater inlet and the maximum length of effluent weir around the tank periphery. The long effluent weir around the outer wall helps to maintain a uniform, low velocity of fluid flow over the top of the sedimentation tank. Circular sedimentation tanks are better able to absorb surges in fluid flow than rectangular sedimentation tanks, making circular sedimentation tanks more desirable than rectangular tanks for small to medium size municipal WWTP that have greater flow variations each day than large municipal WWTP. Square sedimentation tanks are used when space is limited and common wall construction is desired both between tanks and for underground utility tunnels. Square sedimentation tanks are essentially circular sedimentation tanks in a square box.

The most difficult part in final sedimentation tanks is the settled sludge collection system. Once the suspended solids have been collected in the settled sludge, the solids must be removed and returned to the aeration tank as RAS or wasted from the system as WAS. The sludge collection mechanisms originally used in final sedimentation tanks were the same as used in primary sedimentation tanks. Unfortunately, there was a major difference in sludge characteristics between primary sludge and secondary activated sludge that was overlooked for years. Primary sludge has a greater density than secondary activated sludge and can be easily pushed in a fluid environment. Settled activated sludge cannot be pushed in a fluid environment. The sludge scrapers in primary sedimentation tanks push the settled primary sludge to the sludge hopper for removal from the primary sedimentation tanks by positive displacement sludge pumps. The sludge scrapers in secondary activated sludge sedimentation tanks are unable to push the settled activated sludge. The sludge scrapers push through the settled sludge, forcing the sludge to ride up and over the scraping mechanism. In effect, the sludge scrapers help to keep the thickened sludge in a fluidized state that allows the thickened sludge to flow continuously towards the sludge collection hopper for recirculation back to the aeration tank. The sludge recirculation pumps supply the energy to move the thickened sludge through the final sedimentation tanks.

With increased emphasis on activated sludge in the 1950s and 1960s more attention was paid to removing the settled from the final sedimentation basins, as quickly as possible. A hydraulic sludge collection system, employing a pipe manifold, had been around for several decades with limited use compared to the conventional sludge scrapers. The pipe manifold collector moved across the flat bottom of circular sedimentation tanks, moving settled sludge hydraulically into the perforated manifold and back to the aeration tank. Suddenly, design engineers seemed to discover the hydraulic sludge collection system. Other equipment manufacturers quickly developed new versions of the hydraulic sludge collection system and hydraulic sludge collectors became the accepted sludge collection system in the United States. Figure 11-16 shows a typical final sedimentation tank with hydraulic sludge removal tubes and sludge scrapers.

Good solids separation depends upon the activated sludge floc characteristics. The NVSS are important for their contribution to the density of floc particles. Current activated sludge treating primary effluent from domestic wastewater contain about 20% NVSS. Activated sludge produced from municipal wastewater collected in combined sewers contains 25 to 35% NVSS. The activated sludge with the higher NVSS fraction not only settled at a faster rate, but also compacted to a greater sludge concentration after settling, 12,000 to 15,000 mg/L. Activated sludge with lower NVSS fractions can be expected to settle at a slow rate and compact less dense than normal activated sludge. Excessive growths of filamentous bacteria will also slow activated sludge settling. The filamentous bacteria prevent the floc particles from aggregating to their maximum density. With excessive filamentous bacteria growth the settled sludge volume increases significantly. Unless corrective action is taken by the WWTP operator, the final sedimentation tank will fill completely and upset the plant operations. The normal activated sludge environment cannot prevent some growth of filamentous bacteria. The filamentous

Figure 11-16 FINAL SEDIMENTATION TANK WITH HYDRAULIC SLUDGE REMOVAL TUBES AND SLUDGE SCRAPERS

bacteria growth in normal activated sludge systems should not be sufficient to adversely affect the rate of separation and the density of the thickened sludge.

Return Activated Sludge - The RAS is an important part of most activated sludge systems and often determines the success or failure of the activated sludge process. The settled sludge in the secondary sedimentation tanks is collected and pumped continuously back to the aeration tank to maintain an excess of active bacteria at all times. In most activated sludge WWTP the RAS is discharged at the influent end of the aeration tank. True CMAS plants have the RAS discharged under the aerators to insure rapid dispersal throughout the aeration tank. The actual fluid flow through the aeration tank is determined by the sum of the incoming wastewater flow (Q) and the RAS flow (Q_r). The estimated MLSS concentration in the aeration tank is determined by the RAS flow and the concentration of RAS as shown in shown in Equation 11-9.

$$(Q_r)(RAS) = (Q + Q_r)(MLSS) \qquad (10\text{-}5)$$

With Q varying continuously during the day, the WWTP operator must decide if Q_r is to vary or kept constant. Simplicity of operations favors a constant Q_r. A constant Q_r results in shifting MLSS from the aeration tanks to the final sedimentation tanks as Q increases. Since the RAS moves a fixed quantity of sludge from the sedimentation tank to the aeration tank, the MLSS in the aeration tank decreases and the suspended solids accumulate in the final sedimentation tank. As Q decreases, less MLSS are displaced into the final sedimentation tank and the suspended solids in the final sedimentation tanks decrease. At minimum Q the RAS concentrations often decrease as minimum suspended solids remain in the sedimentation tanks. Some operators adjust the Q_r to specific levels each shift to maintain an average MLSS in the aeration tank. During the high flow shift, the Q_r is reduced to match the average Q expected for the shift to maintain the desired MLSS concentrations. As the expected Q decreases on the evening shift, Q_r is increased to keep the MLSS as close to constant as possible. The night shift sees minimum Q and the highest Q_r. Storm water infiltration creates problems for operations since they usually arrive unexpectedly and last for unknown periods of time. Storm water infiltration increases the incoming flow with MLSS displaced into the final sedimentation tanks. Since the incoming organic concentration is reduced by the storm water dilution, Q_r can be reduced to allow the secondary sedimentation tank to absorb the excess solids on a temporary basis. Balancing the flows to keep the MLSS at suitable levels is difficult. It tends to be a trial-and-error procedure for most WWTP operators. It also helps make the SBR plants popular since there is no separate sedimentation tank. The Q_r is currently designed to be varied from 25% Q to 100% Q, giving the WWTP operators considerable flexibility.

One of the problems that WWTP operators tend to create for themselves is using the final sedimentation tanks for storing excess activated sludge. Failure to waste the proper amount of excess activated sludge can allow the excess sludge to accumulate in the final sedimentation tank unless the Q_r is adjusted to move the excess sludge back to the aeration tank. As the thickened sludge layer increases beyond the ability of the sludge collection system to move the sludge through the final sedimentation tank, the upper portion of the sludge layer will become stagnant. If the stagnant sludge is not removed, anaerobic conditions will become well established. Over time the stagnant sludge layer will turn black in color from iron sulfide and other metal sulfides. Carbon dioxide gas may accumulate sufficiently in the stagnant layer to lift a thin layer of black activated sludge to the surface of the tank. The floating layer of septic sludge is quite different from the large clumps of brown activated sludge that rises to the surface as a result of denitrification. Denitrification is a rapid process that creates rising sludge compared to the slow process for septic sludge rising. Both processes are the result of excess retention of activated sludge in the final sedimentation tanks. Both processes can result in deterioration of the final effluent quality. It is important for the operators to recognize that the final sedimentation tank has limited sludge storage capacity. The MLSS entering the final sedimentation tank must be moved through the final sedimentation tank and back to the aeration tank as quickly as possible or problems will occur that will adversely affect the effluent quality. The inability to properly handle the WAS often results in the final sedimentation tanks becoming "temporary" sludge storage tanks that soon become more "permanent" than "temporary".

Activated Sludge Changes - Over the years there have been many changes to activated sludge systems to improve the process. Most of the modifications have been minor and have yielded minimum improvement. It was quickly observed that the placement of aerators along one wall of the aeration tank resulted in good mixing and adequate oxygen transfer with less air than with diffusers covering the entire floor of the aeration tank. The aeration along one wall produced *Spiral Flow* down the aeration tank. Concern over short-circuiting through the center core of the long, narrow aeration tanks resulted in the placement of several transverse baffles across the aeration tank. It was soon learned that the mixing in the aeration tank was adequate to

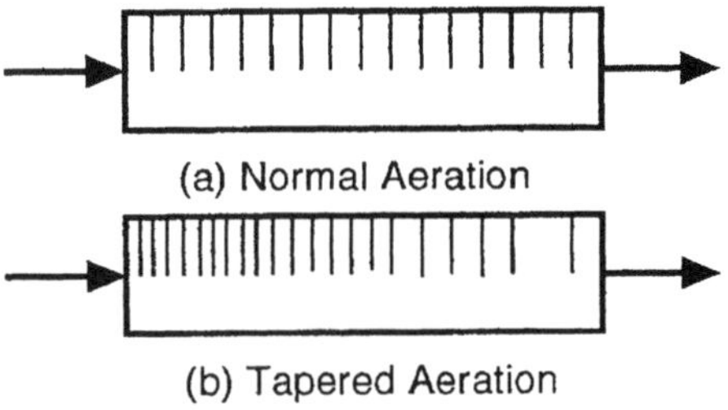

Figure 11-17 LOCATION OF AIR DIFFUSERS IN AERATION TANKS

prevent shortcircuiting and the transverse baffles were eliminated. When it was observed that the DO at the head end of the aeration tank was lower than the DO at the effluent end of the aeration tank, it was suggested that some of the aerators be removed from the effluent end of the aeration tank and placed near the influent end of the aeration tank. The new aeration placement was called *Tapered Aeration*, as shown in Figure 11-17. Tapered aeration provided no improvement in effluent quality. The oxygen demand rate at the head end of the aeration tank still exceeded the oxygen transfer rate. In spite of its limited value, tapered aeration became the standard for aerator placement in the United States.

If oxygen transfer could not be increased sufficiently to meet the oxygen demand, Richard Gould reasoned that the oxygen demand rate could be reduced by distributing the incoming wastewaters at several points along the aeration tank. By spreading the organic load over more of the aeration tank, Gould hoped to be able to treat more wastewater in the same volume of aeration tanks. The new activated sludge process was developed in New York City and was called *Step Aeration.* Gould demonstrated that by adding the incoming wastewater at three or four points along the long narrow aeration tank, it was possible to reduce the oxygen demand rate below the oxygen transfer rate for diffused aeration system. Although the activated sludge modification was called "Step Aeration", it was actually "Step Feeding". The term, Step Aeration, stuck with the process. Step Aeration also permitted the treatment of greater volumes of wastewater in the same aeration tank volume without a loss in effluent quality. The uniqueness of the Step Aeration concept permitted the new treatment process to be patented. Unfortunately, the patent on Step Aeration inhibited its use outside of New York City. Difficulties with the British patent on the basic activated sludge process and a series of legal problems a few years earlier made most cities leery of having to pay royalties on a continuous basis. Figure 11-18 shows a four pass, schematic diagram of the Step Aeration process.

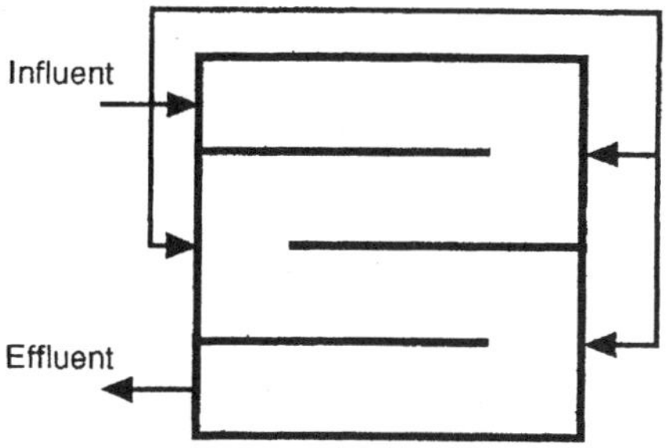

Figure 11-18 SCHEMATIC DIAGRAM OF THE STEP AERATION PROCESS

As municipal wastewater plants became overloaded from expanding populations after World War II, treatment plant operators were forced to look for new ways to modify existing treatment plants to process more wastewater in the same tanks without any loss in treatment efficiency. At Austin, TX, A. H.

Ullrich and M. W. Smith looked at the adsorptive and absorptive characteristics of activated sludge and figured out a new way to use these two properties to remove the contaminants from municipal wastewater. They found that Austin, TX, wastewater contained more suspended contaminants than soluble contaminants, allowing them to take advantage of the adsorptive characteristics of activated sludge. Their research demonstrated that mixing return activated sludge with raw, unsettled wastewater for about 5 minutes, followed by sedimentation, produced a high quality effluent. They regenerated the activated sludge by aerating the RAS before mixing again with fresh wastewater. Ullrich and Smith named their new treatment process, *Biosorption*. Efforts to obtain a patent on Biosorption failed when it was found that the process had been discovered years earlier. W. W. Eckenfelder was also doing research on the same concept for industrial wastes and found that it worked as well as it had for municipal wastewater. The new treatment process soon became widely accepted as the *Contact Stabilization* process. Figure 11-19 illustrates the schematic diagram of the Biosorption process modifications made to the aeration tanks at Austin, TX. The long narrow aeration tanks made excellent bioreactors for the process modifications. The longitudinal mixing in the aeration tanks produced more mixing between the aerated RAS and the incoming wastewater than was recognized in the early studies on the contact stabilization process.

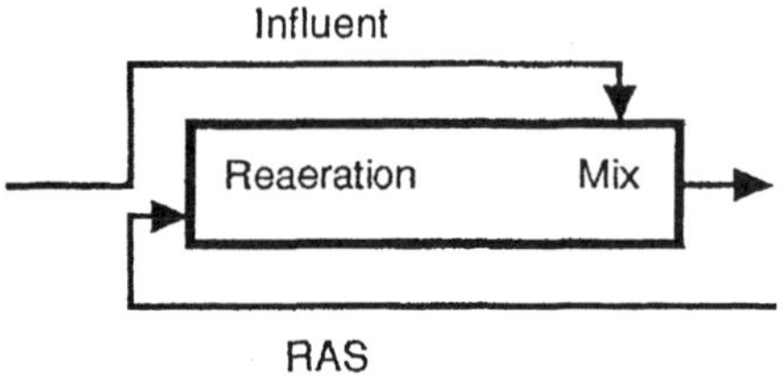

Figure 11-19 SCHEMATIC DIAGRAM OF THE BIOSORPTION PROCESS

Contact stabilization took advantage of the adsorptive and absorptive characteristics of activated sludge to quickly remove the organic compounds by mixing the raw wastewater with activated sludge for a short time period, 5 to 10 minutes, settling the activated sludge and aerating the concentrated sludge prior to its return to the mixing tank. The original design eliminated the primary sedimentation tanks for municipal wastewaters, employed 15 minutes aerated mixing, 2 hrs settling, and 2 hrs reaeration. Unfortunately, the equipment manufacturers, selling the process equipment, the regulatory engineers, responsible for approving the process, and the design engineers, using the process, all added their "safety factors" to the original design and completely changed the process without realizing it. The short term aerated mixing tank became the soluble organic metabolism tank as its size was increased. In some cases, it was the complete metabolism tank. The reaeration tank was increased to 4 - 6 hrs capacity. Essentially, the reaeration tank became an aerobic digestion tank instead of the metabolism tank. Increasing the size of biotreatment units to provide a "safety factor" became an "unsafe factor" as the larger, aerated, mixing tank shifted from an adsorption and absorption unit to a metabolism unit. Even with all the design modifications contact stabilization WWTP have been widely used in municipal wastewater treatment. The inability to patent the contact stabilization process allowed open access for everyone to use this process at no additional cost for royalty payments.

The *Extended Aeration* modification was first recognized in the 1940s and became popular after World War II. The extended aeration process is actually a completely mixed activated sludge system that combines synthesis with aerobic digestion by extending the aeration time to at least 24 hrs. The extended aeration systems gained popularity by treating small quantities of wastewaters generated by isolated subdivisions or small communities with a minimum of operational supervision. The SRTa for extended aeration sludge plants ranges from 30 to 40 days. Minimum excess sludge production occurs with the long sludge aeration times. Originally, it was believed that extended aeration systems would not produce any excess sludge to be removed from the system. The operators of small extended aeration plants quickly learned that excess activated sludge had to be wasted from the system or discharged in the final effluent. The *Oxidation Ditch*, shown in Figure 11-20 is one of the most widely used modifications of the extended aeration activated sludge process. Oxidation ditches are oval shaped aeration tanks that use mechanical surface aerators across the width of the aeration tank to provide oxygen and mixing. Many oxidation ditch plants operate with 36 to 48 hours raw wastewater aeration time. Figure 11-20 shows two horizontal, rotor type aerators across the oxidation ditch. Some designs employ vertically mounted, surface mechanical aerators instead of the horizontal, rotor aerators. Oxidation ditches have aerobic zones immediately after the aerators, followed by anaerobic zones. The size of the

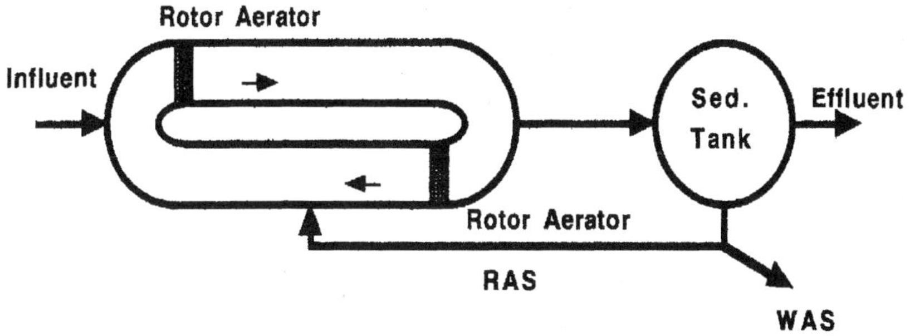

Figure 11-20 SCHEMATIC DIAGRAM OF AN OXIDATION DITCH TYPE EXTENDED AERATION PROCESS

anaerobic zone is a function of the oxygen transferred and the organic loading rate. Mixing in oxidation ditches is definitely different than in conventional aeration tanks. DO and oxygen uptake rate measurements are important in evaluating the aerobic and anaerobic zones in oxidation ditches.

The *Activated Biofiltration* (ABF) modification uses a wood media trickling filter as a high rate synthesis unit ahead of a small aeration tank. The RAS is mixed with primary effluent before being applied to the top of the trickling filter to allow rapid synthesis of new cells and a high rate of oxygen transfer before entering the aeration tank to complete metabolism of biodegradable organic matter in the wastewater. The ABF system has produced mixed results as a result of design engineers not fully understanding the quantitative relationships in each unit. The short contact between the biodegradable organic compounds and the bacteria in the wood media trickling filter limits the amount of bacteria synthesis and the amount of organic compounds metabolized in the trickling filter. The remaining biodegradable organic matter must be metabolized in the aeration tank. Balancing the metabolism between the trickling filter and the aeration tank is critical for complete metabolism of the biodegradable organic compounds in the wastewater being treated. If the aeration tank is large enough, some nitrification will occur. A well designed and well operated ABF WWTP can produce an excellent effluent. For the most part, ABF systems are best applied to space limited treatment plants with adequate land for proper disposing of the excess sludge.

The *Sequencing Batch Reactor* (SBR) process is the modern version of the original batch fed activated sludge system with both aeration and solids separation occurring in the same tank. Current interest in the batch fed activated sludge process appears to have begun in the 1970s by R. L. Irvine. The raw wastewater is discharged to the SBR tank under anaerobic conditions for a short period of time. The anaerobic fill produces the same effect as the anaerobic selector. The air is turned on and the tank fills to the desired level under aerobic conditions. The air is stopped after a specific time interval and the MLSS are allowed to settle under quiescent conditions. After a short settling period, the top layer of the settled supernatant is drawn off through a pipe manifold as the effluent. The effluent pipe manifold is maintained at a fixed depth by floats on the surface of the sedimentation tank. The length of the effluent pipe is quite short compared with the length of the two side weir boxes used in conventional final sedimentation tanks. The length of the effluent pipe in the SBR plants demonstrated that weir length design parameter for conventional final sedimentation tanks is not a valid design concept. The accumulation of scum on the surface of the SBR requires that surface scum removal equipment be a normal part of the equipment in an SBR plant. WAS is removed by a pump from the bottom of the tank, as needed to maintain the desired MLSS. Most SBR plants have higher MLSS than normal activated sludge plants as a result of a longer SRTa and greater endogenous respiration. A microprocessor control system is used to adjust the proper time sequences for the various functions during each operating cycle. The time element for each treatment sequence can be adjusted to fit different wastewater characteristics. Most SBR plants have two or more tanks to allow continuous wastewater treatment in the system on an alternating basis. The simplicity of the SBR systems has made them quite popular in small WWTP. Very small plants can operate with only one tank and still produce a satisfactory effluent.

Although *Nitrification/Denitrification* concepts have been around for years, few treatment plants have been designed for nitrogen reduction. The various forms of inorganic nitrogen in the effluents from activated sludge plants, NH_3-N and NO_x-N, were not considered serious environmental problems. As the nitrogen concentrations rose in specific streams, researchers found that the free NH_3 was toxic to fish at high concentrations. Initial efforts were made to minimize the NH_3 in natural waters by keeping the pH close to 7. Unfortunately, pH control turned out to be quite impractical. Efforts were then turned to nitrification with the conversion of the NH_3-N to NO_3-N. While nitrates were not toxic at the normal concentrations found in natural waters, the concern over NO_3-N in well water carried over to surface waters. In addition, algae were able to use the NO_3-N together with the PO_4 in the activated sludge effluent to produce considerable algal growths in ponds and lakes. The public considered the algal growths as a sign of water pollution and demanded removal of the NO_3-N to prevent excessive algae accumulation in the water environment used for recreation and fishing. The federal EPA began to add nitrogen removal to NPDES permits in critical areas when it was time for NPDES permit renewals. Nitrification/Denitrification plants require larger sized aeration tanks for maximum nitrification. Denitrification requires a source of biodegradable organic matter in an anoxic tank with a mechanical mixer and nitrates from the aeration tank. Current treatment plants use the primary effluent for the source of biodegradable organic matter with nitrates being returned to the anoxic tank in the RAS and in direct MLSS recycling from the aeration tank effluent. Figure 11-21 illustrates a schematic diagram of the nitrification-denitrification wastewater treatment process. Denitrification reduces the amount of oxygen required for total treatment of the wastewaters by allowing part of the oxygen used for nitrate formation to be reused for the oxidation of the incoming organic matter in the primary effluent. The reduction of nitrates in the anoxic tanks also allows the alkalinity to increase, about 3.5 mg/L alkalinity as $CaCO_3$ for each mg/L NO_x-N reduced to N_2. The recovery of half of the required alkalinity for nitrification is very important in soft water areas. Nitrification-denitrification systems are a challenge for the design engineers and an even greater challenge for the WWTP operators.

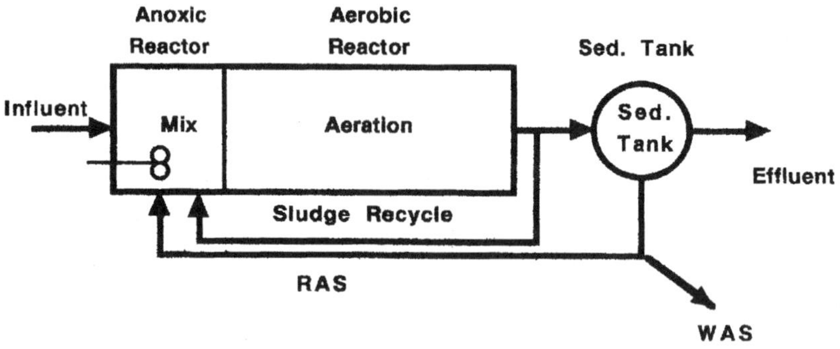

Figure 11-21 SCHEMATIC DIAGRAM OF NITRIFICATION/ DENITRIFICATION SYSTEM

As the number of activated sludge WWTP constructed and placed into operation increased in the United States, a new problem surfaced. The highly purified effluents stimulated excessive growths of algae in the receiving streams, lakes, and reservoirs. Examination of the activated sludge effluents indicated low turbidity, few carbonaceous organic compounds, excess nitrogen, and excess phosphorus. Research quickly pointed to the effluent phosphorus in the form of phosphates as the major stimulant for algae growth. It soon became apparent that phosphates needed to be reduced in certain critical areas. Examination of domestic wastewaters indicated that about half of the phosphates in the wastewaters were from synthetic detergents. Regulatory agencies quickly decided that the greatest impact could be made if the detergent manufacturers could stop adding phosphates to synthetic detergents. Research by the detergent manufacturers eventually found other chemical compounds that could replace phosphates in detergents and not stimulate the growth of algae. In some areas of the country it was necessary to remove the excess phosphorus from the WWTP effluents to minimize the adverse impact of algae growths. Aluminum salts or iron salts were used to precipitate the phosphates from the WWTP effluents. The major problems with chemical precipitation of phosphates are the continuous requirement for chemicals and the additional sludge produced for disposal.

The newest modification of the activated sludge process is the *Nutrient Removal* process that includes Nitrification/Denitrification and Biological Phosphorus removal, *Bio-P*. Figure 11-22 illustrates a schematic diagram of the Nutrient Removal process. The primary effluent is added to an anaerobic cell at the head end of the biotreatment system along with the RAS. The anaerobic cell has a raw wastewater retention time of 1 to 2 hrs and is mechanically mixed to keep the contents in suspension. The facultative bacteria in the RAS metabolize the soluble organic compounds to short chain fatty acids under anaerobic conditions and begin some hydrolysis of colloidal suspended solids. The phosphate bacteria metabolize the short chain fatty acids with the release of phosphate back into the liquid, increasing the overall phosphate concentration in solution. The release of phosphate is part of the energy metabolism by the phosphate-concentrating bacteria, often designated as Bio-P bacteria. The Bio-P bacteria store some of the short chain fatty acids as a poly-β-hydroxybutyrate polymer in their cells as they grow. Denitrification begins in the final clarifier and continues in the RAS pipe. Denitrification will continue in the anaerobic cell. If it is desired to keep

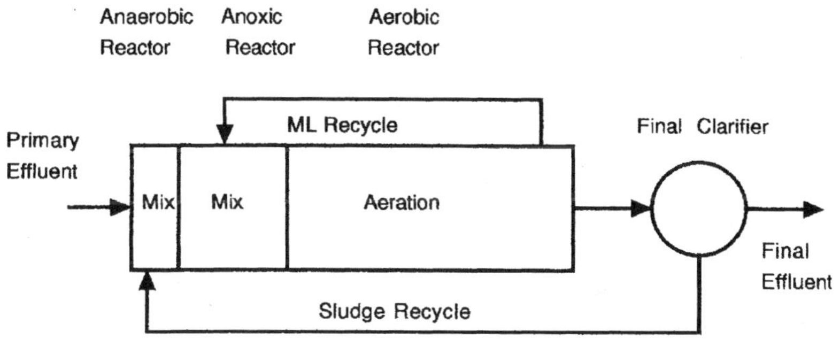

Figure 11-22 SCHEMATIC DIAGRAM OF A NUTRIENT REMOVAL PROCESS

denitrification out of the anaerobic cell, the RAS will be discharged in the anoxic cell instead of the anaerobic cell. Recycle of sludge from the anoxic cell to the anaerobic cell may or may not be used, depending upon the rate of short chain fatty acid production. The wastewater moves from the anaerobic cell to the anoxic cell where MLSS is recycled from the end of the aeration tank. The MLSS recycle stream from the aeration tank will range from 100% to 400% raw wastewater flow, depending on the extent of denitrification desired. The RAS flow from the final clarifier tends to be equal to the raw wastewater flow. The anoxic cell has a raw wastewater retention time between 2 and 4 hrs and is also mechanically mixed to help maintain anaerobic conditions in the anoxic cell for rapid denitrification. The effluent from the anoxic cell enters the aeration cell where the remainder of the biodegradable organic matter is metabolized aerobically. The aeration cell has a raw wastewater retention time between 6 and 10 hrs. The Bio-P bacteria metabolize their stored poly-β-hydroxybutyrate and remove excess phosphates from solution. The excess phosphorus removed from solution is stored as polyphosphate inside the Bio-P cells. Nitrification occurs in the aerobic environment, converting the excess NH_3-N into NO_3-N. The mixed liquor at the end of the aeration cell is split into two streams. One stream is recycled back through the anoxic cell and the aeration cell. The other stream is discharged to the final clarifier for solids separation. The settled solids are largely recycled back to the anaerobic cell with the excess solids wasted from the system. The process of removing WAS results in the quantity of phosphates actually removed from the WWT system. The process of biologically removing phosphates from wastewater has been termed the *Bio-P* process. Needless to say, the Bio-P concept has stimulated considerable research to determine which bacteria are responsible for enhanced phosphate removal and the development of quick methods for their identification using 16S rDNA probes.

These are but a few of the modifications of activated sludge that have been developed over the years. Most of the modifications have been developed to solve specific problems that have arisen at activated sludge plants. As design engineers and operators learn more of the microbiology and the biochemistry of activated sludge together with the fluid characteristics of various types of tanks used in the biotreatment process, other modifications of activated sludge will be developed in future years.

Stabilization Ponds

Stabilization ponds are the simplest WWTP for small communities where low cost land is readily available. Stabilization ponds are simply large, shallow ponds that allow municipal wastewaters to be retained for several months. Stabilization ponds had their American start in California and Texas in the 1920s. The real development of raw wastewater stabilization ponds occurred in the Upper Midwestern states after World War II. The need for improved wastewater treatment at a relatively low cost for construction and operation stimulated the rapid development of stabilization ponds all through the Midwest and Southwest. By 1970 wastewater stabilization ponds were well established for both municipal wastewater and industrial wastewater treatment. For the most part, the design criteria for stabilization ponds were developed by trial-and-error in the field.

Design Concepts – Stabilization ponds require large flat areas close to the source of the wastewater. Most stabilization ponds are constructed with 0.9 - 1.5 m (3 - 5 ft) water depth. In warm, dry climates with a high rate of water evaporation stabilization ponds are constructed with water depths of 3 - 6 m (10 - 20 ft). The ponds have a 0.9 m (3 ft) freeboard above the water surface with a 3:1, horizontal to vertical, slope on all side slopes. The berm around the stabilization pond is constructed of compacted soil removed from the pond excavation. The top of the berm is at least 3 m (10 ft) wide to allow a car or a small truck to drive completely around the pond. The submerged soil surfaces are treated with clay and phosphates to minimize water loss from the pond into the groundwater below the pond. Recently, the pond surfaces have been covered with plastic liners to prevent seepage from the pond and to protect the side slopes from erosion. The volume of the stabilization pond is determined by the volume of wastewater produced each day, the quantity of biodegradable organic material in the wastewater, and the number of days with temperatures below freezing. The treatment of normal municipal wastewater from small communities requires 60 to 120 days raw wastewater retention in the southern half of the United States and 120 to 180 days raw wastewater retention in the northern half of the United States. Currently, stabilization ponds are divided into 2 or 3 cells with a few ponds having even more than 3 cells. The discharge structure from the final cell controls the water depths in the other cells. The discharge overflow pipe in the final cell is baffled to minimize the discharge of algae from the pond system. The intermediate cells are normally connected by a single pipe located at mid-depth between adjacent cells.

Microbiology of Stabilization Ponds - Although there has been limited interest in the microbiology of stabilization ponds, bacteria, algae, protozoa, and higher animals were recognized as being important in the operation of stabilization ponds. The major research in stabilization pond microbiology has centered on the role of algae in the overall stabilization process. Later, research focused on the roles of protozoa, rotifers, and crustaceans found in stabilization ponds. Public health concerns have led to examination of the survival of pathogens in the treated effluent. There is no doubt that the size and the simplicity of stabilization ponds has limited the microbiological research on stabilization ponds.

The first efforts to define the role of the algae in stabilization ponds were carried out at the University of California by H. F. Ludwig *et al* in 1951. Their initial studies examined *Euglena gracilis*, a motile, green algae, and *Chlorella pyrenoidosa*, a non-motile green algae. These algae were very common in wastewater stabilization ponds in California. Their research indicated that the bacteria metabolized the organic compounds in the wastewaters with the production of CO_2 that the algae used for their cell synthesis. Overall, carbon was the limiting element in the growth of algae with N and P both in excess. Their data also indicated that *Chlorella* tended to clump together as the cells aged.

In 1953 P. C. Silva and G. F. Papenfus presented data on the microscopic examination of algae in 8 stabilization ponds for the California State Water Pollution Control Board. Their observations indicated *Euglena, Chlorella, Chlamydomonas, Chlorogonium,* and *Scenedesmus* were the most common green algae. *Oscillatoria, Anabaena,* and *Microcystis* were the most common blue-green algae. The diatom, *Navicula*, was also common in hard water areas.

A study by Gann *et al* showed that the common bacteria in stabilization ponds were the soil bacteria: *Achromobacter, Pseudomonas, Flavobacterium,* and *Bacillus*. As stabilization ponds increased in numbers, efforts were made to determine their effectiveness in removing fecal coliform bacteria and other indicators of fecal contamination. While coliform reductions in municipal wastewater stabilization ponds exceeded 99% reduction, the effluents did not meet coliform requirements without disinfection.

Heavily loaded stabilization ponds tended to produce sulfate reducing bacteria and often have the red-purple, photosynthetic, sulfide oxidizing bacteria growing at the pond surface where there is adequate light. Blue-green algae will also be found in stabilization ponds having sulfide production. Environmental conditions within each pond or within each cell of a multi-cell stabilization pond determine the growth and predomination of each group of microorganisms. The accumulation of biodegradable solids on the bottom of the stabilization pond will result in the production of anaerobic and facultative bacteria in the solids layer. The upper volume of water in the pond may be either aerobic or anaerobic. It is quite normal for the bottom of the pond to be anaerobic and the top of the pond to be aerobic.

The growth of bacteria and *Chlorella* will stimulate the growth of free-swimming ciliated protozoa: *Paramecium, Glaucoma, and Colpidium*, under aerobic conditions. The crawling protozoa, *Euplotes*, and the stalked ciliated protozoa, *Vorticella*, may also be found in large numbers in stabilization ponds. The rotifers: *Epiphanes, Philodina,* and *Proales*, may also grow in this aerobic environment. The crustaceans, *Moina* and *Daphnia*, can occur in very large numbers in the spring and may completely remove all the common green algae from the ponds. *Diaptomus* and *Cyclops* are also found in varying numbers. The algae populations vary with the seasons of the year and reflect the effect of temperature changes, as well as, the predator populations. *Chlorella* grow rapidly in the early spring as the temperature increases and are quickly eaten by the higher animal forms. Loss of the *Chlorella* allows *Scendesmus* to grow and predominate. The sharp spikes on *Scenedesmus* prevent their being rapidly eaten by the crustaceans, allowing them to survive during the summer and fall. When the temperature declines and the predator populations decline, the *Chlorella* return and displace the less efficient *Scenedesmus* until the next spring. There is no doubt that the microbiology of stabilization ponds is the most dynamic of all the wastewater treatment systems.

Biochemistry of Stabilization Ponds - The biochemistry of stabilization ponds did not attract the attention that the biochemistry of activated sludge did. The complexity of the microbial populations in stabilization ponds and the limited understanding of mixed microbial populations in general discouraged researchers. Most researchers chose to look at specific microorganisms under carefully controlled conditions. Unfortunately, mixed microbial populations respond differently than pure cultures in carefully controlled environments. Even laboratory studies with mixed microbial populations fail to provide the environment required for comparable evaluation with field units, unlike activated sludge systems. The biochemistry of stabilization ponds is best studied through examination of many ponds. Long term studies of individual stabilization ponds are necessary to produce the information required for a general understanding of the biochemistry of stabilization ponds in the United States.

The size of stabilization ponds allows the incoming wastewater flow velocity to slow quickly, allowing the heavy suspended solids to settle to the bottom of the pond. The soluble solids in the incoming wastewater move steadily from the discharge of the influent pipe towards the effluent pipe in the first cell. Liquid mixing within the stabilization pond is dependent upon the turbulence generated by wind currents over the pond surface. In the absence of wind action on the pond surface, mixing will be limited to fluid movement from the influent pipe to the effluent pipe with some molecular diffusion. Mixing can range from limited to considerable, depending upon the velocity of the wind and its steadiness over time. Density currents are not sufficient to affect mixing either positively or negatively. This means that the soluble waste contaminants travel as a stream from the influent to the effluent, with the flowing stream increasing in size by dispersion. The settled biodegradable solids undergo both aerobic and anaerobic metabolism. The wind induced motion of the pond liquid brings oxygenated water from the pond surface to the top of the settled solids, allowing aerobic metabolism in a thin surface layer, similar to a trickling filter. Below the aerobic surface layer of solids, facultative bacteria create anaerobic conditions. Over time, methane bacteria should grow in the sludge layer and produce some rising bubbles of methane gas. The production of methane gas and its loss to the atmosphere above the stabilization pond is one of the ways carbon is removed from the wastewater being treated. If the carriage water is high in sulfates, sulfate-reducing bacteria will grow and produce hydrogen sulfide that collects in the gas bubbles and moves to the pond surface. Although the loss of hydrogen sulfide can be detected as an obnoxious odor, very little sulfur will actually be lost from the treated wastewater. During low temperature periods the rate of metabolism of the settled solids slows and the suspended solids accumulate on the pond bottom. With the onset of warm weather the metabolism rates increase rapidly. Sufficient gas can be produced to lift the settled solids to the surface of the pond. Obnoxious odors may be released to the atmosphere when the rising solids reach the pond surface. Increased concentrations of sulfides in the water often stimulate the growth of blue-green algae. *Oscillatoria*, a blue-green algae that oxidizes sulfides to sulfates in the presence of sunlight, will find the sulfide environment suitable for growth near the pond surface. Excessive growths of blue-green algae tend to produce thin mats of cells that cover the pond surface and stop wind mixing. The blue-green algae mats prevent oxygen transfer into the water and create stronger anaerobic conditions. Limited growth of *Oscillatoria* is normal and minimizes the release of sulfides to the atmosphere during the warm months. The dispersed bacteria metabolize the soluble organic contaminants under aerobic conditions, as the wastewaters flow through the first cell. The bacteria growth follows the normal growth pattern, based on the nutrient concentrations available to the bacteria. Endogenous respiration occurs continuously, but has limited impact until all the readily biodegradable organic compounds have been metabolized. The free-swimming ciliated protozoa help remove the dispersed bacteria the same as in activated sludge systems. With a long fluid retention time the bacteria population decreases significantly and the rate of oxygen uptake also decreases. The nitrifying bacteria grow slower than the carbonaceous bacteria and often do not reach high enough levels to provide measurable nitrites or nitrates. Competition with algae for CO_2 further limits the growth of nitrifying bacteria. The net result is that nitrifying bacteria have little impact on the effluent quality from stabilization ponds. Pathogenic bacteria find the stabilization ponds an unsuitable environment for growth. Some of the pathogenic bacteria are metabolized by the protozoa. Spore forming pathogenic bacteria form spores for survival until the spores are placed in a suitable environment for growth again. With limited wind mixing the bacteria spores will slowly settle to the bottom of the oxidation pond

Algae find the stabilization pond an excellent environment for growth. The relatively clear liquid allows light energy to penetrate into the liquid. In the presence of CO_2, NH_3-N, P, and trace metals, the algae are able to grow until the concentration of one of the nutrient parameters or light becomes limiting for growth. Municipal wastewaters supply excess NH_3-N, P, and trace metals for algae growth. Carbon dioxide is the growth limiting chemical parameter in many stabilization ponds. As the algae remove CO_2 for their cellular carbon and use water for their cellular hydrogen and cellular oxygen, they release dissolved oxygen back into the water, keeping the pond surface quite aerobic during daylight hours. Wind mixing pushes the oxygenated water across the top of the pond to the edge and then down to and across the bottom of the pond, bringing the anaerobic liquid on the top of the settled sludge layer with its nutrients to the pond surface for increased algae growth. Wind action generates waves on the surface of the pond that result in oxygen being transferred into the atmosphere above the pond or into the pond, depending on the DO of the surface water. The algae undergo endogenous respiration on a continuous basis, the same as bacteria. In the daylight hours the nutrients released into the water by the algae are immediately assimilated back into new cells. Only the dead cell residue remains as a slowly settling suspended solid with its nutrients. If the stabilization pond has sufficient retention without much wind mixing, much of the dead algae mass settles to the bottom and accumulates over time. If the pond is well mixed by wind, the dead cell residue remains suspended and can be carried out as effluent TSS. At night the algae utilize DO in the pond water for endogenous respiration and are unable to assimilate the nutrients released by endogenous respiration. These nutrients accumulate in the water until daylight returns and are quickly assimilated back into new algae cells. The ability of algae to recycle nutrients allows them to produce more organic mass than originally enters the stabilization pond in the raw wastewaters. The algae move towards the water surface to obtain the light they need for synthesis energy during the daylight hours. In the first cell the algae mass will often increase to the point where light becomes the limiting factor in algae growth, allowing the excess nutrients to pass into the second cell. By using baffled discharge pipes or submerged outlets to collect the effluent from each cell, the algae can be kept mostly in the first two cells. In an aerobic environment crustaceans will feed on the dispersed algae and remove many of the remaining algae. The final effluent from multi-cell stabilization ponds will normally contain TSS and BOD_5 related to algae rather than to raw wastewater TSS and BOD_5. Failure to contain the algae within the stabilization pond system will result in a poor quality effluent.

The growth of algae has a pronounced impact on the stabilization pond environment. Removal of CO_2 for cell growth causes a shift in the form of alkalinity and an increase in pH. Some of the bicarbonate alkalinity in the pond is changed to carbonate alkalinity with a shift in pH from pH 7 to pH 9 and higher. As the alkalinity shifts to carbonate alkalinity, calcium carbonate can be precipitated out in hard water areas, causing the total alkalinity to decrease. It is also possible to precipitate calcium hydroxylphosphate, in addition to the calcium carbonate precipitation, as the pH increases to pH 9.5 or higher. As the alkalinity shifts from bicarbonate alkalinity to carbonate alkalinity, the carbon dioxide available for algae metabolism decreases. In effect, the algae create an environment that limits further algae growth.

Operating Characteristics - For the most part stabilization ponds develop their own operating characteristics. No two stabilization ponds have exactly the same operating characteristics. The same stabilization pond will not have the same biochemical characteristics from year to year as the environment within the stabilization pond changes continuously. Yet, it is possible to make some generalization about what to expect with regard to stabilization pond operations.

The theoretical hydraulic detention time in any oxidation pond will always be greater than the actual hydraulic detention time. The actual hydraulic detention time will be a function of the extent of wind mixing. In the absence of wind mixing the incoming wastewater creates a slow moving submerged stream from the influent discharge point to the effluent pipe. The energy in the incoming flow of wastewater is absorbed by the mass of water near the influent pipe. The use of several cells in series will still result in a direct flow from the inlet to the outlet in each cell. Tracer studies can show how the actual retention time compares to the theoretical retention time. Even with wind mixing, few stabilization ponds will show an actual retention time close to the theoretical retention time. Internal baffles can increase the actual retention time in the different cells. Fortunately, the retention time is sufficiently long that the effluent quality is satisfactory until the system is overloaded from an organic standpoint. It is important to recognize that the flow velocity slows sufficiently to allow the suspended solids in the raw wastewater to settle out near the influent pipe. Only the soluble organic compounds in the flowing wastewater will be affected by the shortened hydraulic retention time. There should be adequate time for the bacteria to aerobically stabilize the biodegradable organic compounds. Stabilization of the organic compounds in municipal wastewater stabilization ponds should never be a problem unless they contain large quantities of industrial wastes. The filtered effluent from a municipal wastewater stabilization pond system should show less than 2 mg/L soluble BCOD. Microscopic examination of the effluent from municipal wastewater stabilization ponds will show that most of the effluent suspended solids are not related to the raw wastewater suspended solids. The effluent suspended solids are related primarily to algae and higher animals that feed on the algae. There will be some bacteria solids that have grown up from decomposing algae, the waste products of higher aquatic animals, and the waste products of birds and other small animals in the area. The light weight dead cell mass that has not settled out will also be part of the effluent suspended solids.

The aerobic conditions in the stabilization pond allow the bacteria to grow aerobically. Endogenous respiration by the bacteria and predation by protozoa produces complete metabolism of the biodegradable contaminants in normal municipal wastewater within 20 days fluid retention at 20°C and 40 days fluid retention at 10°C. Vegetative pathogenic bacteria die off naturally and only spores survive to be carried out in the effluent. Pathogenic protozoa survive only by cyst formation. Protozoa cysts tend to settle out in the stabilization ponds unless well mixed by wind action. The final effluent from stabilization ponds will definitely show a significant reduction in all pathogenic microorganisms in proportion to the actual fluid retention time.

The large surface area exposed to sunlight and adequate nutrients stimulate the growth of algae as the predominant group of microorganisms in stabilization ponds. Since only a small amount of the light energy reaching the pond surface will be used by the algae, cloud cover has little effect on the overall algae growth. The availability of light with the changing seasons of the year has a major impact on the algae growth. Greater algae growth will occur during the long summer days and less algae growth will occur during the shorter winter days. Cold temperatures during the winter months slow the algae growth even further. Yet, algae have the ability to grow under ice cover in proportion to the light penetration and water temperature under the ice layer. The cold temperatures and short periods of daylight during winter months require that stabilization ponds in northern climates be constructed with longer hydraulic retention periods to hold the municipal wastewater during the winter period. When warm temperatures return, microbial growth will be rapid enough to create anaerobic conditions within the stabilization pond with the release of obnoxious odors to the atmosphere above the pond surface. As soon as the bacteria metabolize the excess organic matter in the pond, the algae will begin to grow in the upper layer of pond water and aerobic conditions will return. Maximum concentrations of algae occur at the end of the warm summer months. As the daylight hours become shorter during the fall months, the demand for oxygen by the algae and the bacteria eventually exceeds the ability of the algae and the wind to maintain aerobic conditions. Once again anaerobic conditions may become sufficient to allow obnoxious odors to occur.

Stabilization ponds are the simplest system for wastewater treatment in current use. The primary problems with stabilization ponds are excessive concentrations of algae in the final effluent during the warm summer months and obnoxious odor production in the spring and fall seasons. Unfortunately, there are no simple methods for solving these operational problems. The large land area required for stabilization ponds may be viewed by some people as a problem, but not for rural areas.

Aerated Lagoons

The successful operation of stabilization ponds resulted in the steady increase in the organic loads to determine the maximum loading that could be applied and not produce either odor nuisances or poor effluent quality on a continuous basis. Eventually, most stabilization ponds were overloaded, requiring immediate action to resolve the overloaded plant condition. Increasing the size of the stabilization ponds was never considered as an immediate solution to the overloaded condition. Building new ponds was too slow and too expensive. Since the problem was simply too much organic matter in the stabilization pond, it was determined that adding aeration equipment to the first cell should provide the oxygen required to shift the environment from anaerobic to aerobic. One of the first aeration systems employed in an overloaded stabilization pond utilized a small air compressor connected to perforated plastic hose, placed at regular intervals across the bottom of the cell. The plastic hose was attached to metal stakes and was weighted along the bottom of the plastic hose to prevent it from floating to the water surface. Unfortunately, the efficiency of oxygen transfer in the shallow stabilization ponds was very low. The primary impact of the diffused aeration in the first cell was to increase the rate of fluid mixing though the entire cell. The rising air bubbles generated a fluid flow pattern around each plastic hose that allowed the oxygen produced by algae at the water surface and/or the oxygen transferred from the atmosphere to be dispersed throughout the cell. The net effect was improved aerobic conditions in the aerated cell and improved effluent quality. In hard water areas $CaCO_3$ deposits built up on the perforations in the plastic hose and stopped the flow of air. It was necessary to periodically add HCl gas through the air distribution system to keep the perforations in the plastic hose open for normal aeration. Over time, the plastic hose deteriorated and broke. Some of the hoses broke loose from their anchors and floated to the surface. The initial efforts at diffused aeration in shallow stabilization ponds ultimately were impractical.

Studies with mechanical, surface aerators at the end of the 1950s indicated that these aerators had potential use in aerated lagoons. The surface aerators consisted of three parts: 1. an electric motor, 2. a gear reducer, and 3. a large diameter mixing blade. The electric motor supplied the power to drive the surface aerator. The gear reducer converted the speed of the electric motor shaft to a slower speed of the output shaft. The slow speed output shaft allowed the use of a relatively large diameter mixing blade that was capable of pumping a large volume of water into the air and transferring oxygen to water as it was pumped into the air. The mechanical surface aerator was mounted on a platform with Styrofoam floats to provide the proper water submergence on the mixing blade. In 1960 the University of Kansas made one of the early detailed studies on aerated lagoons. Howard Edde studied the Red Bridge aerated lagoon in Kansas City, MO. Odor problems with a second stabilization pond in the Red Bridge Subdivision resulted in the construction of a second aerated lagoon ahead of the stabilization pond. The second aerated lagoon utilized a fixed platform mounting in the center of a square lagoon for the mechanical aerator rather than utilizing a floating platform. A study by Henry Benjes, Jr. showed that the aerator did not have sufficient power to keep the suspended solids from settling out in the aerated lagoon. Periodically, the settled solids rose to the surface of the aerated lagoon and released obnoxious odors to the atmosphere. It became apparent that the aerated lagoon should have sufficient mixing to be completely mixed to prevent the obnoxious odor problems.

Evaluation of the oxygen demand characteristics of municipal wastewater and the mixing energy required to maintain the MLSS completely mixed indicated that a 24 hrs retention tank would be the proper size for an aerated lagoon ahead of an overloaded oxidation pond. The aerated lagoon would allow rapid bacteria metabolism of the biodegradable organic compounds with the stabilization pond providing settling of the suspended solids and stabilization of the active bacteria produced in the aerated lagoon. Unfortunately, the available nutrients in the stabilization pond allowed the algae to grow the same as before the aerated lagoon was constructed. Without removing the algae, the effluent quality from the aerated lagoon-stabilization pond combination was almost the same as from the stabilization pond by itself. The only real difference was elimination of obnoxious odors and floating scum on the surface of stabilization pond.

Competition between equipment manufacturers resulted in the development of a floating mechanical surface aerator with a direct connection between the electric motor and the mixing blade. Elimination of the gear reducer lowered the cost and allowed the use of a smaller mixing blade. The floating surface aerators were placed near the influent of the overloaded stabilization pond to stimulate rapid metabolism of the incoming organic waste materials. In effect, the surface aerators reduced the organic load on the rest of the stabilization pond, allowing normal biological organisms to develop. Further increases in the organic load resulted in more floating surface aerators being added to increase oxygen transfer. Examination of the large area of the first cell with the surface aerators and the limited mixing capacity of the surface aerators indicated that the simple addition of more floating surface aerators was not the best method for supplying oxygen to aerated lagoons. Evaluation of aerated lagoons indicated that the construction of a small, completely mixed, aeration cell with a slow speed, fixed mounted or floating, mechanical, surface aerator, ahead of the overloaded stabilization pond, offered the best use of the equipment and space and produced the best quality effluent.

It quickly became obvious that the aerated cell did not need to be the same depth as the stabilization pond. Deep aeration cells, 3 - 5 m (10 - 16 ft), permitted the use of more efficient aeration equipment. The deep aeration cells were normally made of reinforced concrete, either square or round. Diffused aeration equipment was more efficient in the deep cells than in the shallow stabilization ponds, providing both better oxygen transfer and better mixing. In small aerated cells a single mechanical surface aerator with a draft tube could provide the desired oxygen transfer and mixing. As the aeration cell size increased, diffused aeration proved more efficient than mechanical aerators. Both aeration systems provided improved flocculation of the suspended solids. The shallow stabilization pond was still needed after the aerated cell to remove dispersed bacteria and the settling of the suspended solids.

There is nothing unique about the microbiology of aerated lagoon systems. The bacteria that grow in the aeration cell are mostly facultative bacteria found in the soil. The bacteria best able to metabolize the soluble biodegradable organic compounds in the wastewater within the aeration retention time will predominate. If the environment is sufficiently aerobic, the bacteria will stimulate zooflagellated protozoa and free-swimming ciliated protozoa growth in the aeration cell. The protozoa will metabolize the bacteria in proportion to their ability to grow within the retention time in the aeration cell. Higher animals will not be found to any significant extent in the aeration cell because of the short aeration period. Nitrifying bacteria will also not grow to any significant extent in most aerated lagoon systems. Turbidity in the aeration cell will prevent algae growth except at the wall-water interface. Some algae will grow on the wetted concrete wall where there is adequate sunlight. Aerated lagoons are simply a dispersed growth, activated sludge system.

Disinfection

The original basis for municipal wastewater treatment was for the protection of public health. Polluted streams and rivers allowed pathogenic bacteria to be spread far and wide. The original slow sand filters produced a very high quality effluent from a biological point of view and effectively provided a major method to stop the spread of enteric diseases in natural waters. The development of trickling filters allowed widespread application of municipal wastewater treatment to major cities. While trickling filters were not as effective as slow sand filters in removing pathogenic bacteria, trickling filter plants removed most of the pathogenic microorganisms and greatly reduced the spread of enteric diseases. Dilution of the treated effluent in the receiving water, followed by natural die-off and predation by protozoa and other microscopic animals, allowed the numbers of pathogenic bacteria to decrease significantly.

The advent of chlorine as a disinfectant in water treatment plants was a major step forward in reducing enteric pathogens from polluted water sources. It was recognized that chlorine could also be used as a disinfectant in WWTP effluents. Unfortunately, the chlorine demand in treated wastewaters was much higher than in municipal water treatment plants. In treated wastewaters chlorine not only reacted with bacteria and other microorganisms, chlorine also reacted with organic matter and with NH_3-N. Since water treatment plants obtained excellent disinfection with small amounts of chlorine, regulatory agencies did not require WWTP effluents to be disinfected unless there was a potential public health problem in the receiving stream just below the effluent discharge pipe from the wastewater treatment plant. As a net result, very few WWTP had disinfection requirements. The development of activated sludge WWTP resulted in improved effluent quality and a greater reduction in pathogenic bacteria than in trickling filter WWTP. The number of cases of enteric disease, related to polluted water, decreased in the industrial areas of the world with the expanded use of activated sludge and other wastewater treatment processes Water treatment and wastewater treatment demonstrated their value in reducing the spread of enteric pathogens through water.

In recent years the Federal EPA has responded to public concerns over potential enteric pathogens in the treated effluents from municipal wastewater treatment plants by requiring the addition of disinfection facilities in many plants when their NPDES permits were renewed. It is expected that all wastewater treatment plants in the United States will eventually have effluent disinfection facilities. The widespread use of chlorination in water treatment plants made chlorination the disinfectant of choice in wastewater treatment plants. It was not long before the regulatory agencies began to require dechlorination after chlorination in wastewater effluents as a result of concerns raised by aquatic biologists as to the adverse effect that chlorine residuals might have on the aquatic life in the receiving streams. The complexity of the chlorination-dechlorination system stimulated research on the use of ultra-violet (UV) light for disinfection of WWTP effluents. UV light disinfects by damaging the bacteria cells and preventing bacteria replication. Since suspended solids absorb UV light, UV disinfection works best on very clear effluents with very few bacteria. Plant scale research is currently being conducted to determine the life of the UV bulbs and the light energy required for good disinfection under real world conditions. One of the positive aspects of UV disinfection is the lack of a residual in the receiving water. Concerns have been raised about potential side reactions created between the stabilized organic compounds in solution and UV light. If the chemical side reactions are minor, UV disinfection may well become required in all wastewater treatment plants until some other new method for disinfection of wastewater effluent is developed.

INDUSTRIAL WASTEWATER TREATMENT

Industrial waste treatment systems are as highly varied as the wastewaters themselves. There are no uniform criteria for the design of industrial WWTP as there are for the design of municipal WWTP. Each industrial plant has its own specific wastewater characteristics that must be carefully evaluated before designing the wastewater treatment facilities. While industries that manufacture the same products produce similar wastewaters, each plant will still have its own wastewater characteristics. All industrial plants produce a domestic wastewater stream that can be discharged to the municipal sewerage system, can be treated the same as municipal wastewaters, or can be combined with the industrial process wastewater for treatment. The problem with industrial wastewater lies with the process wastewaters that are generated during product manufacturing. It starts with the raw materials that are spilled, leaked or drained into the process sewers. Intermediates are often manufactured as the products move from the raw materials to the finished goods. Intermediates are spilled, leaked, and washed into the process sewers. Even the final products can be spilled, leaked, or washed into the product sewer before being packaged and shipped from the manufacturing plant. There are far too many factors involved in the different industrial processes for duplicate industrial plants to produce identical wastewaters.

Industrial process wastewaters, containing mostly biodegradable organic compounds, are often discharged to a municipal sewer for treatment in a municipal WWTP. Process wastewaters, deficient in N, P, or trace metals, may have to be supplemented with the deficient elements before being discharged to the municipal sewer. The combined industrial-municipal WWTP requires a balanced nutrient mixture at the municipal wastewater treatment plant for proper biotreatment. If the process wastewaters have a high pH or a low pH, it will be necessary to adjust the industrial wastewater pH between 6 and 9 before discharging the industrial wastewater to the municipal sewer. High concentrations of insoluble oils and greases will have to be removed by pretreatment before discharge to the municipal sewer. Toxic chemicals will have to be reduced below the toxic limits before being discharged to the municipal sewer.

The regulatory requirements for industrial pretreatment before discharge to the municipal sewer have forced many industrial plants to look at constructing and operating their own process WWTP. The tendency for municipal officials to blame municipal wastewater treatment problems on industrial waste discharges to the municipal sewers also has forced many industrial plants to build their own treatment plants within the municipal sewer district. Industrial plants that are located outside of the municipal sewer service area are required to obtain their own NPDES permit and to have their own WWTP. The regulatory agencies set the effluent requirements and the required monitoring schedule for each of the separate industrial WWTP. Each industrial plant is responsible for designing, constructing and operating their own WWTP to meet the required effluent criteria and to properly process the residual solids for return to the land environment. Many of the industrial WWTP utilize the same treatment processes as municipal WWTP. The important thing with industrial wastewater treatment design is recognizing how to modify standard treatment processes to fit specific industrial wastewater criteria and still meet the effluent criteria.

AEROBIC BIOTREATMENT

Aerobic biotreatment for industrial wastewaters is best suited to dilute organic wastewater; but has been used for small volumes of strong industrial wastewater. Trickling filters, activated sludge, stabilization ponds, and aerated lagoons have all been part of treating industrial wastewater. Industrial wastewater treatment systems are often constructed as simple as possible because of the possible need to move the wastewater treatment system as the industrial plant expands or as the treatment process needs to be replaced. It is important to recognize that the funding of industrial WWTP is quite different than funding for municipal WWTP. An industrial WWTP has a lower capital cost and a greater operating cost than the equivalent municipal WWTP. Industries try to finance the capital costs for wastewater treatment over a 5 year period while municipalities use general obligation bonds over a 20 year period. This difference in economics results in some differences in the design and operation of industrial WWTP when compared to municipal WWTP design and operation. Fundamentally, the industrial biotreatment systems utilize the same microbiological concepts as municipal biotreatment systems. The key to good industrial WWTP design is the proper application of the biological concepts to the specific wastewater characteristics.

Plastic Media Trickling Filters

Plastic media was developed in 1954 by the Dow Chemical Co. to replace rock media in trickling filters for industrial wastewater treatment. Although E. H. Bryan at Dow Chemical Co. did extensive research on different types of industrial wastewater to demonstrate the value of plastic media, Dow Chemical Co. chose not to market the plastic trickling media. Dow Chemical Co. decided to market the plastic to companies that would market the plastic trickling filter media. Over the years many different types of plastic media were made available for trickling filters. Efforts were directed to maximizing the surface area to volume ratio. It was believed that the greater surface area provided greater microbial growth for the organic stabilization.

Plastic media allowed the construction of tall trickling filters, 3 – 12 m (10 – 40 ft) rather than shallow depth rock trickling filters. The tall, plastic media trickling filter allowed the use of a smaller base to support the media than that required for rock trickling filters. Plastic media trickling filters are designed on the basis of organic loading rates and hydraulic loading rates. Balancing the organic loading rates and the hydraulic loading rates requires adjustment of the effluent recirculation rate to dilute the organic concentration applied for the desired hydraulic flow rate. The high hydraulic flow rates in plastic media trickling filters produces a high shear rate next to the plastic media, keeping the microbial layer thin compared to the rock media trickling filter. Plastic media trickling filters have about 95% void spaces, allowing the air to move easily through the trickling filter. The thin film of water passing over the plastic media, allows rapid oxygen transfer from the air to the water. The oxygen transfer within the plastic media trickling filter can be measured by the difference between the COD of the influent to the filter and the COD of the effluent from the filter, once equilibrium conditions have been reached.

Plastic media trickling filters operate as high synthesis systems. The soluble biodegradable organic matter in the wastewater is quickly metabolized to bacteria cell mass. The suspended solids in the wastewater will largely pass through plastic media trickling filters without change. While some suspended solids will be adsorbed onto the biomass on the plastic media, the hydraulic shear forces will remove the suspended solids before significant metabolism can occur. The microbial solids produced by metabolism and the suspended solids passing through the trickling filters will have to be removed in the final sedimentation tanks. The settled sludge solids removed by the final sedimentation tanks will have to be stabilized further before being placed onto the land. The sludge solids can be treated either aerobically or anaerobically.

It is possible to place a short term aeration tanks between the plastic media trickling filters and the final sedimentation tanks. Recycling settled sludge from the final sedimentation tanks back to the aeration tanks will allow activated sludge to buildup in the aeration tank to complete metabolism of the biodegradable organic matter in the wastewater. Recycling settled sludge over the top of the plastic media trickling filter allows active microbial solids to metabolize the organic matter in the liquid phase, as well as, in the attached phase. Plastic media trickling filters ahead of aeration tanks are often called *roughing filters*. It is not surprising that the use of plastic media trickling filters have been limited in recent years with the higher effluent requirements in the United States.

Activated Sludge

Activated sludge systems can be used to treat biodegradable industrial wastewaters with the production of a high quality effluent for return to the environment. Because many industrial wastewaters contain unusual organic compounds, it is essential to use an acclimated microbial seed for accurate determination of the BOD_5 data. Improper BOD_5 data on the raw wastewater characteristics will result in a poor treatment plant design. Variations in wastewater characteristics are also critical in industrial wastewater systems. Biotreatment plants cannot produce a uniform quality effluent with a highly variable wastewater influent. Industrial plants that produce batch dumps or periodic cleanups should give major consideration to the construction of a suitable surge tank ahead of the activated sludge units. If the wastewater BOD_5 is relatively uniform and the flow is the major variable, a simple holding tank will allow the wastewaters to be discharged to the WWTP at a constant flow rate. If the BOD_5 is highly variable, a well mixed surge tank will be needed to level out the major organic variations. An industrial plant that operates on a single shift each day and a 5 day week will need a surge tank of sufficient size to allow the WWTP to receive wastewater 24 hrs a day, 7 days a week. If the waste organic materials are highly putrescible, the surge tank will have to be aerated to keep odors under control. Surge tanks will not completely eliminate the variations in wastewater characteristics; but surge tanks will greatly reduce the variations on the treatment tanks that follow.

The treatment of high BOD_5 industrial wastewaters can create problems in the design of activated sludge systems. Experience has shown that completely mixed activated sludge (CMAS) systems work best for high BOD_5 wastewaters. By using the entire volume of the aeration tank for dilution, the high BOD_5 in the influent wastewaters is quickly reduced to a low BOD_5 level that does not exceed the oxygen transfer capability in the aeration tank. In order to produce the same effluent quality with the high BOD_5 industrial wastewaters, it will be necessary to use aeration tanks with longer raw wastewater hydraulic retention periods than are used in municipal WWTP design. The same basic concepts work for high BOD_5 industrial wastewaters as for municipal wastewater. About one-third of the BCOD metabolized will be oxidized for energy for bacteria synthesis. The living microorganisms will undergo endogenous respiration at about 2%/hr, based on the living microbial mass in the aeration tank. Nitrification can create an additional oxygen demand in the aeration tank. The total oxygen demand will be the sum of the synthesis oxygen demand, endogenous respiration, and nitrification.

Shock organic loadings from sudden spills within the industrial plant create the largest problems for industrial WWTP operators. Surge tanks can help absorb the impact of spills to a limited extent. It is important to recognize that the spills add a definite quantity of wastes to the normal daily load that the WWTP must handle. There will be an immediate microbial response to the additional organic load. The oxygen resources will shift from normal operations to meeting the additional load. There may well be a reduction in nitrification until the organic load from the spill has been absorbed. If the organic load is large enough, nitrification will cease and endogenous respiration will be reduced. The bacteria synthesis reactions will predominate. As long as there is sufficient oxygen transfer available to meet the synthesis oxygen demand, the soluble BCOD in the treated effluent will only rise in proportion to the increased organic load. As soon as the organic matter from the spill has been metabolized, the effluent soluble BCOD level will return back to normal. The real problem from spills occurs when the demand for oxygen exceeds the oxygen transfer rate. The unmetabolized soluble organic matter will pass through the aeration tank and out in the final effluent, creating a violation in the industrial plant NPDES permit. The seriousness of NPDES permit violations for industrial plants cannot be over emphasized. Plant management must recognize that the industrial WWTP operations will determine if the industrial plant will be allowed to continue to operate. Like it or not, the industrial WWTP operations will control how the industrial plant is allowed to operate in the United States.

Successful absorption of shock organic loads by the industrial WWTP means a sudden increase in suspended solids in the total system and the need to properly process the extra suspended solids load. If the final sedimentation tanks can absorb the extra suspended solids, the effluent TSS will not change significantly. If the final sedimentation tank cannot absorb the extra suspended solids, the final effluent will exceed the NPDES permit and create a violation. The increase in bacteria solids in the aeration tank will be followed by an increase in suspended solids accumulating in the sedimentation tank. The return sludge (RAS) from the final sedimentation tank to the aeration tank will determine the ability of the suspended solids to move through the final sedimentation tank. Normally, a definite increase in waste sludge (WAS) will be needed to remove the excess suspended solids from the system.

Industrial wastewaters containing high concentrations of suspended solids create special problems for activated sludge. If the suspended solids are largely biodegradable, sufficient aeration time should be provided to allow the biodegradable suspended solids to be metabolized to bacteria solids. Non-biodegradable suspended solids should be reduced to a minimum before the wastewaters are added to the activated sludge system. A gravity sedimentation tank or an air flotation tank can be useful in removing most of the non-biodegradable suspended solids. The inert suspended solids accumulate in the activated sludge in proportion to the sludge age in the treatment system. High concentrations of inert suspended solids in the industrial wastewaters fill up the aeration tank and the sedimentation tank with accumulated inert suspended solids and make it difficult for the bacteria to accumulate to the desired concentrations. Normally, it is considered good practice to remove the suspended solids prior to activated sludge treatment and to process the concentrated sludge solids separately.

Temperature is an important factor in the treatment of industrial wastewaters. Many industrial wastewaters are warm and some are quite hot. The rate of bacteria metabolism increases as the aeration tank temperature increases. In the mesophilic temperature range, 5°C to 40°C, the rate of metabolism doubles for each 10°C temperature increase. The rate of metabolism drops in half for each 10°C temperature decrease. As the temperature increases above 40°C, the mesophilic bacteria begin to die at a rapid rate and are replaced by thermophilic bacteria that can live up to 70°C. Very little is known about thermophilic activated sludge except that the normal protozoa die off at temperatures much above 40°C. It appears that the microorganisms do not like to be in the transition zone between mesophilic operations and thermophilic operations. It is very difficult to maintain stable operations if the aeration temperatures slide back and forth between 38°C and

45°C. Oxygen transfer rates increase with increasing temperature; but the DO saturation values decrease. Since the effluent quality deteriorates without an active protozoa population, efforts are normally made to keep the aeration tank temperature below 40°C by removing the excess heat with heat exchangers prior to treatment.

With the development of improved aeration equipment, efforts have been made to reduce the size of activated sludge systems. Increased oxygen transfer rates allow higher organic loading rates in the aeration tank. Unfortunately, high organic loading rates produce poorer quality effluents. It is possible to design the first aeration tank as a high bacteria synthesis tank, operating at a high oxygen demand rate to meet the high oxygen transfer rate. A second aeration tank is placed in series with the synthesis aeration tank and operated at a low oxygen transfer rate to permit endogenous respiration with growth of protozoa and higher animals and the production of a high quality effluent. Placing more than two aeration tanks in series does not appear to offer any advantages.

High nitrogen wastewaters lend themselves to nitrification-denitrification for both nitrogen removal and more effective use of oxygen resources. The concepts of nitrification-denitrification systems for industrial wastewaters are the same as for municipal wastewater.

Sludge processing is the most difficult aspect of industrial wastewater treatment. There is no simple solution for the excess sludge. If an industrial plant has large furnaces, the sludge can be dewatered and burned in the furnaces. The furnace gas discharges will have to meet the Federal EPA air pollution criteria. If land is readily available, the sludge can be dewatered and placed in a sanitary landfill or plowed into the surface of the soil for further treatment. Sludge that is plowed into the surface of the soil is stabilized much faster than sludge placed into a sanitary landfill. It is possible to place the sludge in deep lagoons to allow the solids to separate and concentrate over time. Eventually, solids will have to be removed from the lagoon for the operation to continue. If sludge is to be placed into a deep lagoon, it is important to stabilize the sludge aerobically before it is placed in the lagoon to prevent the production of nuisance odors. One of the hardest lessons for industrial plant managers to learn is that they started with a liquid waste and end up with having to return not only the treated wastewaters to the environment but also have to treat the microbial solids and return them back to the environment as a solid waste.

Stabilization Ponds

Stabilization ponds have been widely used for treating large volume, dilute, organic wastewaters. The design criteria for industrial wastewater stabilization ponds are the same as for municipal wastewater stabilization ponds. Where suspended solids are a primary factor in the wastewater characteristics, industrial wastewater stabilization ponds tend to be deeper than municipal wastewater stabilization ponds. The additional pond depth is used to store the suspended solids that settle out and concentrate on the pond bottom. Eventually, the concentrated solids will have to be removed and placed on the land or the pond will have to be closed for further use. Multi-cell stabilization ponds are quite common as new cells have been added to expand existing systems. It is not uncommon for industrial stabilization pond systems to have 10 or more cells in various combinations of series and parallel cells. Industrial stabilization pond systems grow to fit the existing terrain. One of the major differences in the design criteria between industrial ponds and municipal ponds is sizing the ponds for future wastewater loads. Municipal stabilization ponds are designed on the basis of 10 to 20 years future loads. The variability of industrial production and the rapidly changing technology makes it impossible to predict future wastewater production. Industrial economics limits the construction of stabilization ponds to the maximum wastewater loads for the current industrial plant size. Periodic changes in industrial production will produce definite changes in wastewater production rates and characteristics that will not overload the stabilization pond system. Major increases in the industrial plant production will require additional plant facilities, allowing time to plan for stabilization pond expansion. Industrial plants are able to expand their wastewater treatment facilities faster than municipalities, allowing industries to operate closer to existing wastewater production.

Some industrial wastewaters are deficient in nitrogen, phosphorus, and/or trace metals that the bacteria need for their metabolism of the waste organic materials. The nutrient deficient wastewaters will need the addition of sufficient nutrients to meet the microbial needs. The long wastewater retention periods in stabilization ponds will allow recycling of nutrients within the pond system, minimizing the amount of nutrients required to be added. The microbial populations in industrial stabilization ponds are more varied than in municipal stabilization ponds. The predominant bacteria will depend on the industrial chemicals in the wastewaters. The bacteria best able to metabolize the industrial organic compounds will predominate initially. As the acclimated bacteria grow and die, a secondary group of bacteria will grow in the industrial stabilization ponds, the same as in municipal stabilization ponds. The protozoa, rotifers, and higher animal life will develop the same as in municipal stabilization ponds unless toxic compounds are in the industrial wastewaters. The higher animals are more sensitive to toxic compounds than either bacteria or protozoa. It may well be that toxic organic compounds will be metabolized in the initial cells by acclimated bacteria, allowing the higher animals to develop in later cells within the stabilization pond system. Regular microscopic examination of the water in each cell will provide basic information on microbial activity that can be confirmed from the chemical analyses. Algae may or may not grow in industrial stabilization ponds. The chemical characteristics of the wastewater in the stabilization ponds will determine the extent of algae growth and photosynthetic bacteria. Highly colored wastewaters may prevent or reduce the growth of algae by absorption of the light energy. Excessive turbidity remaining in suspension in the stabilization pond cells will also absorb light energy and prevent or reduce algae in industrial stabilization ponds.

Aerated Lagoons

Industrial aerated lagoons developed as an alternative to constructing more cells for existing stabilization ponds. As industrial loads increased and obnoxious odor production also increased, the need for a simple source of oxygen became obvious. Floating surface aerators were proposed by various equipment manufacturers as the best method to solve the lack of DO and the odor problems. The low cost, floating surface aerators fit industrial economics perfectly. The capital cost was low and the operating cost was part of the normal industrial plant operations. Unfortunately, the small, floating surface aerators were not very efficient in the initial large cell of the industrial stabilization pond system. Each aerator had too large a volume to mix and to aerate. The solution became the addition of more aerators at regular intervals. The bacteria stabilized the wastewater organic compounds around each aerator and helped flocculate the suspended solids. The suspended solids tended to settle out in the spaces just beyond the mixing zone of the aerators. Some of the wastewater passed through the aerated lagoon portion of the stabilization pond without significant biological treatment.

The best solution to aeration of industrial wastewaters in stabilization ponds appears to be the construction of a completely mixed aeration cell ahead of the stabilization pond. The aeration cell needs to be sized on the basis of the wastewater characteristics to allow for complete metabolism of the biodegradable organic compounds in the industrial wastewaters. Best results will be obtained by providing adequate oxygen for synthesis plus endogenous respiration for the retention time in the aeration cell. Balancing the oxygen transfer and mixing characteristics of the aeration equipment in the aerated volume will determine the success of the aerated lagoon part of the industrial wastewater treatment system. The microbial solids produced in the aeration cell will have to be removed in the stabilization ponds that follow the aerated lagoon. The primary purpose of aerated lagoons is to convert the wastewater organic matter to a new chemical form that is easier to remove and is less hazardous to the environment.

ANAEROBIC BIOTREATMENT

Anaerobic biotreatment of industrial wastewaters has been used for years with limited success. In the past few decades research has yielded a better understanding of the anaerobic processes and has produced a new form of biotreatment that is gaining popularity for processing concentrated, biodegradable wastewaters. More research will be carried out over the next few decades on the design and operation of various anaerobic treatment systems to improve treatment plant designs. The key to anaerobic biotreatment systems lies in understanding the microbiology and biochemistry of the anaerobic bacteria together with the engineering of the bioreactors. Overall, the microbiology of anaerobic biotreatment systems is the same as the anaerobic microbiology of municipal sludge digestion. Facultative bacteria metabolize the complex organic compounds to short chain fatty acids that are metabolized by methane forming bacteria to methane gas, carbon dioxide, and water. Some of the organic compounds will be metabolized by sulfate reducing bacteria, depending on the concentration of sulfates in the industrial wastewaters. The metabolism by the sulfate reducing bacteria will reduce the organic matter for the methane bacteria to metabolize. The facultative bacteria can also reduce the organic matter available for methane bacteria if the industrial wastewaters contain nitrates or nitrites. The presence of sulfates in the wastewaters will stimulate sulfate reducing bacteria. Each group of bacteria must work together in the anaerobic environment to produce the desired treatment results.

Anaerobic Lagoons

Anaerobic lagoons are the simplest form of anaerobic biotreatment to achieve widespread use in the United States. The primary use of anaerobic lagoons has been with industrial wastewaters containing high concentrations of biodegradable suspended solids. Over the years the anaerobic lagoon has been changed from a simple hole in the ground to an engineered hole in the ground with a plastic liner and a plastic cover with a gas collection line. The plastic liner is designed to prevent leakage into ground water. Where the ground water table is close to the surface, regulatory agencies may require two liners with a layer of sand and liquid collection pipes between the two liners. The plastic cover is designed to prevent the discharge of obnoxious odors into the atmosphere and to collect the methane gas for reuse as an energy source. The liquid volumes of anaerobic lagoons range from about 3 days to 300 or more days wastewater retention. The short retention time anaerobic lagoons require mechanical mixing and removal of the excess suspended solids in the treated effluent. The larger anaerobic lagoons provide sludge storage for several months and trade mechanical mixing for mechanical mixing.

Anaerobic lagoons can remove 60% to 90% or more of the biodegradable organic compounds in wastewater. Since the BCOD removed in anaerobic systems is converted to methane gas and new microbial cell mass, measurement of the methane gas generated on a daily basis is the best indicator of organic matter removed through the anaerobic lagoon. Even with 90% BCOD removal from the industrial wastewaters, the treated effluent contains considerable organic compounds that must be treated aerobically to produce a high quality effluent. Essentially, the anaerobic lagoon is just the first stage of a complete treatment system. Another item of concern is the inert suspended solids in the wastewaters. The inert suspended solids either accumulate at the bottom of the anaerobic lagoon or are discharged in the lagoon effluent. As the concentrated suspended solids accumulate on the bottom of the anaerobic lagoon, the active lagoon volume is reduced proportionally. Eventually, sludge will have to be removed to keep the anaerobic lagoon in operation. Grease and oils in the industrial wastewaters can also create problems. The methane gas can lift particles containing grease to the lagoon surface. Since only the water-grease interface provides a proper environment for bacteria metabolism of the organic grease materials, metabolism is slow. Provision should be made to keep floating materials to a minimum for good operations. The small fraction of BCOD converted to microbial cells for most industrial wastewaters will either become part of the settled solids or will be discharged in the anaerobic effluent. The simplicity of anaerobic lagoons will continue their use for decades to come for specific industrial wastes.

High Rate Anaerobic Treatment

Industrial wastewaters having a high daily flow rate and a high BOD_5 concentration, greater than 1,000 mg/L BOD_5, were not easily treated aerobically. The success of anaerobic digestion in treating concentrated municipal sludge stimulated researchers to investigate the suitability of anaerobic digestion for concentrated industrial wastewaters. Researchers for the meat packing plants in Minnesota sought to use the concepts of anaerobic sludge digestion with the concepts of activated sludge by settling the anaerobic sludge in a final sedimentation tank and returning much of the settled sludge to the anaerobic digester. The success of pilot plant studies in the meat packing industry in the 1950s led to the construction and operation of full-scale anaerobic treatment plants. By the end of the 1950s the *anaerobic contact* process had been placed into full operation in Minnesota. While the anaerobic contact process for treating meat packing wastewaters proved that high rate anaerobic industrial wastewater treatment was possible, it was too complex for most industrial plants. The desire for better anaerobic treatment systems also stimulated research at MIT in the 1950s on the biochemistry of anaerobic treatment. It was believed that better anaerobic wastewater treatment systems would be developed only when engineers had a better understanding of the fundamental mechanisms that were responsible for determining the limits of the anaerobic bacteria in treating industrial wastewaters. Much remained to be learned before high rate anaerobic treatment was to become a reality for industrial wastewater treatment.

R. R. Dague *et al* published a report in 1966 on their research at the University of Kansas dealing with the flocculation and settling characteristics of anaerobic bacteria fed a synthetic wastewater. It was found that when most of the organic nutrients had been metabolized, the anaerobic bacteria flocculated in the same fashion as aerobic bacteria in activated sludge. When the anaerobic floc settled, the supernatant was not as clear as with aerobic floc. Microscopic examination of the anaerobic floc particles showed some flagellated protozoa and some free-swimming ciliated protozoa. The anaerobic floc had less protozoa than aerobic floc found in activated sludge systems. It quickly became apparent that the numbers and types of protozoa were important in determining the clarity of the supernatant in anaerobic systems the same as in aerobic systems.

In 1962 P. L. McCarty moved his anaerobic research from MIT to Stanford and established himself as one of the leading researchers on anaerobic treatment in the United States. In 1967 J. C. Young and P. L. McCarty published the results of their study on the use of a submerged rock filter to treat wastewaters containing concentrated soluble organic contaminants. Their research results demonstrated that a submerged rock filter could accumulate sufficient numbers of bacteria to metabolize pure soluble organic compounds at a high organic loading rate and a short hydraulic retention time. The simulated wastewater moved upward through the rock filter, producing an effluent low in suspended solids and BOD_5. A 1969 report by Young and McCarty indicated that the microbial growth flocculated and accumulated in the spaces between the rocks and became granular over time. Microscopic examination of the microbial masses within the rock filter showed some free-swimming ciliated protozoa and some amoeba.

Young and McCarty's research stimulated studies on specific wastewater treatment by anaerobic filters. In spite of the success with laboratory studies, industries were not ready to invest in a full-scale anaerobic plant. The anaerobic treatment picture changed when Schonheit *et al* published the results of their research in 1979, showing that Ni, Co, and Mo were required trace metals for the growth of methane bacteria. The requirement for Ni by the methane bacteria surprised most researchers, since Ni was considered to be a toxic heavy metal. The addition of Ni to anaerobic units allowed anaerobic treatment systems to attain the microbial stability that had been missing in previous long-term anaerobic studies. Most anaerobic units treating pure organic compounds began with a large seed of anaerobic sludge from a municipal sludge digester. As the anaerobic seed was washed out, the treatment efficiency would slowly decline over time. If additional seed sludge was not added to the anaerobic reactor, the treatment efficiency would continue to decline. The treatment efficiency would return to normal with the addition of a fresh sample of digesting municipal sludge. It appeared that the digesting municipal sludge provided the anaerobic bacteria with the necessary trace metals for good growth. While iron had long been recognized as an important trace metal in anaerobic bacteria metabolism, the need for Ni as an essential trace metal had escaped everyone.

This information on trace elements came as G. Lettinga at the Agricultural University at Wageningen in the Netherlands was developing an upflow anaerobic sludge blanket (UASB) reactor for the treatment of beet sugar wastewater. A critical part of the UASB reactor was the formation of dense, microbial granules that allowed the development of the upflow blanket and the retention of the microbial granules in the anaerobic reactor. The UASB eliminated the need for rocks to retain the anaerobic bacteria in the treatment system. The UASB reactors became the most widely used high-rate anaerobic treatment system throughout the world.

Further research on anaerobic systems quickly followed with the development of numerous types of anaerobic reactors. Pilot plants led to full-scale plants for treating different industrial wastewaters. By 1990 high-rate anaerobic treatment was the accepted method for treating concentrated organic wastewaters having a minimum of inert suspended solids. The problem with inert suspended solids is their accumulation with the active bacteria. Too many inert suspended solids will bloc the normal fluid flow through the bioreactor and will result in less stabilization of the incoming organic waste materials. The low synthesis mass produced by the anaerobic bacteria can be quickly overwhelmed by a high concentration of inert suspended solids entering the bioreactor. In 1996 R. E. Speece of Vanderbilt University published an excellent book summarizing much of the published research information on high rate anaerobic treatment of industrial wastewaters.

The microbiology of high rate anaerobic treatment is similar to that found in municipal anaerobic sludge digesters. Various facultative bacteria from the soil are responsible for the metabolism of complex organic compounds to simple organic acids that must be neutralized with alkalinity to maintain a pH above 6.5. The sulfate-reducing bacteria produce sulfides from sulfates while metabolizing the short chain fatty acids produced by the facultative bacteria. The sulfate-reducing bacteria create a strongly reduced environment that stimulates the growth of the methane bacteria. The sulfides produced by the sulfate reducing bacteria can react with heavy metals in the industrial wastewaters and produce metal-sulfide precipitates that prevent heavy metal toxicity to the bacteria. The hydrogen-metabolizing methane bacteria compete with the acetogenic bacteria for carbon dioxide and hydrogen with both groups of bacteria showing some growth. The acetate-utilizing methane bacteria convert acetate to methane and bicarbonate alkalinity. Since the acetate-utilizing methane bacteria obtain very little energy from their metabolism of acetate, this group of bacteria becomes the key group in the overall metabolic reactions. If the acetate-utilizing methane bacteria fail to metabolize the acetate, acetate accumulates in the system and methane production is reduced proportionally. When the methane bacteria metabolize acetate normally, methane, carbon dioxide, and new bacteria are the end products. The poor solubility of methane gas in water results in the loss of methane with its COD from the liquid phase. Carbon dioxide gas and hydrogen sulfide gas are also partially stripped from solution with the methane gas.

The energy contained in the organic matter fed to anaerobic bioreactors is converted to energy in the methane gas, the hydrogen sulfide gas, new cell mass, and partially degraded organic intermediates. Since the majority of energy will be transferred to methane gas, continuous measurement of the methane gas production is the best parameter to use in determining the microbial health of high rate anaerobic bioreactors. A decrease in gas production for the same organic load usually means problems with the acetate utilizing methane bacteria. A check of volatile acids can be used to confirm that the acetic acid concentration is increasing. Next, checks of the pH and the alkalinity should be made to insure that the environment is satisfactory for these methane bacteria. Trace metal deficiencies can also adversely affect the methane bacteria. It is essential that the cause of the methane gas decrease be established and corrected or the anaerobic treatment system will continue to decline in treatment efficiency until it fails completely. When properly treated, the methane bacteria will always produce the best results possible.

Rapid growth of the methane bacteria results in the production of considerable gas and the generation of turbulence in the bioreactor as the gas bubbles rise through the liquid to the surface. The induced turbulence helps to produce floc particles the same as occur in activated sludge systems. The floc particles have a definite surface area/mass ratio that affects the floc particles reaction to rising gas bubbles. When the floc particles become heavy enough, they will form a blanket of floc particles in the anaerobic reactor, while the dispersed bacteria will be discharged in the treated effluent unless they are eaten first by the protozoa. Over time, the floc particles will increase in density as more bacteria are incorporated into the floc and the floc particles will become granules. Granule formation is a natural progression of mixed, anaerobic growth in stable systems. Iron sulfide and other inorganic precipitates contribute to the mass of the granules.

Distribution of incoming wastewaters is a real problem in field scale treatment plants. Most anaerobic bioreactors have been designed for upflow operations. The major problem with upflow feeding is in the wastewater distribution. It is very difficult to achieve uniform distribution in large diameter tanks. Collection of the treated effluent at the surface of the upflow bioreactor will also affect the hydraulic flow through the anaerobic bioreactor. Upflow distribution works best in small diameter, tall bioreactors. Good downflow distribution of incoming wastewaters can be achieved with several sprays above the liquid. The use of recirculation pumps in both upflow and downflow systems can be helpful in mixing the contents of the bioreactor to provide a more uniform dispersion of the bacteria and the wastewaters being treated.

High rate anaerobic bacteria are not faced with an oxygen limitation that aerobic bacteria are faced with. The primary limitations in anaerobic systems are the bacterial population, the ability of the different groups of bacteria to work together for successful metabolism of the organic, and the environment in the anaerobic bioreactor. The production of organic acids from complex neutral compounds, such as carbohydrates, alcohols, aldehydes, ketones, and complex hydrocarbons, requires sufficient alkalinity to neutralize the organic acids as fast as they are formed. High carbon dioxide concentrations in the anaerobic liquid will also create a demand for strong alkalinity to produce bicarbonate alkalinity. Care must be taken with the addition of strong alkalinity to the anaerobic system. Direct addition of strong alkalinity to the anaerobic bacteria will result in killing the bacteria by the high pH or the high salt concentration of the strong alkalinity solution before it is dispersed into the system. As the bacteria metabolize the neutralized acetic acid, alkalinity is reformed. Unfortunately, the alkalinity will be lost in treated effluent, requiring that fresh alkalinity be added continuously. Failure to add sufficient alkalinity will result in failure of the system to provide maximum treatment. Industrial wastewaters that contain organic nitrogen will produce ammonium bicarbonate alkalinity to meet part or all of the alkalinity requirements. Since alkalinity and carbon dioxide combine to affect the pH of the liquid in the bioreactor, the pH must be maintained between 6.5 and 8.5 for optimum growth of the mixture of anaerobic bacteria required for high rate anaerobic treatment.

The current loading rates for BCOD in high rate anaerobic bioreactors range from 10 g BCOD/d/L tank volume to 50 g BCOD/d/L tank volume. The effluent quality from the anaerobic bioreactors will show a 98% - 99% BCOD reduction at the lower loading rate. The 10 g BCOD/d/L tank volume loading rate appears to be a reasonable loading rate for normal operations of high rate anaerobic bioreactors. The higher loading rates will produce 90% BCOD metabolism and require more careful operational control that the lower loaded tanks to keep the system working properly. The BCOD concentration in the industrial wastewaters will determine the hydraulic retention time (HRT) within the anaerobic bioreactor. Industrial wastewaters with a high BCOD concentration are easier to treat than dilute BCOD wastewater. The high BCOD concentration wastewaters will have a long HRT in the anaerobic bioreactor. A long HRT in the anaerobic bioreactor will provide sufficient time for the bacteria to metabolize the BCOD. Dilute BCOD wastewaters produce short HRT values and tend to wash the bacteria out of the anaerobic bioreactor before all of the BCOD has been metabolized.

An industrial wastewater containing 1,000 mg/L BCOD would have a HRT of 0.1 day at a 10 g BCOD/d/L tank volume loading rate. This system would require careful operational control to maintain a suitable microbial population. A small increase in wastewater flow could create a serious problem from loss of the bacteria in the treated effluent. Variations in hydraulic loading rates are as important as organic loading rates in controlling high rate anaerobic bioreactors. Relatively uniform organic loading rates and flow rates will produce the best operating results. Short HRT periods will be possible only if the bacteria are in granules or are attached to fixed surfaces, providing the bacteria with sufficient solids retention time (SRT) to maintain satisfactory growth within the anaerobic bioreactor. The biggest problem for anaerobic systems is during startup. It is essential that sufficient time is allowed for the bacteria to adapt to the various organic compounds in the industrial wastewater and to maximize bacteria growth. Batch feeding for one or two displacement periods may be needed to produce sufficient bacteria growth to allow a slow, continuous feed rate. Anaerobic systems will normally require several months to a year to accumulate sufficient bacteria mass to produce the best effluent quality. Operators skilled with the rapid changes that occur with activated sludge systems will be totally frustrated by the slow pace of change in anaerobic systems. The microbial mass in anaerobic systems cannot change as rapidly as the microbial mass in activated sludge systems. Patience is required in the startup and continued operation of successful anaerobic systems.

Trace metals play an important role in anaerobic systems. Iron (Fe) appears in almost all energy transfer enzymes within the bacteria, making it the most important trace metal. Nickel (Ni) is the next most important trace metal, followed by cobalt (Co), zinc (Zn), molybdenum (Mo), and copper (Cu). Manganese (Mn) and Selenium (Se) have also been indicated as important in anaerobic systems. Industrial wastewaters with high sulfates create special problems for the trace metals. The sulfate reducing bacteria will compete with the methane bacteria and reduce the amount of methane gas in proportion to the sulfates reduced. The sulfides produced as end products of the sulfate reducing bacteria will precipitate the trace metals. It appears that the bacteria are able to find sufficient trace metals in spite of the insolubility of the metal sulfides. Industrial wastewaters should contain at least 1 mg/L Fe and 0.1 mg/L of the other important trace metals. If the wastewaters do not contain sufficient trace metals, it will be necessary to add them continuously to the wastewaters prior to treatment to obtain proper growth of the desired bacteria.

Temperature also has an impact on the high rate anaerobic bioreactors. As previously indicated, the bacteria metabolism is affected temperature. Anaerobic bioreactors can operate at *mesophilic* temperatures, 5°C to 40°C, and at *thermophilic* temperatures, 40°C to 70°C. Mesophilic bacteria cover the normal range of natural temperatures, including warm blooded animal temperatures. As the temperature rises above 40°C, the mesophilic bacteria are killed, leaving the thermophilic bacteria to survive at the higher temperatures. The thermophilic bacteria grow at mesophilic temperatures, but grow at a slower rate than the mesophilic bacteria at mesophilic temperatures. The rate of metabolism of thermophilic bacteria will increase as the temperatures increase. The faster rates of metabolism of the thermophilic bacteria make them very hard to control unless the environment is kept very constant. Slight variations in the BCOD loading rates, the HRT, or the temperature will cause serious upsets in operations. Thermophilic treatment should only be considered if the temperature of the wastewaters falls in the thermophilic range and cannot be easily reduced.

The main objective of high rate anaerobic treatment is the conversion of organic matter into CH_4 gas and microbial cell mass. Anaerobic bacteria maximize the conversion of organic matter into CH_4 gas and minimize the production of cell mass. There is no simple method for evaluating anaerobic relationships like there is for aerobic systems. The problem with anaerobic systems lies in the limited metabolism of different bacteria. The CH_4 gas production comes from two groups of methane bacteria. The acetate utilizing bacteria obtain very little energy from the metabolism of acetate and produce large quantities of CH_4 gas and small quantities of cell mass. It appears that 98% of the acetate energy, measured as COD, is metabolized to CH_4 gas with 2% of the acetate energy, also measured as COD, transferred to cell mass COD. The hydrogen utilizing methane bacteria obtain more energy from metabolizing hydrogen than the acetate utilizing bacteria metabolizing acetate. About 84% of the hydrogen COD metabolized will be converted to methane with 16% of the hydrogen COD metabolized converted to cell mass. Carbohydrates are metabolized by the facultative bacteria to acetate and hydrogen before being metabolized to CH_4 gas. The net result is that carbohydrate metabolism converts about 84% of the COD metabolized into CH_4 gas and 16% of the COD metabolized into cell mass COD. While the overall metabolism of carbohydrates and hydrogen appear similar, the cell mass production is quite different. The carbohydrate energy is split between three major groups of bacteria with the hydrogen energy being used by only one group of bacteria. Each bacteria group has many different species helping with the overall metabolism. Proteins and fats result in considerably less energy for new cells than carbohydrates. Proteins are hydrolyzed to amino acids that yield about 4% cell mass COD and 96% CH_4 when metabolized anaerobically. Aromatic ring compounds are metabolized to about 7% cell mass and 93% CH_4 The rapid rate of endogenous respiration of the bacteria at 37°C will result in the reduction in active bacteria mass and the accumulation of large amounts of dead cells in anaerobic systems having SRT periods of 10 days or greater. Any inert suspended solids in the industrial wastewater will add to the overall suspended solids accumulation in the anaerobic bioreactor. The energy-synthesis reactions in complex anaerobic systems will be a fertile area of research for years to come. The biggest problem to date has been the inability to accurately measure all the environmental parameters in high rate anaerobic systems. The new cell mass production is often masked by the mass of solids in the system. It suffices that the metabolism of most biodegradable organic compounds found in industrial wastewaters will only yield between 2% and 7% conversion of the BCOD to cell mass COD with the remaining 93% to 98% BCOD converted to CH_4 gas. Only carbohydrate containing wastewaters will show a greater cell mass COD production, about 16%. It will be interesting to watch future research in this area of anaerobic energy transformations.

Industrial wastewater treatment is the most interesting type of wastewater treatment. Industrial wastewaters show the greatest variations in different organic compounds to be treated. Each industrial wastewater is unique and needs to be carefully evaluated to produce the optimum wastewater treatment system that can be easily operated under widely varying conditions. Because industrial production is subject to the demands of the market places, industrial wastewater flows show wide swings over periods that range from a few months to several years. These factors combine to make industrial wastewater treatment the major environmental challenge for both engineers and WWTP operators.

THINGS TO REMEMBER

1. There are three different types of wastewaters: storm water, domestic wastewater, and industrial wastewater.

2. Evaluation of wastewaters starts with determination of the primary wastewater characteristics.

3. Wastewater treatment systems are designed to remove the contaminants from wastewater for safe return of both the water and the contaminants to the environment.

4. Biotreatment systems utilize natural microorganisms to remove the biodegradable contaminants in wastewaters.

5. Anaerobic biotreatment systems utilize bacteria to convert the concentrated organic contaminants to methane, carbon dioxide, and water plus more bacteria.

6. Aerobic biotreatment systems utilize bacteria to convert dilute organic contaminants to carbon dioxide, water, and more bacteria.

7. Protozoa and rotifers also grow in aerobic biotreatment systems, eating the dispersed bacteria and helping to produce a clarified effluent after suspended solids separation.

8. Preliminary treatment devices include screens and grit removal equipment.

9. Primary treatment includes gravity sedimentation with anaerobic digestion of the settled sludge.

10. Secondary treatment includes trickling filters, activated sludge, and secondary sedimentation with settled sludge being treated by aerobic or anaerobic digestion prior to discharge to the land.

11. Tertiary treatment includes all treatment units following secondary treatment.

12. Biological nutrient removal systems include nitrification-denitrification and phosphorus removal.

13. Secondary sedimentation tanks must remove at least 99% of the entering MLSS in activated sludge plants to meet current effluent criteria.

14. The concentrated suspended solids must be continuously removed from the bottom of secondary sedimentation and either returned to the aeration tank or wasted from the system.

15. The excess activated sludge generated each day must be removed from the system to maintain biological equilibrium.

16. Stabilization ponds consist of a series of shallow ponds that utilize bacteria, algae, protozoa, and higher animals to produce a suitable effluent for discharge back into the environment.

17. Aerated lagoons have been used to stabilize the excess organic load in overloaded stabilization ponds.

18. Disinfection of treated effluents is concerned with destruction of pathogenic microorganisms.

19. Chlorination and ultra violet-light are used for wastewater effluent disinfection in the United States.

20. The concentration of pathogenic microorganisms in domestic wastewater in the United States has been steadily decreasing with the increased use of biotreatment systems.

21. Aerobic biotreatment systems have been used extensively to treat dilute organic contaminants in industrial wastewaters.

22. High rate anaerobic treatment systems are currently being used to treat concentrated organic contaminants in industrial wastewaters.

REFERENCES

Adamse, A. D. (1966) *Bacteriological Studies on Dairy Waste Activated Sludge*, PhD Thesis, H. Veenman en Zonen N. V., Wageningen .

Allen, L. A. (1944) The Bacteriology of Activated Sludge, *Jour. of Hygiene*, **43**, 424.

Al-Shahwani, S. M. and Horan, N. J. (1991) The Use of Protozoa to Indicate Changes in the Performance of Activated Sludge Plants, *Wat. Research*, **25**, 633.

Ardern, E. and Lockett, W. T. (1914) Experiments on the Oxidation of Sewage Without the Aid of Filters, *Jour. Soc. Chem. Indust.*, **33**, 523.

Barak, Y. and van Rijn, J. (2000) Atypical Polyphosphate Accumulation by the Denitrifying Bacterium *Paracoccus denitrificans*, *Appl. Environ. Microbiol.*, **66**, 1209.

Barclay, D. L. (1979) *Aerobic Digestion of Waste Activated Sludge*, MS Thesis, University of Kansas.

Berg, L. V. den, Lentz, C. P. and Armstrong, D. W. (1980) Anaerobic Waste Treatment Efficiency Comparisons Between Fixed Film Reactors, Contact Digesters and Fully Mixed, Continuously Fed Digesters, *Proc. 35th Ind. Waste Conf., Purdue Univ.*, 788.

Bond, P. L., Erhart, R., Wagner, M., Keller, J., and Blackall, L. L. (1999) Identification of Some of the Major Groups of Bacteria in Efficient and Nonefficient Biological Phosphorus Removal Activated Sludge Systems, *Appl. Environ. Microbiol.*, **65**, 4077.

Boyko, B. I. (1968) *Mixing Studies on a Full-Scale Aeration Tank*, Ontario Water Resources Commission Research Publication, v. **19**

Bryan, E. H. (1955) Molded Polystyrene Media for Trickling Filters, *Proc. 10th Indust. Waste Conf., Purdue University*, 164.

Burkhead, C. E. (1966) *Energy Relationships in Aerobic Microbial Metabolism*, PhD Thesis, University of Kansas.

Buswell, A. M. and Long, H. L. (1923) Microbiology and Theory of Activated Sludge, *Activated Sludge Studies, 1920-1921*, Illinois Water Survey Bulletin, **18**, 82.

Butterfield, C. T. (1935) A Zooglea Forming Bacterium Isolated from Activated Sludge, *Pub. Health Rpts.*, **50**, 671.

Butterfield, C. T. and Wattie, E. (1941) Studies on Sewage Purification XV. Effective Bacteria in Purification of Trickling Filters, *Sew. Works Jour.*, **13**, 639.

Calaway, W. T. (1963) Nematodes in Wastewater Treatment, *Jour. Wat. Poll. Cont. Fed.*, **35**, 1006.

Calaway, W. T. (1968) The Metazoa of Waste Treatment Processes - Rotifers, *Jour. Wat. Poll. Cont. Fed.*, **40**, R412.

Cawley, W. A. (1955) *The Metabolism of Single Carbon Organic Compounds by Activated Sludge*, SM Thesis, MIT.

Crabtree, K., Boyle, W., McCoy, E. and Rohlich, G. A. (1966) Mechanism of Floc Formation by *Zoogloea ramigera*, *Jour. Wat. Poll. Cont. Fed.*, **38**, 1968.

Cramer, R. (1931) The Role of Protozoa in Activated Sludge, *Ind. & Engr. Chem.*, **23**, 309.

Crocetti, G. R., Hugenholtz, P., Bond, P. L., Schuler, A., Keller, J., Jenkins, D. and Blackall, L. (2000) Identification of Polyphosphate-Accumulating Organisms and Design of 16S rRNA-Directed Probes for Their Detection and Quantitation, *Appl. Environ. Microbiol.*, **66**, 1175.

Curds, C. R. and Cockburn, A. (1970) Protozoa in Biological Sewage Treatment Plants II. Protozoa as Indicators in the Activated Sludge Process, *Wat. Research*, **4**, 225.

Dawson, P. S. S. and Jenkins, S. H. (1949) The Oxygen Requirements of Activated Sludge Determined by Manometric Methods, *Sew. Works Jour.*, **21**, 643.

Deinema, M. H. (1972) Bacterial Flocculation and Production of Poly-β-Hydroxybutyrate, *Appl. Microbiol.*, **24**, 857.

Dias, F. F. and Bhat, J. V. (1964) Microbial Ecology of Activated Sludge, I. Dominant Bacteria, *Appl. Microbiol.*, **12**, 412.

Eikelboom, D.H. (1975) Filamentous Organisms Observed in Activated Sludge, *Wat. Research*, **9**, 365.

Engelbrecht, R. S. (1954) *Energy Relationships in the Activated Sludge Process*, PhD Thesis, MIT.

Frye, W.W. and Becker, E. R. (1929) "he Fauna of an Experimental Trickling Filter, *Sew. Works Jour.*, **1**, 286.

Fuhs, G. W. and Chen, M. (1975) Microbiological Basis of Phosphate Removal in the Activated Sludge Process for the Treatment of Wastewater, *Microb. Ecol.*, **2**, 119.

Gann, J. D., Collier, R.E. and Lawrence, C. H. (1968) Aerobic Bacteriology of Waste Stabilization Ponds, *Jour. Wat. Poll. Cont. Fed.*, **40**, 185.

Garber, W. F. (1954) Plant Scale Studies of Thermophilic Digestion at Los Angeles, *Sew. & Ind. Wastes*, **26**, 1202.

Gee, C.S., Pfeffer J. T. and Suidan, M. T. (1990) *Nitrosomonas* and *Nitrobacter* Interactions in Biological Nitrification, *Jour. Environ. Engr., ASCE*, **116**, 4.

Gray, N. F. (1989) *Biology of Wastewater Treatment*, Oxford Univ. Press, Oxford, England.

Greenberg, A. E., Klein, G., and Kaufman, W. J. (1955) Effect of Phosphorus Removal on the Activated Sludge Process, *Sew. & Ind. Waste*, **27**, 227.

Heukelekian, H. and Littman, M. L. (1939) Carbon and Nitrogen Transformations in the Purification of Sewage by the Activated Sludge Process II. Morphological and Biochemical Studies of Zoogleal Organisms, *Sew. Works Jour.*, **11**, 752.

Heukelekian, H., Orford, H. E., and Manganelli, R. (1951) Factors Affecting the Quantity of Sludge Production in the Activated Sludge Process, *Sew. & Ind. Wastes*, **23**, 945.

Holtje, R. H. (1943) The Biology of Sewage Sprinkling Filters, *Sew. Works Jour.*, **15**, 14.

Hoover, S. R. and Porges, N. (1952) Assimilation of Dairy Wastes by Activated Sludge - II. The Equation of Synthesis and Rate of Oxygen Utilization, *Sew. & Ind. Wastes*, **24**, 306.

Hotchkiss, M. (1923) *Studies on the Biology of Sewage Disposal*, New Jersey Ag. Expt. Sta., Bulletin **390**.

Jeris, J. S. (1954) *The Metabolism of Simple Organic Compounds by Activated Sludge*, SM Thesis, MIT.

Jordan, E. O. and Richards, E. H. (1890) Investigations Upon Nitrification and the Nitrifying Organism, *Report on Water Supply and Sewerage, Part II*, Mass. State Bd. of Health, 865.

Kinner, N. E. and Curds, C. R. (1987) Development of Protozoa and Metazoa Communities in Rotating Biological Contactor Biofilms, *Wat. Research*, **21**, 481.

Koelsch, R. and Shapiro, C. (1997) Estimating Manure Nutrients Production (Part 1), *Manure Matters*, University of Nebraska-Lincoln, **3**, No. 2.

Lettinga, G. and Vinken, J. N. (1980) Feasibility of the Upflow Anaerobic Sludge Blanket (UASB) Process for the Treatment of Low-Strength Wastes, *Proc. 35th Ind. Waste Conf., Purdue Univ.*, 625.

Levin, G. V. and Shapiro, J. (1965) Metabolic Uptake of Phosphorus by

Wastewater Organisms, *Jour. Wat. Poll. Cont. Fed.*, **37**, 800.

Levine, M., Burke, G. W., and Watkins, J. H. (1929) *Removal of Milk Constituents by Filtration*, Iowa Engr. Expt. Sta. Bulletin, **95**.

Lighthart, B. and Oglesby, R. T. (1969) Bacteriology of an Activated Sludge Wastewater Treatment Plant - A Guide to Methodology, *Jour. Wat. Poll. Cont. Fed.*, **41**, R267.

Lighthart, B. and Loew, G. A. (1972) Identification Key for Bacteria Clusters from an Activated Sludge Plant, *Jour. Wat. Poll. Cont. Fed.*, **44**, 2078.

Ludwig, H. F., Oswald, W. J., Gotaas, H. B., and Lynch, V. (1951) Algae Symbiosis in Oxidation Ponds I. Growth Characteristics of *Euglena gracilis* Cultured in Sewage, *Sew. & Ind. Wastes*, **23**, 1337.

McCarty, P. L. (1964) Anaerobic Waste Treatment Fundamentals - I. Chemistry and Microbiology, *Pub. Works*, **95**: 9, 107.

McCarty, P. L. (1964) Anaerobic Waste Treatment Fundamentals - II. Environmental Requirements and Control, *Pub. Works*, **95**: 10, 123.

McCarty, P. L. and McKinney, R. E. (1961) Salt Toxicity in Anaerobic Digestion, *Jour. Wat. Poll. Cont. Fed.*, **33**, 399.

McKinney, R. E. and Horwood, M. P. (1952) Fundamental Approach to the Activated Sludge Process - I. Floc-Producing Bacteria, *Sew. & Ind. Wastes*, **24**, 117.

McKinney, R. E. (1951) *Biology and Biochemistry of the Micro-organisms in Activated Sludge*, ScD Thesis, MIT.

McKinney, R. E. (1952) Fundamental Approach to the Activated Sludge Process - II. A Proposed Theory of Floc Formation, *Sew. & Ind. Wastes*, **24**, 280.

McKinney, R. E. (1977) *Performance Evaluation of an Existing Lagoon System at Eudora, Kansas*, EPA 600/2-77-167.

Martin, A. J. (1927) *The Activated Sludge Process*, MacDonald and Evans, London.

Mass. State Board of Health (1890) *Report on Water Supply and Sewerage*, 2 Volumes.

Morgan, E. H. and Beck, A. J. (1928) Carbohydrate Wastes Stimulate Growth of Undesirable Filamentous Organism in Activated Sludge, *Sew. Works Jour.*, **1**, 46.

Neave, S. L. and Buswell, A. M. (1928) Biological Data on the Sprinkling Filter, *Part IV, Illinois State Water Survey*, Bulletin **26** , 95.

Nelson, E. W. (1964) *Manometric Observations of Algal Endogenous Respiration*, MS Thesis, Kansas University.

Nelson, F. B. (1975) *Varying Biological Solids Concentrations and Substrate Concentrations in the Warburg Respirometer*, MS Thesis, Kansas University.

Oberton, A. C. E. (1955) *Stabilization of Isopropyl Alcohol by Activated Sludge*, SB Thesis, MIT.

O'Brien, J. E. (1958) *Activated Sludge Processes and the Warburg Respirometer*, SM Thesis, MIT.

Parker, C. D. (1962) Microbiological Aspects of Lagoon Treatment, *Jour. Wat. Poll. Cont. Fed.*, **34**, 149.

Parsons, A. B. and Dugan, P. R. (1971) Production of Extracellular Polysaccharide Matrix by *Zoogloea ramigera, Appl. Microbiol.,* **21**, 657.

Phaup, J. D. (1968) The Biology of *Sphaerotilus* Species, *Wat. Research*, **2**, 597.

Placak, O. R. and Ruchhoft, C. C. (1947) Studies of Sewage Purification - XVII. The Utilization of Organic Substrates by Activated Sludge, *Public Health Reports*, **62**, 697.

Rao, V. C., Metcalf, T. G. and Melnick, J. L. (1987) Removal of Indigenous Rotaviruses During Primary Settling and Activated sludge Treatment of Raw Sewage, *Wat. Research*, **21**, 171.

Richards, T., Nungesser, P. and Jones, C. (1990) Solution of *Nocardia* Foaming Problems, *Res. Jour. Wat. Poll. Cont. Fed.*, **62,** 915.

Rossello-Mora, R. A., Wagner, M., Amann, R. and Schleifer, K-H. (1995) The Abundance of *Zoogloea ramigera* in Sewage Treatment Plants, *Appl. Environ. Microbiol.*, **61**, 702.

Sandino, J. (1993) *Development of a Rational Model for the Design and*

Performance Evaluation of the ABF/AS Process, PhD Thesis, Kansas University.

Sawyer, C. N. (1956) Bacterial Nutrition and Synthesis, *Biological Treatment of Sewage and Industrial Wastes*, **1**, 3.

Schonheit, P., Moll, J., and Thauer, R. K. (1979) Nickel, Cobalt, and Molybdenum Requirement for Growth of *Methanobacterium thermoautotrophicum*, *Arch. Microbiol.*, **123**, 105.

Sezgin, M., Jenkins, D. and Parker, D. S. (1978) A Unified Theory of Filamentous Activated Sludge Bulking, *Jour. Wat. Poll. Cont. Fed.*, **50**, 362.

Silva, P. C. and Papenfuss, G. F. (1953) *A Systematic Study of the Algae of Sewage Oxidation Ponds*, Calif. State Wat. Poll. Cont. Bd., Publ. No. 7.

Smart, P. and McKinney, R. E. (1970) Odorless Pork Production: From Conception to Market, *Proc. 25th Ind. Waste Conf., Purdue Univ.*, 757

Smolders, G. J. F., van der Meij, J., van Loosdrecht, M. C. M. and Heijnen, J. J. (1994) Stoichiometric Model of the Aerobic Metabolism of the Biological Phosphorus Removal Process, *Biotech. & Bioengr.*, **44**, 837.

Speece, R. E. (1988) A Survey of Municipal Anaerobic Sludge Digesters and Diagnostic Activity Assays, *Wat. Research*, **22**, 365.

Speece, R. E. (1996) *Anaerobic Biotechnology For Industrial Wastewaters*, Archae Press, Nashville, TN.

Stoltenberg, D. H. (1962) *Algal Metabolism as Related to the Theory of Oxidation Ponds*, MS Thesis, Kansas University.

Stone, A. R., Platt, H. M. and Khalil, L. F. (Editors) (1983) *Concepts in Nematode Systematics*, Academic Press, New York.

Strom, P.F. and Jenkins, D. (1984) Identification and Significance of Filamentous Microorganisms in Activated Sludge, *Jour. Wat. Poll. Cont. Fed.*, **50**, 449.

Theios, E. P., Morris, J. G., Rosenbaum, M. J. and Baker, A. G. (1967) Effect of Sewage Treatment on Recovery of Poliovirus Following Mass Oral Immunization, *Amer. Jour. Pub. Health*, **57**, 295.

Tomlinson, H. D. (1955) *The Metabolism of Certain Aromatic Hydrocarbons by Activated Sludge*, SM Thesis, MIT.

Towner, K. J., Bergogne-Berezin, E. and Fewson, C. A. (Editors) (1991) *The Biology of Acinetobacter*, Plenum Press, New York.

Ullrich, A. H. (1957) Experiences with the Austin, Texas, Biosorption Plant, *Water & Sew. Wks.*, **104**, 23.

US EPA (1987) *Design Manual – Phosphorus Removal,* EPA/625/1-87/001.

van Gils, H. W. (1964) *Bacteriology of Activated Sludge*, IG-TNO Report No. **32**, Research Institute for Public Health Engineering, Hague, Holland.

van Rensburg, J.E.J. (1983) *Anaerobic Digestion of Waste Activated Sludge*, MS Thesis, University of Kansas.

Wagner, M., Erhart, R., Manz, W., Amann, R., Lemmer, H., Wedi, D., and Schleifer, K. (1994) Development of an rRNA-Tagged Oligonucleotide Probe Specific for the Genus *Actinobacter* and Its Application for In Situ Monitoring in Activated Sludge, *Appl. Environ. Microbiol.*, **60**, 792.

Wattie, E. (1942) Cultural Characteristics of Zooglea-Forming Bacteria Isolated From Activated Sludge and Trickling Filters, *Pub. Health Rpts.*, **57**, 1519.

Wilcox, R. L. (1955) *Metabolism of Phenolic Compounds by Activated Sludge*, SM Thesis, MIT.

Williams, A. E. (1976) *Solids Retention Time and the Metabolism of Acetate by Activated Sludge*, MS Thesis, Kansas University.

Wrigley, T. J. and Toerien, D. F. (1990) Limnological Aspects of Small Sewage Ponds, *Wat. Research*, **24**, 83.

Young, J. C. and McCarty, P. L. (1969) The Anaerobic Filter for Waste Treatment, *Jour. Wat. Poll. Cont. Fed.,* **41**, R160.

Chapter 12

AIR MICROBIOLOGY

Air microbiology has long fascinated a few environmental microbiologists. Primary concern has been with the spread of pathogenic microbes through the air. Originally, people believed that diseases were transmitted through the air. Once it was demonstrated that most bacterial pathogens could not survive very long, much less increase in the air, many microbiologist lost interest in air microbiology. There is no doubt that air microbiology is unique, requiring more imagination and creativity than water or soil microbiology. Recently, interest in air microbiology has been stimulated with the construction of relatively air-tight houses and buildings. Heating and air conditioning systems have created environments that allow various microorganisms to survive and adversely affect the health of people. Emphasis on reuse of wastewaters in spray irrigation has raised concern over the potential spread of pathogenic organisms through the air in fine mists. Even wastewater treatment plants are being examined as potential sources of airborne pathogens. Recently, the concern has been raised that terrorists could disperse extreme pathogens through the air. As populations increase and greater demands are made on the environment, there will be additional concerns about air microbiology. For these reasons it is important that environmental microbiologists understand air microbiology, as well as, water microbiology and soil microbiology.

THE AIR ENVIRONMENT

Air is composed primarily of nitrogen and oxygen, 78% by volume nitrogen and 21% by volume oxygen. Air also contains water vapor in varying amounts.

Water vapor can help the microbes survive in the unfavorable air environment. There are also many tiny suspended particles in the air. Microorganisms are part of those suspended particles dispersed in the air. Gravity is the major force affecting the removal of suspended particles from air. Large particles quickly settle onto solid surfaces; while small particles settle slowly, and very small particles settle very slowly. Wind currents determine the ability of suspended particles to remain in suspension. Natural wind currents can move suspended particles for hundreds of miles. Not only can wind currents carry suspended particles long distance, but wind currents can also lift tiny suspended particles from the land surfaces. Movement of cars, trucks, and people in urban areas help generate micro-currents that combine with natural wind currents to keep particles in motion. Industrial operations often result in the production of gaseous wastes that contain particulates. The velocity of the discharge gases can carry the particulates high into the air, allowing wind currents to carry tiny particulates great distances.

People sneezing and coughing discharge microorganisms into the adjacent air environment. In close quarters people can breathe in some of the suspended microorganisms before the discharged particulates settle to the ground. Splashing water also discharges microbes into the air environment in tiny droplets. As the water evaporates, the active microorganisms lose their protective environment and quickly die. Spore forming microorganisms have the ability to survive as spores in the air environment, since the spore coating protects the nuclear material from drying out. The microbes not only suffer from a loss of moisture, but also from ultra-violet light in natural sunlight. Ultra-violet light can kill vegetative cells. Only the dense spore cells are partially protected from ultra-violet light. Long-term exposure to sunlight can even kill microbial spores and cysts. The oxygen in air adversely affects anaerobic bacteria that are discharged into the air environment. The hostile air environment has greatly limited the spread of diseases except for closely confined populations. Schools, hospitals, factories, and offices provide environments that allow some pathogenic microorganisms to be spread from person to person through the air.

The decomposition of crop materials left on agricultural fields by actinomycetes and fungi results in the production of large quantities of aerial spores. Natural wind currents can carry the fungi spores for long distances. Most of the spores settle out onto surfaces that are not suitable for their germination. Still, a sufficient number of spores find a suitable environment for growth and continued survival. There is no doubt that the survival of spores in the air has helped transmit specific microorganisms to different parts of the world.

SAMPLING TECHNIQUES

Initially, the exposure of solid biological media to air provided the air microorganisms for study. Sterile nutrient media plates were exposed to the air by lifting the glass cover and exposing the media surface to air for a given period of time. Spores and a few vegetative cells, which settled out on the media surface, found the media suitable for rapid growth, allowing them to be examined in detail. While this technique is reasonable for a qualitative measure of the microbes in the immediate vicinity, it was not suitable for quantitative evaluation of microbes in the air. In 1931 W. F. Wells at Harvard University developed an air sampler that allowed quantitative measurement of the microorganisms in air. The Wells air sampler was a combination centrifuge that sucked in air at a measured flow rate and forced the suspended particles onto the surface of biological media that coated the surface of clear glass centrifuge tubes. As the centrifugal action of the sampler pushed the microbial particles against the media surface, the microbes found an environment for rapid growth. Based on initial studies on the survival of bacteria in the air, Wells and Stone found that bacteria could survive for a sufficient period to allow pathogenic bacteria to be transmitted from person to person through the air.

As interest increased in collecting data on all types of air contaminants, the impingement type bubbler sampler was developed. The impingement sampler used a small air compressor to pull air through a series of bubbler tubes to collect the air contaminants in water tubes. Different size orifices in the air tubing controlled the airflow rate through the sampler. The glass tube inside the sampler was drawn to form a nozzle, allowing the entering air to create maximum mixing when air bubbles were formed in the water. Rapid mixing insured that the contaminants in the air were quickly transferred to the water in the sampler. The use of three bubbler tubes in series provided good collection of most of contaminants. In 1970 David Armstrong reported on an impingement method for collecting samples from incinerator stacks. Instead of pulling the air through the liquid, he placed the inlet tube just above a phosphate buffer collection fluid. The velocity of airflow through the inlet tube caused the microbial particles to impinge on the phosphate buffer solution and to be captured for later transfer to normal microbiological media. The key was in the mass of the bacteria and the velocity of airflow. The surface of the phosphate buffer solution was placed about 1.27 cm below the exit from the sampling tube. Currently, the AGI-30 impingement sampler is considered the standard air sampler for liquid collection of microbial samples.

After World War II, concern over the use of pathogenic organisms for biological warfare led to the development of the membrane filter for collecting air samples for microbial growth. Membrane filters were cellulose acetate filters of specific pore sizes to allow retention of different size particles. Measured volumes of air were drawn through the membrane filters to capture all particles of a certain size. By using a series of different filters a spectrum of particle sizes could be measured. After passing the desired amount of air through the membrane filter, the membrane filter was placed on top of a porous pad containing specific microbiological media to stimulate the growth of the desired microbes. Eventually, the membrane filters were made available for general use in air and water microbiology. The primary concern with membrane filters is damage to viable cells during collection. High concentrations of suspended particulates in the air sample can suppress microbial growth by completely covering the microbes and not allowing them to obtain nutrients for growth. The use of staged filters can minimize the impact of suspended solids by removing the large suspended particles ahead of the microorganisms. Polycarbonate membrane filters eventually became more popular than cellulose acetate membrane filters in air microbiology since the polycarbonate membrane filters had a more uniform pore size than the cellulose acetate membrane filters.

The Anderson cascade impactor has become one of the more popular microbiological air samplers. The Anderson sampler utilizes a series of stacked agar plates in which the air impacts on the surface of agar media plates similar to the Wells air sampler. Anderson samplers have either two stages or six stages. A series of plates with specific size holes separate the different stages in the Anderson sampler. The size of the holes decreases as the air moves through the sampler. In effect, different particle size organisms are removed at the different levels. The major differences between the Wells sampler and the Anderson sampler are the lack of centrifuging by the Anderson sampler and its multi-staging. Recently, increased interest in air microbiology has stimulated the development of new air samplers and evaluation of the old samplers. In 1994 Juozaitis *et al* reported on their research using several commercially available air samplers, as well as, two new samplers for the capture of *Pseudomonas fluorescens* from air in a bioaerosol chamber. They found that a new, agar slide impinger was the most efficient sampler, capturing 26% of the viable cells expected in the air sample. The Anderson two-stage impactor recovered 1.25% of the viable cells. The Anderson six-stage compactor captured 16.8% of the *Ps. fluorescens* in the air sample. These data indicate the limitations of the current air sampling equipment. The lack of interest in air microbiology has prevented the development of more efficient sampling equipment. Air sampling equipment is highly specialized with a limited market. As better air sampling equipment is developed, there should be an increase in air microbiology research.

CENTRAL HEATING AND AIR CONDITIONING

The development of central heating and air conditioning systems has created problems that are stimulating new interest in air microbiology. Houses and office buildings are being constructed tighter to prevent significant energy losses. Heated air in the winter and cool air in the summer are recirculated on a semi-continuous to continuous basis for maximum efficiency. Various types of filters are placed ahead of the circulation fans to remove particulates from the recycled air. Most of the simple filters trap large particles. A few specialized filters will remove small particles, such as bacteria. In large buildings electrostatic precipitators are used to remove microbial particles from the air. Electrostatic

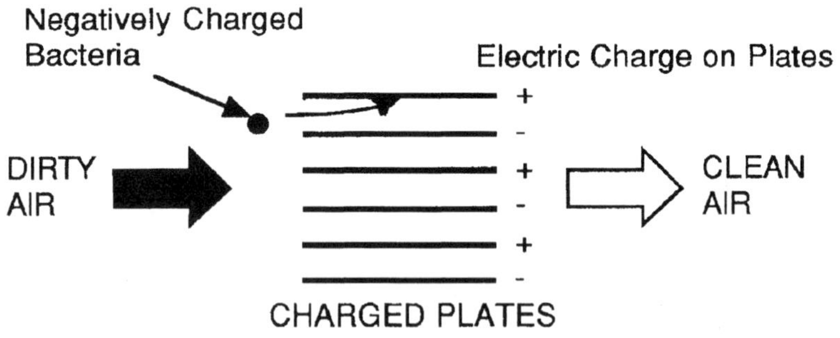

Figure 12-1 SCHEMATIC DIAGRAM OF AN ELECTROSTATIC PRECIPITATOR

precipitators use high voltage, about 50,000 volts, across a series of flat metal plates to pull the tiny charged particles onto the plate surfaces. Figure 12-1 illustrates the removal of bacteria by an electrostatic precipitator. The velocity of airflow, the spacing between plates, the area of the plates, and the number of plates are important design parameters. Periodically, the metal plates are cleaned to remove the attached particles. Since bacteria and colloidal particles in the air tend to be negatively charged, all of the tiny particles are removed together onto the positively charged plates. Electrostatic precipitators can be very efficient in removing microorganisms from the air

The use of humidifiers to adjust the moisture content of internal air can stimulate microbial growth that can be dispersed through central ventilation ducts to all parts of the building. Most humidifiers pick up moisture by passing the air through a mesh screen coated with water. The air removes some of the water from the screen; but the water picks up various contaminants from the air. Over time, microorganisms are able to grow on the wetted mesh screen. Any splashing allows the microbes to be transferred from the water to the air and circulated through the ventilation ducts. Dust and suspended particles also tend to accumulate in the ventilation ducts, especially around bends and at junctions of several ducts. The moist air and the nutrients in the settled particles can allow fungi to grow and produce aerial mycelia with discharge of large numbers of spores into the flowing air. It is normal to see fungi and some bacteria circulated through household air ducts, especially during the warm weather when air conditioning is widely used. People bring microbes into the house as they come and go. The central ventilation system moves the indoor air around the house. The settling of dust on furniture is indicative of the increase in particulates in the home environment. The normal household vacuum cleaner collects the large dirt particles and suspends the tiny microbial particles in the adjacent air. Recently, some vacuum cleaner manufacturers have added fine filters to minimize the return of small particles to the adjacent environment. The use of air filters in the return air ducts removes the larger suspended particles with limited removal of microorganisms. By removing the larger particles of dust and dirt, the microorganisms have less opportunity to grow in the air ducts. The overall effect is reduced spore production and distribution through the central ventilation system even though the air filters do not remove many microorganisms, if any. Some air filters use finer materials to remove small particles. These filters increase resistance to airflow, providing less air circulation. The finer material filters also clog more quickly, requiring greater maintenance.

The discovery of the transmission of *Legionella* in 1976 through the central ventilation system in a Philadelphia hotel focused attention on the importance of internal air contamination in the transmission of disease. *Legionella* was considered to be a non-pathogenic bacteria that normally lived in water. It became pathogenic when introduced into the lungs of certain individuals through contaminated aerosols. *Legionella* appears to be a secondary pathogen, affecting people with reduced immune systems. While *Legionella* has been found in normal water systems and in hot tubs, it is easily controlled by maintaining a chlorine residual in the water system. As primary pathogens are reduced in the environment, allowing more people to live longer, secondary pathogens that have not been significant, will move to the level of primary pathogens. Non-pathogenic bacteria suddenly become pathogenic, creating undue concern among the public. The problem is simply one of bacteria moving into an available niche when opportunities exist for increased growth. New pathogens will arise at regular intervals to replace the old pathogens brought under control by modern science and technology. This is part of our natural environment and it should be accepted as such without fear or panic. It simply means that science will never fully eliminate microbial pathogens.

Spores from bacteria, fungi, and actinomycetes, as well as, protozoa cysts, are the primary microbial forms circulated through central ventilation systems. Spores have the ability to remain dormant until they fall onto sites where they can grow into vegetative cells. Since moisture is a major factor affecting microbial growth, it is not surprising that most microbial spores do not germinate. Fungi and actinomycetes tend to grow on paper, cloth or leather objects in damp basements where the moisture level is sufficient. The spores produced by the fungi and actinomycetes in basements are easily picked up by ventilation systems and circulated through the house. While the spores may not cause disease, they adversely affect people who are allergic to microbial proteins. Allergies are more of a problem with central ventilation systems than pathogens. Microbial spores can also prove to be contaminants in industrial products unless care is taken to eliminate them. Many industrial plants have clean rooms with very fine air filters and ultra-violet lights to kill any vegetative cells. Multiple doors and negative pressure vents are used to remove contaminants before people enter the clean process rooms. With reasonable care, indoor microbial contamination can be controlled.

SEWAGE IRRIGATION AND SLUDGE APPLICATION

The environmental movement favored simple concepts for wastewater disposal over modern technology. They suggested that municipal wastewaters could be used for irrigation. This would allow reuse of the nutrients contained in the wastewater and reduce the demand on the use of clean water for agriculture. This positive approach to solving the sewage disposal problem had its corresponding negative side. Concerns were raised about the potential spread of pathogenic microorganisms, as well as, the accumulation of toxic materials in agricultural soils. It is not surprising that wastewater irrigation research studies were approached from two different points of view. One group wanted to show that wastewater irrigation represented a definite health hazard and should be stopped. The other group wanted to demonstrate that wastewater irrigation was a sound agricultural practice that did not create a health hazard, even to the workers in the immediate vicinity of the irrigation projects. Much of the initial research focused on spray irrigation systems. A study in Israel in 1978 showed that coliform bacteria and enteric viruses were found downwind from a spray irrigation system using treated wastewater. This research was followed in 1980 by a more detailed study that showed that low levels of *Salmonella*, coliform bacteria, and enteroviruses could be detected in mists sampled from 40m (131 ft) to 100m (328 ft) downwind from the spray nozzles discharging treated wastewater. The data showed that coliform bacteria were contained in aerosols generated by spray irrigation. Fortunately, the aerosols did not travel very far from the spray nozzles. None of the data showed that pathogenic microorganisms survived in sufficient numbers to create measurable cases of enteric diseases. Since the research was carried out in areas where few pathogens existed in the wastewaters, the results were not surprising and could be misleading for developing areas of the world where wastewaters contain large numbers of pathogenic organisms. The 1978 study carried out in Israel indicated that survival of coliform bacteria in aerosols was related to the relative humidity of the air. Here again, the moisture in the air helped the coliform bacteria survive. Survival was also greater at night than during the daylight hours. Part of the increased survival was related to the increased relative humidity and part was related to the lack of sunlight with its lethal uv radiation and its dehydration effect. Ultraviolet radiation definitely reduced the survival of coliforms as did increased temperatures. There is no doubt that spray irrigation of treated wastewaters poses a risk, based on the potential pathogenic microorganisms in the wastewater being used. With the reduction in pathogens in wastewaters before and after treatment, the risk of spreading contagious diseases is quite low. Unfortunately the public perception of the risk of spreading diseases remains quite high and will continue to remain high for some time in the future.

In many locations wastewater sludge is applied to the land as a source of nutrients and for ultimate disposal back onto the land environment. When the liquid sludge is applied to land from trucks, aerosols may be formed. Application of dewatered sludge does not normally produce aerosols. As the dewatered sludge dries out, it breaks down into smaller particles. Wind action can pick up and carry small particles great distances, causing concern that pathogens might be transported within the tiny particles. Studies have shown that wastewater sludge applied to land quickly lose their fecal coliform bacteria. Air sampling did not indicate that fecal coliform bacteria were present in the air particles. One of the problems with agricultural sampling lies with normal bacteria from the soil and from animals in the vicinity. Again, it should be emphasized that wastewater sludge in the United States does not contain as many pathogens as sludge in some of the developing countries where enteric diseases are still endemic. The probability of American wastewater and wastewater sludge having significant numbers of pathogens is quite low even though many environmentalists and microbiologists do not want to recognize that simple fact. It has required a large number of federally funded research projects to confirm that the application of sewage sludge to agricultural lands does not pose a health hazard from pathogens in the air. One of the major problems with the current university education is that knowledge is only accepted when the individual raising the questions carries out the experimental research. Existing knowledge has no real value. Common sense and logic are not considered of importance in evaluating research problems. As a net result, the same research is carried out over and over again with varying conclusions, depending upon the individual researcher's objectives.

AERATION TANKS

Aerosol generation is common in modern wastewater treatment plants. Aerosols are formed where wastewaters undergo turbulent mixing by normal hydraulic currents or by mechanical agitation. The prime sources of aerosols are the aeration tanks used for grit removal and for organic stabilization in activated sludge plants. Trickling filters also produce aerosols as the wastewaters splash on the rock media from the distribution arms. Primary and secondary sedimentation tanks produce aerosols when the treated effluent is collected around the periphery of the sedimentation tanks and drops into the pipe inlet box. Most of the aerosol research to date in municipal wastewater treatment plants has been carried out on aerosols created in activated sludge aeration tanks.

The basic problem with aeration tanks lies in their potential for creating of large quantities of aerosols. Moisture around the aeration tanks and brown colored particulates coating nearby surfaces indicate where the majority of the aerosols are deposited. Only the tiny droplets move any significant distance from the aeration tanks. One of the problems in collecting aerosol data from aeration tanks is the large number of heterotrophic bacteria growing in activated sludge. The heterotrophic bacteria easily overgrow the fecal coliform bacteria, making data collection and evaluation difficult.

The Chicago Metropolitan Water Reclamation District has carried out studies on the hazards of aerosols from activated sludge aeration. Their research demonstrated that the air in the residential environment before their WWTP was constructed, contained enteric microorganisms of unknown origin. As expected, the microbial populations were highest at night when the relative humidity was the highest and there was no ultra-violet radiation. After the WWTP was constructed and placed into operation, the microbial populations in aerosol samples increased on the downwind side of the WWTP, as would be expected. Enteric viruses were isolated from the downwind samples. It appeared that there was a potential for pathogenic microbes to be produced in the aeration tank aerosols. Unfortunately, the lack of microbial discharge data from the aeration tank made it difficult to evaluate the degree of reduction in the number of bacteria as the distance from the aeration tank increased. In 1993 researchers at Chicago reported on their study to measure the bacteria discharged into the air above an aeration tank. They used a six stage, Anderson, aerosol sampler in a smokestack tower to eliminate extraneous bacteria from other parts of the aeration tank. The sampler was located approximately 15 cm above the surface of the aeration tank. Their data indicated a standard plate count between 0.66 and 2.65 bacteria/m^2/sec with between 0.02 and 0.40 coliform bacteria/m^2/sec. Since the Anderson sampler had an air flow rate of 0.47 l/s and the aeration tank air flow rate up the tower was 0.33 l/s, it was assumed that most of the air from the aeration tank passed through the sampler, giving reasonable bacteria capture from the aeration tank.

Activated sludge plants have been in operation in the United States for about 80 years. Many of the WWTP are located in large cities with people living immediately adjacent to the plants. The people living adjacent to the activated sludge plants and the plant operators form the base population of individuals who could have been potentially affected by the aerosols from the aeration tanks. No data have been produced to date to show that either the WWTP operators or significant numbers of the people living adjacent to the WWTP have suffered from enteric diseases or other diseases that might have been transmitted by pathogenic microorganisms carried in wastewater aerosols. The number of potential pathogens in American municipal wastewater is simply not great enough to cause serious concern for the spread of enteric diseases by aeration tank aerosols. Yet, this lack of serious danger does not mean that operators do not have to be concerned about basic sanitation and personal hygiene. As already indicated, common bacteria can become pathogenic under the right circumstances. The important aspect of the studies to date is that there have been no serious health hazards from activated sludge tank aerosols. The public living next to a wastewater treatment plant should not have to worry about potential health problems from the treatment plant, provided the plant is properly designed and operated.

The 1965 design of an aeration only, activated sludge system below the floor in a confined hog building led to the production of aerosols around the aerator. Additional designs in larger, finishing buildings resulted in several aerators located directly adjacent to the hog pens. While some experts predicted that the aeration equipment would quickly produce aerosols that would spread disease throughout the confined hog population within the buildings, there was never any disease transmission related to the aeration aerosols. The potential for serious contamination in a confined animal building is far greater than in open aeration tanks in wastewater treatment plants. Not only were the animals healthy in the confined buildings, the growth data indicated normal weight gains. The aerosols were not sufficient to adversely affect normal metabolism. If the aerosols had been greater, animal health could have been adversely affected. The magnitude of the danger in the confined hog buildings turned out to be lower than the experts expected.

URBAN COMMUNITIES

Simple, gravity exposure of microbial media plates has demonstrated that the air in urban areas is more contaminated than the air in rural areas as far as bacteria, yeast, actinomycetes, and fungi are concerned. The difference between the air quality in urban areas and rural areas is largely related to increased activities in urban areas, as contrasted to rural areas, creating a greater opportunity for microbial dispersion. A 2 year study on the bacteria in the air at the University of Colorado in Boulder, published in 1978, indicated that 41% of the bacteria were *Micrococcus*; 11% were *Staphylococcus*; and 8% *Aerococcus*. It was apparent that cocci survived better than other bacteria. Since the cocci bacteria did not predominate in the soil around the sampling stations, it was not clear where the cocci originated. On the other hand, a study in Sweden, reported in 1978, indicated that bacterial spores of the common soil bacteria, *Bacillus*, traveled from a sandstorm north of the Black Sea to Sweden by air currents. While sandstorms are unusual events, it showed how microbial spores could travel considerable distances between different areas of the world. If the spores land on favorable sites, the microbes will begin to grow and will become part of the local environment.

The problems in studying urban air microbiology lie in the sources of the specific organisms and the airflow patterns. Air movement in urban areas is not uniform. The density and the shape of buildings, the human activities around the area being studied, and wind velocities affect the localized air currents. Samples collected on any given day will show wide variations in numbers and types of microorganisms. Samples taken at different times of the year will also show wide variations. Viable bacteria cannot survive very long in the air environment; but will survive longer at night than in the daylight. Spores will germinate in the proper environment. Once the microorganisms are cultured on nutrient media, it cannot be determined if the growth started with a viable cell or a dormant spore. The types of media used for microbial isolation and counting will determine which microorganisms are observed. The isolation of specific organisms from air samples simply indicates their presence when the sample was collected and analyzed. Healthy people come into almost continuous contact with many different spores and vegetative organisms without any negative consequences.

The most important aerial contacts appear to come from close contact. Large numbers of microorganisms are released into the immediate environment with every sneeze and cough. Individuals in the nearby vicinity can easily come into contact with the microbes contained in these aerosols. Under the proper circumstances oral diseases can be passed from person to person without much difficulty. The personal types of urban aerosols are much more dangerous than WWTP aerosols. Unfortunately, they receive much less attention by researchers and the public media.

A study was made to characterize the air microbiology of a suburban area in Washington, D. C. over a two-year period, September 1978 to December 1980,. A total of 379 samples were taken at 17 sampling locations, using a two stage Anderson sampler. The data were highly variable and had to be subjected to extensive statistical analyses for reasonable evaluation. These data reflect a basic problem with air microbiology. The collection of air samples is simply a measurement of a small fraction of the total air volume at a specific location at a definite time. Air is not a uniform medium, carrying a homogenous mixture of microbes. Air is a continuously fluctuating fluid that cannot be accurately sampled and defined from a contamination point of view. It is possible to collect a large number of data over a long time period and obtain general numbers of microbial populations. As time changes, the numbers and activities of all living creatures in the area change. The air microbiology will also change. A second long-term study will produce different data. Any similarities will be more due to random chance than to significant consistency in the environment. This is the real dilemma for all air microbiology sampling carried on in the outdoor environment. Fundamental concepts indicate that vegetative cells have short survival times; while spores have longer survival times. Air pathogens for humans are transmitted primarily through the lungs. Healthy lungs have the ability to repel the spores; while unhealthy lungs may or may not be as successful. Viruses tend to have short survival times in the air. Respiratory bacteria and viruses are infectious only if transmitted over short distances from person to person. This is why children in schools and people in the same office space tend to share the same common respiratory infections. In spite of all the problems with sampling and analysis, air microbiology will continue to be an interesting area of environmental microbiology. Concern for the unknown and a desire for specific answers will stimulate various individuals and groups to examine the microbiology in the air around us. Organisms will be identified and counted. Efforts will be made to determine the real significance of the numbers and names of the different organisms. Care will have to be exercised or misinformation will be generated instead of information. This will be a challenge for the air microbiologists.

THINGS TO REMEMBER

1. Air is a very light fluid that can carry small particles with wind energy against the force of gravity.

2. The ability of vegetative microbial cells to survive in air is limited.

3. Spore cells have the ability to survive for long periods in the atmosphere.

4. Sunlight provides uv radiation and heat to limit the survival of microbes in air.

5. Suitable equipment is available for collecting air microbiology samples on a quantitative basis.

6. Current houses with central air conditioning and heating systems tend to create localized problems with microbial growths in the air ducts.

7. Air filters can remove the air contaminants that build up in air ducts and provide for the growth media for microorganisms.

8. *Legionella* is an airborne pathogen that normally lives in water and can be transmitted by air.

9. Irrigation with treated wastewaters creates aerosols containing microorganisms in direct proportion to the microorganisms in the treated wastewaters.

10. Spray irrigation aerosols do not travel very far from their point of release.

11. Aeration tanks in WWTP generate aerosols in the vicinity of the aeration tanks.

12. Aeration tank aerosols also have limited distances of travel.

13. The limited number of pathogenic bacteria in domestic wastewaters in the United States helps to limit the potential transfer of pathogen bacteria in wastewater aerosols.

14. Urban communities have a more diverse microbial population than rural areas.

15. Human generated aerosols in enclosed spaces with large numbers of people are the most dangerous aerosols as far as people are concerned.

REFERENCES

Armstrong, D. H. (1970) Portable Sampler for Microorganisms in Incinerator Stack Emissions, *Appl. Microbiol.*, **19**, 204.

Bausum, H. T., Schaub, S. A., Bates, R. E., McKim, H. L., Schumacher, P. W. and Brockett, B. E. (1983) Microbiological Aerosols from a Field Source Wastewater Irrigation System, *Jour. Wat. Poll. Cont. Fed.*, **55**, 65.

Bovallius, A., Bucht, B., Roffey, R. and Anas, P. (1978) Long Range Air Transmission of Bacteria, *Appl. Environ. Microbiol.*, **35**, 1231.

Cannon, R. E. (1983) Aerosol Release of Cyanophages and Coliforms from Activated Sludge Basins, *Jour. Wat. Poll. Cont. Fed.*, **55**, 1070.

Cox, C. S. (1987) *The Aerobiological Pathway of Microorganisms*, John Wiley & Sons, NYC.

Cox, C. S. and Wather, C. M. (Editors) (1995) *Bioaerosols Handbook*, CRC Press, Boca Raton, Florida.

Cronholm, L. S. (1980) Potential Health Hazards from Microbial Aerosols in Densely Populated Urban Areas, *Appl. Environ. Microbiol.*, **39**, 6.

Dennis, P. J. L. (1990) An Unnecessary Risk: Legionaires Disease, *Biological Contaminants in Indoor Air*, ASTM STP 1071.

Edmonds, R. L. (Editor) (1979) *Aerobiology: The Ecological Systems Approach*, US/IBP Synthesis Series 10, Institute of Ecology.

Fannin, K. F., Vana, S. C. and Jakubowski, W. (1985) Effect of an Activated Sludge Wastewater Treatment Plant on Ambient Air Densities of Aerosols Containing Bacteria and Viruses, *Appl. Environ. Microbiol.*, **49**, 1191.

Goetz, A. and Tsuneishi, N. (1951) Application of Molecular Filter Membranes to the Bacteriological Analysis of Water, *Jour. Amer. Water Works Assoc.*, **43**, 943.

Jones, B. L. and Cookson, J. T. (1983) Natural Atmospheric Microbial Conditions in a Typical Suburban Area, *Appl. Environ. Microbiol.*, **45**, 919.

Juozaitis, A., Willeke, K., Grinshpun, S. A. and Donnelly, J. (1994) Impaction onto a Glass Slide or Agar Versus Impingement into a Liquid for the Collection and Recovery of Airborne Microorganisms, *Appl. Environ. Microbiol.*, **60**, 861.

Mancinelli, R. L. and Shulls, W. A. (1978) Airborne Bacteria in an Urban Environment, *Appl. Environ. Microbiol.*, **35**, 1095.

Pillai, S. D., Widmer, K. W., Dowd, S. E. and Ricke, S. C. (1996) Occurrence of Airborne Bacteria and Pathogen Indicators during Land Application of Sewage Sludge, *Appl. Environ. Microbiol.*, **62**, 296.

Sawyer, B., Elenbogen, G., Rao, K. C., O'Brien, P., Zenz, D. R. and Lue-Hing, C. (1993) Bacterial Aerosol Emission Rates from Municipal Wastewater Aeration Tanks, *Appl. Environ. Microbiol.*, **59**, 3183.

Sorber, C. A., Bausum, H. T., Schaub, S. A. and Small, M. J. (1976) A Study of Bacterial Aerosols at a Wastewater Irrigation Site, *Jour. Wat. Poll. Cont. Fed.*, **48**, 2367.

Sorber, C. A., Moore, B. E., Johnson, D. E., Harding, H. J. and Thomas, R. E. (1984) Microbiological Aerosols from the Application of Liquid Sludge to the Land, *Jour. Wat. Poll. Cont. Fed.*, **56**, 830.

Stetzenbach, L. D. (1992) Airborne Microorganisms, *Encyclopedia of Microbiol.*, **1**, 53.

Teltsch, B. and Katzenelson, E. (1978) Airborne Enteric Bacteria and Viruses from Spray Irrigation with Wastewater, *Appl. Environ. Microbiol.*, **35**, 290.

Teltsch, B., Kedmi, S., Bonnet, L., Borenzstajn-Rotem, Y., and Katzenelson, E. (1980) Isolation and Identification of Pathogenic Microorganisms at Wastewater-Irrigated Fields: Ratios in Air and Wastewater, *Appl. Environ. Microbiol.*, **39**, 1183.

Wells, W. F. (1933) Apparatus for Study of the Bacterial Behavior of Air, *Amer. Jour. Pub. Health*, **23**, No. 1.

Chapter 13

SOLID WASTES

Solid wastes are one of the more interesting environmental contaminants to deal with. The primary characteristic of importance for solid wastes lies in the fact that they are solid. Unlike gaseous wastes that flow into the vast atmosphere around us or liquid wastes that flow downhill until they ultimately reach the ocean, solid wastes tend to stay put until a major effort is made to move them to another spot. Solid wastes tend to remain as such until someone finds a use for the solid wastes and converts them into something of value. Suddenly, the solid wastes disappear until the new product loses its value and becomes solid wastes once again. The lack of value of solid wastes has resulted in limited information on their interaction with various microorganisms. Few people bothered to study solid wastes until the Federal government provided the funds for studies. It is not surprising that most of our understanding of solid wastes has come in the last 40 years.

SOLID WASTE CHARACTERISTICS

One of the frustrating problems confronting environmental scientists and environmental engineers lies in trying to determine the characteristics of solid wastes. Solid wastes are simply solid materials that have lost value for their owner and are discarded. It does not mean that the solid wastes being discarded have no value. It simply means that the solid wastes have no value for the current owner. It may well have value for another owner. If the solid wastes have value for a new owner, these materials are no longer solid wastes, but rather raw materials for further use with renewed value until the new owner decides to discard them as solid wastes. All material goods that society makes and uses will become solid wastes in time.

Solid wastes cannot be characterized by their chemical composition or their size or their weight alone. Solid wastes are characterized by many parameters. Chemical composition is an important parameter, along with size and weight. Bio-stability is also an important parameter for solid wastes that has largely been ignored. Bio-stability is the parameter that defines how the solid waste materials react to microorganisms and the rate of that reaction. Most solid wastes currently being produced are largely bio-stable, showing little to no reaction with microorganisms. Food wastes are the least bio-stable solid wastes, undergoing rapid reaction with microorganisms. Grass clippings and leaves are seasonal solid wastes that are not very bio-stable. From a practical point of view environmental microbiologists are only interested in the solid wastes having the least bio-stability.

There are many ways to characterize solid wastes. No one classification is perfect. Classifications tend to start with the major sources of solid wastes.

1. Residential SW – all solid wastes produced by people living in residences within the classification area. Residences include single-family houses, duplexes, and apartments with multiple family units.
2. Commercial SW – all solid wastes produced by commercial establishments within the classification area.
3. Industrial SW – all non-hazardous solid wastes produced by industrial manufacturing plants within the classification area.
4. Construction and Demolition SW – all solid wastes generated during the normal construction of houses, apartments, commercial establishments, and industrial factories or from the destruction of houses, apartments, commercial establishments, or industrial factories.
5. Street Sweepings SW – all solid wastes collected by street sweepers operating in urban communities.
6. Water & Wastewater Treatment Plant Sludge SW – all sludge solid wastes produced by water and wastewater treatment plants within the classification area.
7. Automotive SW – all solid wastes generated when automobiles

and trucks are junked.

8. Bulky SW – all large solid waste items from residences and commercial establishments that require special collection and handling. Bulky solid wastes include washing machines, refrigerators, stoves, sofas, etc.
9. Trees – large trees that die or trees that are cleared for construction projects require special handling.
10. Agricultural SW – all solid wastes produced from farming operations.
11. Mining SW – all residues remaining from mining and mineral processing.

These broad classifications of solid wastes have value for looking at the total solid waste problem at the national level. Regional classifications normally look at municipal SW, rural SW, agricultural SW, and mining SW. Prior to 1965 data on solid waste characteristics were quite limited. Few people were concerned enough about solid wastes to determine their characteristics. The major problem was simply that it was too difficult to take representative samples of solid wastes and make complete analyses. Although every person was concerned with the generation and disposal of solid wastes, no one cared enough to characterize those solid wastes. No one knew what to do with the limited data that appeared in the literature. When Congress passed the first solid waste legislation at the federal level, it became immediately apparent that no one had any real data on the magnitude of the solid waste problem. All available data was largely extrapolation of a limited database that dealt with the weight and volume of solid wastes generated by municipalities. Much of the early efforts of the new federal Office of Solid Wastes dealt with gathering more reliable data. Contracts were given to various engineering firms and research organizations to determine data on solid waste characteristics across the country. The data generated from these studies is the basis for most of the solid waste characterization.

MUNICIPAL SOLID WASTES

Municipal solid wastes include the residential SW, the commercial SW, and limited industrial SW generated within the jurisdiction of the municipality. The latest EPA published data on municipal solid wastes indicated that the United States produced 232 million tons in 2002. The per capita SW generation averaged 4.5 lbs/d. The composition of municipal solid wastes in the United States and the percentage of each group by weight are given in Table 13-1. Examination of the different groups of municipal solid wastes indicates that glass and plastics are the most biostable with metals being relatively biostable. Paper, wood, rubber, leather and textiles are slowly biodegradable under specific conditions. Food wastes and yard wastes are biodegradable. One of the key

Table 13-1 COMPOSITION OF MUNICIPAL SOLID WASTES IN 2000

Paper	37.4%
Yard Waste	12.0%
Plastics	10.7%
Food Wastes	11.2%
Metals	7.8%
Rubber, Leather, & Textiles	6.7%
Glass	5.5%
Wood	5.5%
Other	3.2%

parameters for biodegradability is moisture content. Food wastes contain up to 80% moisture, averaging about 70% moisture. Yard wastes average about 60% moisture with grass clippings having the most moisture and dead branches having the least moisture. Paper and cardboard normally contain less than 10% moisture. Wood can contain up to 40% moisture. Rubber, leather, and textiles usually contain less than 10% moisture. These materials are not biodegradable without the addition of water to increase their moisture level. Dry metal objects are bio-stable except in the presence of water. Microorganisms can slowly react with wet metal products, if the end products of the microbial-metal reactions are not toxic to the microorganisms. Glass and plastics have very low moisture content and are bio-stable even when immersed in water. For the most part municipal solid wastes reflect the lifestyle of the people within the municipal jurisdiction and the changes in the local economy.

INDUSTRIAL SOLID WASTES

Industrial solid wastes consist of hazardous solid wastes and non-hazardous solid wastes. The hazardous solid wastes are dangerous for the environment and are controlled by separate federal legislation than the non-hazardous solid wastes. The characteristics of industrial solid wastes are highly variable, depending upon the specific industrial processes and the workers within specific plants. Industrial solid wastes range from inorganic materials to organic materials. The EPA's estimate of non-hazardous industrial solid wastes in 1996 was 7.6 billion tons/yr.

CONSTRUCTION AND DEMOLITION SOLID WASTES

The construction of new houses and new commercial buildings results in the production of construction solid wastes in direct proportion to the materials employed in the construction project. The characteristics of the construction solid wastes will vary with each project. Normally, construction wastes will include the residual materials that could not be incorporated into the finished building. The construction of houses will include pieces of wood, wallboard fragments, partial shingles, and excess roofing paper. If the house has a cinder block foundation and/or a brick front, broken cinder blocks and broken bricks will also be included. The amounts of construction wastes produced are directly proportional to the number of houses and buildings constructed. As a net result, the construction waste quantities and characteristics are highly variable and not easily estimated.

Eventually, every house and every building will become solid wastes when their value is too low and the maintenance costs are too high. Under good economic conditions, old buildings are bulldozed to allow new replacement structures on the same site. Demolition of old structures results in complete destruction of the old structures. Demolition debris will consist of all the materials remaining in the structure. These materials are simply bulldozed and placed in a truck for removal from the site. The entire weight of all the materials in the old structure comprises the demolition wastes. Major urban renewal projects create large amounts of demolition wastes over a specific time period. The amount and characteristics of demolition wastes require separate evaluation of every project and cannot be generalized.

STREET SWEEPINGS

Many cities in the United States use street sweepers to collect dirt and trash that accumulates on streets in the major commercial and industrial areas. Occasionally, the residential streets will also be swept. The shape of road surfaces allows the dirt and trash to accumulate next to the curb and the road. For this reason street sweepers normally clean the area of the street next to the curb. The quantity of street sweepings collected is a function of the size of the city and the rainfall characteristics. Large cities produce more street sweepings than small cities. Large cities tend to attract dirty industries that can operate with a minimum of complaints. Cities tend to keep the dirty industries together, making it easier to keep the streets reasonably clean. Industrial areas tend to have large impervious surface areas, allowing maximum runoff during and after every rain event. The rapid runoff of rainwater carries the contamination on the streets into storm sewers for collection and removal. Areas with heavy rains will find that the streets are cleaned of all but the heaviest solid materials. Areas with light rains, occurring frequently, will find only the lightweight materials removed by the runoff water.

Examination of street sweepings indicates they are composed of dirt, sand, grit, paper products of various types, plastic materials, broken glass, rubber particles, and miscellaneous materials too difficult to identify and measure. Most of the street sweepings are relatively inert material that can be used as fill or placed into a sanitary landfill. One of the problems with street sweepers is their tendency to suspend fine particles around the street sweeper rather than capturing the fine particles. There are no uniform analyses for street sweepings. Every community must determine its own characteristics of street sweeping wastes, both as to quantities produced and their chemical characteristics.

WATER AND WASTEWATER SLUDGES

Every community that has a water treatment plant and/or a wastewater treatment plant will have water treatment sludge and/or wastewater treatment sludge. Emphasis on water treatment plants and wastewater treatment plants is on the removal of pollutants from the water being treated. The treatment processes produce a clean water and waste sludge. The water treatment plants generate alum sludge and calcium carbonate sludge, if the treatment plant softens the water. The clean water, produced in the water treatment plant, is distributed to every user in the community through a complex pipe network. Once the clean water has been used, it is returned as wastewater. The wastewater collection system is also a network of complex pipes that services every house and building in the community. The wastewaters are collected and discharged into a wastewater treatment plant. The pollutants in wastewater are classified as suspended pollutants and soluble pollutants. The majority of the suspended pollutants are removed by gravity sedimentation. The remaining suspended pollutants and the soluble pollutants are treated biologically with the pollutants converted into suspended solids that form settleable floc particles that are removed by gravity sedimentation. The two types of wastewater sludge are normally mixed together and treated in an anaerobic digester. The anaerobic digester destroys the readily biodegradable suspended solids, leaving a relatively inert residue for disposal on land.

The characteristics of water treatment plant sludge are quite similar from plant to plant as far as chemical characteristics are concerned. The quantity of sludge produced varies with the magnitude of treatment required to remove the contaminants. Alum treatment of surface water produces an aluminum hydroxide floc that removes the suspended solids in the surface water. River water will contain both organic and inorganic particles. During periods of heavy rainfall and runoff in rural areas the river water will contain soil particles and any materials washed from the soil surface. The net result is for wide variations in alum use and in sludge production. Lake water contains colloidal suspended solids and various microorganisms. Understanding the source of the water being treated is essential to knowing the type and magnitudes of contaminants in the water plant sludge. Water plants that soften the water will produce calcium carbonate sludge. A few plants may produce magnesium hydroxide in addition to the calcium carbonate. Softening sludge is an inorganic sludge. Each water treatment plant must determine its own sludge production rates as they are variable from plant to plant.

Wastewater treatment plants produce screenings and grit in addition to the sludge generated every day. Screenings are large solid wastes that are collected on bar screens at the head of the treatment plant. Most plants collect the screenings and dispose of them with the other municipal solid wastes in sanitary landfills. A few plants grind the screenings and return them to the liquid stream. Logic indicates that it is simpler to handle the screenings as solid wastes once they have been collected. Grinding the screenings and returning them to the liquid wastes means that the screenings have to be removed twice from the wastewater. Grit is largely sand and dense organic particles that could damage mechanical equipment in the units that follow the grit chamber. Grit is usually washed as it is collected and used as fill material. The primary sludge and the biological sludge are combined and anaerobically digested before being dewatered and placed on the land. Most municipal wastewater treatment plants will produce about 0.14 kg/m^3 (1,200 lbs TS/MG) wastewater treated. If 0.02 kg/m^3 (200 lbs TSS/MG) are lost in the treated effluent, 0.12 kg/m^3 (1,000 lbs TSS/MG) will be returned to the land environment. Wastewater treatment plants handling large quantities of industrial wastes will produce even more excess sludge.

AUTOMOTIVE SOLID WASTES

Every automobile sold in the United States will eventually become solid waste. Old automobiles are processed by automobile scrap dealers. With an average life of 7 to 8 years the number of automobiles scraped each year can be estimated by looking at the production data 7 to 8 years ago. With increased emphasis on waste recycling the automobile manufacturers have tried to use materials that could be processed and reused with a minimum of effort. The automobile scrap dealers remove all the parts of value before shipping the residue to a metal grinder and separator to recover the metal for reuse. Automobile tires pose one of the major processing problems for recycling.

BULKY SOLID WASTES

Every community produces bulky solid wastes from time to time. Bulky solid wastes are all large household items that are discarded. Stoves, refrigerators, washing machines, sofas, large chairs, beds, mattresses, and other large items require special collection. Usually, the person discarding the item is responsible for seeing that the bulky waste is collected and delivered to the proper disposal site. Often, when a large item is replaced, the dealer selling the new item collects the old item and delivers it to the disposal site. Reclamation dealers remove any useful parts before sending the residue for recycling or for burial in a sanitary landfill.

TREES

As communities age, there will be a steady production of dead trees and trees removed for residential or commercial expansion. The dead trees are usually ground up and used for mulch in the community. Living trees removed from private or public property will be used commercially, where possible. Large trees can be cut for lumber; while small trees are mulched for use in paper production or in parks and gardens. Every effort is made to reuse the wood rather than burying it in sanitary landfills.

AGRICULTURAL SOLID WASTES

Farms produce large quantities of solid wastes. The quantities of solid wastes produced are a function of the specific crops grown on the farm. Farmers have long recognized that manure from farm animals and crop residues after harvest must be returned to the land to help maintain the soil quality for future crop production. Farmers are among the oldest of the recyclers in our society. The major problems in recent years have come from the large confined animal farms. The manure from the confined animal buildings is handled as a liquid waste rather than as a solid waste. Operators of the large confined animal farms have had to learn new methods of handling the liquid manure to minimize environmental pollution. The basic problem is processing the liquid manure for return to the land environment. It is not surprising that there are no simple solutions for processing the liquid manure as previously indicated.

MINING SOLID WASTES

The mining industry produces large quantities of solid wastes that must also be returned to the land environment. Mining results in the removal of non-valuable materials along with the valuable materials. The failure of the mining industry to properly handle the non-valuable solid materials at the same time they handled the valuable materials has produced localized environmental damages and created a negative image for the mining industry. Too often mining companies simply closed down and walked away from the mined area when the mining operation stopped being profitable. Some of the mining residues contain hazardous chemicals that slowly wash into nearby streams, creating even more environmental damage. The net effect has been for the mining industry to have created a very negative impression on the public at large. It will take decades of major effort by the existing mining industry to overcome the current negative image and to correct the environmental damage that has been done over the years.

PROCESSING SOLID WASTES

Normal processing of solid wastes has always followed the path of least resistance. Solid wastes have always been handled with the least effort and the lowest cost. Since our interests are focused on the microbiological side of environmental pollution control, we will only examine those treatment processes that significantly deal with microorganisms. Our focus will be primarily on municipal solid wastes and agricultural solid wastes and how microorganisms can be used to stabilize these solid wastes. We will briefly examine how bacteria react with mining solid wastes to extract mineral elements.

Biological treatment of solid wastes falls into three general categories: sanitary landfills, composting, and soil stabilization. Although sanitary landfills have been extensively used in the United States, the lack of understanding of how microorganisms react in sanitary landfills has created problems in both design and operation of sanitary landfills. Composting is another biological process that has not been fully evaluated from a microbiological point of view. Soil stabilization of solid wastes has long been used for treating agricultural solid wastes and has potential for treating any readily biodegradable solid waste. Since the production of solid waste will continue to be related to existing populations, we can expect to see increased production of solid wastes. It is essential that we understand how the microorganisms metabolize the biodegradable materials in solid wastes and what we can expect from biological treatment systems currently in operation.

SANITARY LANDFILLS

By definition sanitary landfills are engineered burial of solid wastes. Unfortunately, sanitary landfills grew up at a time when municipal government paid little attention to solid waste processing. Although municipal government is the primary beneficiary of well-designed and well-operated sanitary landfills, municipal government must accept full responsibility for all the problems related to sanitary landfills. Municipal officials put solid waste processing at the bottom of the problem pile until the environmentalists forced the federal government to make municipal officials face up to years of neglect in the area of solid waste processing.

Sanitary landfills should be carefully designed for maximum operating efficiency with a minimum of discomfort and objections from the citizens living around the landfill. Fundamentally, all sanitary landfills should be located within the boundaries of the community producing the solid wastes. This location allows the citizens of the community to continuously observe how their elected officials manage one of the most important municipal functions. It is completely possible to design and operate a sanitary landfill within the municipality without creating problems for the citizens or the municipal officials. Unfortunately, few communities are willing to make the effort that is required to construct and operate a sanitary landfill correctly. Most communities choose to establish their sanitary landfill far enough outside the community boundaries that few citizens know where the sanitary landfill is located or how it is being operated.

Basic Concepts

A sanitary landfill should be constructed in dry soil on flat terrain capable of holding at least 20 to 30 years production of solid waste from the contributing population. The natural water drainage should be away from the landfill site or should be diverted around the landfill site to prevent excess runoff from crossing the landfill and eroding the cover soil. Because of concerns about leachate formation and its effect on groundwater below the landfill, the bottom of the landfill should be sealed with clay or an impermeable membrane. A series of perforated pipes should be placed at intervals across the bottom of the landfill to collect the leachate and to convey it to a concrete sump where the leachate can collect and be pumped to a leachate storage facility prior to treatment. A layer of coarse sand is placed over and between the leachate pipes to prevent possible damage to the collection pipes by the tractor or tractors used for compacting the solid wastes in the landfill. The solid wastes should be placed first at the far end of the landfill. Bulldozers push the solid wastes into a corner and compact the solid wastes to their maximum concentration. By moving back and forth over the solid wastes the bulldozer is able to compact the solid wastes to a concentration close to 593 kg/m^3 (1000 lbs/cy). Figure 13-1 shows a bulldozer compacting solid waste in a sanitary landfill. At the end of each day the compacted solid waste is covered with about 0.15 m (6 inches) of topsoil to act as a temporary cover. As more solid wastes are added to the landfill, the landfill slowly fills to the design level. As the landfill fills with solid waste, perforated gas collection pipes are placed in the upper part of the landfill to collect the gas produced in the landfill. The landfill is covered with a layer of clay or an impermeable membrane to minimize infiltration of water. The top of the impermeable layer is then covered with at least 0.61 m (2 ft) of topsoil. The topsoil is sloped to allow the natural storm water runoff to flow away from the

Figure 13-1 BULLDOZER COMPACTING SOLID WASTES AT THE LAWRENCE, KS SANITARY LANDFILL

landfill into catch basins and a storm water collection sewerage system to minimize the infiltration of rainwater into the landfill. Sanitary landfills can be constructed as a series of trenches or as areas, depending on the specific terrain. Trench landfills are built into the soil surface while area landfills are built against a hill. The compaction and burial process for the solid wastes continues until the landfill has been filled and the landfill site has reached its design capacity for solid wastes. A new site must be designed and developed before the old landfill site is filled. The production and processing of solid wastes is a never-ending process that must be continuously handled.

Biological Activity

The municipal solid wastes placed into the sanitary landfill contain materials that are readily biodegradable. In fact, biological activity is well underway by the time the solid wastes are collected and transported to the sanitary landfill. The food wastes that contain over 60% moisture will allow bacteria to metabolize the organic matter in the food waste materials. Fungi will be able to metabolize organic matter with only 40% moisture content, provided oxygen and other required nutrient elements are available for aerobic metabolism. Even though solid wastes have been compacted to a density of 593 kg/m^3 (1,000 lbs/cy), there are still plenty of void spaces in the compacted landfill. Both bacteria and fungi are able to grow on the moist food wastes initially. The microbes grow directly on the surface of the food materials being metabolized. The metabolic reactions are aerobic with carbon dioxide, water, new cell mass, and heat as the end products. As the initial microbial layer spreads across the organic surface, some microbes are forced to grow on top of the existing microbial layer. Soon a second layer of microbes becomes a third layer and then a fourth layer. Metabolism of the food material shifts from being aerobic to being oxygen limiting. The microbes at the surface of the food material are unable to obtain sufficient oxygen for aerobic metabolism. The bacteria and the fungi continue to metabolize to the best of their ability. As oxygen becomes limiting, the microbes slow their rate of metabolism and organic end products increase in solution. Soon oxygen becomes exhausted and only the facultative bacteria are able to continue metabolism at the food surface. Fungi and strict aerobic bacteria on the food surfaces cease growing. The bacterial layer next to the food surfaces slow their growth as metabolism shifts to the production of organic acids and other partially oxidized end products rather than carbon dioxide and water. The pH in the liquid around the bacteria decreases, as the organic acids accumulate. Further metabolism causes the pH to drop sufficiently low that metabolism stops and the microbes die.

If the food waste is in contact with a metal can, the organic acids will cause the metal to dissolve and neutralize some of the organic acids. The presence of alkaline materials in the vicinity of the food wastes will also neutralize the organic acids, raise the pH, and allow metabolism to continue. As the neutralized organic acids accumulate around the organic matter being metabolized, methane bacteria will begin to metabolize the neutralized organic acids, producing methane gas and carbon dioxide as primary end products. Metabolism of the neutralized organic acids releases the cations, allowing the cations to form bicarbonate alkalinity in the water that can neutralize more organic acids. Slowly, the methane bacteria metabolize the neutralized organic acids, producing methane, carbon dioxide, water, new cell mass, and heat. Sulfate reducing bacteria will also grow under anaerobic conditions where there are sulfates in the solid wastes. The sulfate reducing bacteria can compete with the methane bacteria for organic nutrients in the right environment and may reduce the methane gas production. The metabolic process is a slow process since there is no mixing of the organic matter and the microorganisms in the sanitary landfill. The localized environment determines what microbes can grow and to what extent they grow. Temperature, moisture, and alkaline materials are the three major factors that determine the rate of biodegradability of the organic matter in the landfill. A rising temperature will increase the rate of metabolism. Moisture within the landfill will determine how far the microorganism will be able to move to find new food for metabolism. The presence of alkaline materials will determine the pH of the water around the bacteria and the food wastes being metabolized. The production of water forms a thin liquid film for the microorganisms to move to a new location for metabolism. For the most part, microbial movement inside the sanitary landfill is restricted to the immediate environment around the rapidly biodegradable organic matter. The gas movement in the sanitary landfill depends upon pressure differentials and available void spaces. The heat released by metabolism is quickly absorbed by the materials in the sanitary landfill. Overall, microbial metabolism within a sanitary landfill is limited, the environment within the sanitary landfill is simply not conducive to rapid microbial metabolism..

There are few organic compounds in municipal solid wastes that bacteria can metabolize under the best of circumstances. Most paper in solid wastes contains lignin as well as cellulose. Yard wastes also have a considerable amount of lignin with the cellulose. The lignin prevents bacteria from metabolizing the cellulose, allowing the waste paper and much of the yard wastes to remain untouched once the oxygen in the void spaces has been removed. Fungi have the ability to metabolize lignin as well as cellulose; but the fungi must have dissolved oxygen available for metabolism. Bacteria are able to metabolize the cellulose materials that do not have the cellulose combined with the lignin. Completely sealing the solid wastes in the sanitary landfill, as currently recommended, creates an environment that limits the overall metabolism to 10% to 15% of the municipal solid wastes no matter how long the solid wastes are contained in the sealed sanitary landfill. It also means that 85% to 90% of the municipal solid wastes placed into a completely sealed sanitary landfill will be available for reuse whenever conditions favor recycling municipal solid wastes.

Stimulating Bacterial Activity

It is important for everyone to recognize that completely sealing a sanitary landfill, as currently recommended to minimize groundwater contamination, limits biodegradation within the landfill. The simplest method to stimulate biodegradation inside the sanitary landfill is to pass water through the landfill at periodic intervals to allow the bacteria to migrate throughout the landfill and find all the readily biodegradable materials. Metabolism in the landfill will continue to be anaerobic with the production of methane gas. Recirculation of liquid around the sanitary landfill on a continuous basis has been recommended to speed the stabilization of the biodegradable organic matter. The problem with recycling the leachate from the bottom of the landfill around the landfill is the accumulation of soluble, non-biodegradable contaminants in the leachate. The soluble non-biodegradable materials accumulate in the recycled leachate on each pass through the solid wastes in the landfill. Eventually, the accumulated non-biodegradable materials may adversely affect metabolism of the biodegradable organic compounds.

Under the anaerobic environment inside the sanitary landfill iron and other metallic ions, released by reaction with organic acids, tend to remain partly soluble. Some of the iron and metallic ions may form insoluble sulfide salts that move with the leachate as colloidal precipitates. When the leachate is exposed to air, oxygen in the air reacts with the ferrous iron to form insoluble ferric oxide. Some of the other metallic elements also form insoluble metallic oxides. The organic compounds are unaffected by the changes in the metallic elements and remain in solution. Leachate tends to be on the acidic side and may need additional alkalinity to provide a suitable pH, pH 7 to 8, when the leachate is recycled back to the landfill surface. It is important to maintain a suitable environment within the landfill for maximum stabilization of the organic matter. Recycling of leachate will also remove any salts that are soluble in water, increasing the mineral content of the leachate with time. Maintaining a good environment within the sanitary landfill is not an easy task. The solid waste materials are fixed within the landfill and cannot move to produce a better opportunity for biodegradation. The recycled leachate always takes the path of least resistance and flows through the void spaces between the solid particles by gravity to the bottom of the landfill for collection.

Continuous recirculation of leachate should result in moving the bacteria throughout the entire volume of the sanitary landfill and the maintenance of anaerobic conditions within the sanitary landfill. Care must be taken that the recycled leachate is not applied by spraying in the air. Spaying the leachate will create small droplets and allow oxygen in the air to enter the leachate before it reenters the sanitary landfill. A high rate of continuous recirculation of leachate will be a waste of energy will produce a limited increase in treatment efficiency. The basic problems in the biological degradation of solid wastes lie in the large mass/surface area ratio and the complexity of the chemical composition of the different solid waste components.

Periodic application of leachate to the landfill surface, followed by sufficient time to allow the free water to drain through the sanitary landfill should produce adequate dispersion of the microbes throughout the sanitary landfill without slowing the metabolic reactions with the solid waste. As the upper layers of organic material are stabilized, the upper environment will shift from anaerobic to aerobic, allowing the fungi and the actinomycetes to metabolize the complex paper products in proportion to the availability of oxygen to the microorganisms on the surface of the solid wastes. Slowly, aerobic conditions will be established throughout the entire volume of the sanitary landfill.

The lack of nitrogen, phosphorus, and trace elements can also limit both the rate of biodegradation and the total amount of biodegradation. Remember that the metabolism of organic matter by microorganisms results in the production of new cell mass. Without the synthesis of new cell mass, the microorganisms will produce very limited metabolism, only endogenous respiration and the synthesis of new cells from the nutrients released. Adding treated domestic wastewater to the sanitary landfill on a periodic basis is one method that can be used to add water and to increase the nitrogen, phosphorus, and trace metals that the microorganisms can use in their metabolism of the biodegradable materials in the sanitary landfill. Storm water runoff can be collected and stored for use as needed during the dry season in water limited regions. There is no single method for stimulating microbial growth in a sanitary landfill. It is possible to design the top of the landfill to allow natural rainfall to enter the landfill as the primary water source while still capturing the gas and the leachate. In every sanitary landfill there is a fixed amount of biodegradable materials that can be metabolized. Normally, there will be a slow increase in biological activity to a maximum, followed by a slowing rate of increase to a plateau and a slowing rate of metabolism that eventually ceases to be significant. It normally takes several years for the biodegradable materials in a sanitary landfill to be metabolized.

Gas Production

In recent years efforts have been made to collect the gas from sanitary landfills for power generation. The concept of gas production and energy generation has attracted considerable attention from landfill owners and landfill operators as it is a method for recovering some of the operating costs of the sanitary landfill. While sanitary landfills handling biodegradable solid wastes will produce methane gas that can be collected and burned for energy, there is a fixed amount of potential energy available that can be produced. The actual energy recovery will be something less than the potential energy available from the solid waste analyses. The value of the landfill gas must be sufficient to defray the cost of the extraction equipment and the energy conversion equipment along with the operating costs for the energy system. The landfill gas will contain methane, carbon dioxide, and low concentrations of hydrogen sulfide, the same as contained in gas from an anaerobic wastewater sludge digester. The methane fraction is important in determining the extent of biodegradation and in energy recovery. The amount of methane gas that can potentially be produced is related to the biodegradable COD (BCOD) of the organic matter metabolized. If the BCOD of solid wastes in the sanitary landfill averages 1.2 times the weight of the biodegradable organic fraction of solid wastes, the maximum amount of methane that could be expected from a municipal solid waste sanitary landfill would be about 40 m^3 CH_4/m^3 (1,080 cf CH_4/cy) solid waste. The theoretical energy yield from the biodegradable organic compounds in the solid wastes contained in the sanitary landfill should be about 1.5×10^6 kJ/m^3 (1.1×10^6

Btu/cy). The actual energy capture will depend on the efficiency of the gas collection system. As gas pressure builds in the encapsulated landfill, punctures or cracks in the liner system will allow loss of the landfill gas to the outer environment. The value of the gas produced from biodegradation of municipal solid wastes is not sufficient to warrant complex gas detection equipment to detect all gas leaks from the sanitary landfill. Most of the gas production will occur within the first 3 to 5 years after the solid waste has been placed in the sanitary landfill. Gas production will slow and continue to be produced for a number of years.

Metabolism of the solid wastes in the sanitary landfill will result in loss of solid materials and the creation of increased void spaces. The weight of soil and solid wastes above the metabolized organic matter will often cause the surface of the sanitary landfill to settle over time. Since the surface settling is not uniform, it will be necessary to add soil to the landfill surface to maintain the desired surface for proper rainfall runoff. Since sanitary landfills may contain useful materials for future generations, it is important to keep the surface of finished sanitary landfills free of complex structures that would have to be removed before the solid wastes could be mined for reuse. Completed sanitary landfills should be designed as neighborhood parks when the communities expand. Park areas provide suitable open spaces for the people to enjoy in their residential areas. Too often the finished sanitary landfill areas are used for industrial buildings, commercial buildings, and dense residential structures. These structures will find the slowly settling sanitary landfill unsuitable as a solid foundation and will retard future recovery of the solid wastes. Proper urban planning can insure these potential problems do not occur.

Nutrient Deficient

One of the problems with biological treatment of municipal solid wastes lies with the deficiency of nutrient elements for proper metabolism. Not only are municipal solid wastes deficient in nitrogen and phosphorus, they are also deficient in trace metals needed for proper enzyme development. The daily soil cover added to the compacted solid wastes in the sanitary landfill is the primary source of nutrient elements for many of the bacteria. If leachate is not recycled through the surface soil, the nutrient elements in the soil will not be available for the bacteria. Even the nutrient elements in food wastes will not be readily available unless leachate is recycled or infiltration of surface water is allowed to enter the landfill. The food wastes with excess nutrient elements will undergo metabolism first and will be degraded to the greatest extent. The lack of sufficient nitrogen and phosphorus, as well as, trace metals will seriously limit the degradation of cellulose materials in the sanitary landfill. Metabolism of organic matter in the solid wastes requires sufficient bacteria, acclimated to the organic matter being metabolized, together with water, nutrient elements, and trace metals.

Sanitary Landfill Research

In 1970 the City of Lawrence, KS decided to locate their new sanitary landfill in the floodplain of the Kansas River, north of the City. The sanitary landfill lay on a flat area between the Kansas River and the North Lawrence flood levees, constructed by the U. S. Corp of Engineers. Although concerns had been raised about the sanitary landfill being located in a floodplain, the flow in the Kansas River was considered sufficient to handle any materials leached from the landfill. Corp of Engineers data indicated that the sanitary landfill site could be flooded every few years for a few days. The velocity of flood flow over the sanitary landfill was not considered sufficient to scour the solid wastes in the landfill. It was expected that materials would be deposited on the landfill surface. The first trench was dug at this site and municipal solid wastes were placed into the trench on July 20, 1970.

Lawrence municipal solid wastes were added to this site for three weeks before the Corp of Engineers indicated that it could not approve the City's permit to operate the sanitary landfill in the Kansas River floodplain. The City modified its permit request to allow the site to be used for research on solid wastes disposal. The sanitary landfill site was completed, covered, and left for research by graduate students at the University of Kansas under a grant from the EPA Office of Solid Waste.

The first step in the research project was the placement of a series of sampling wells at various depths within the landfill, below the landfill, upstream, and downstream of the landfill. A total of 11 shallow wells were drilled for this study. Each well was designed to collect water through a stainless steel well screen at a specific depth. Three temperature thermistors were buried at different depths within the landfill to provide direct temperature measurements. A one year detailed study was made on the microbiological and the chemical characteristics of the water samples collected on a biweekly basis for a year, February, 1971 to February, 1972.

During this one year study the water table only reached a level between the bottom of the landfill and 1.2 m (4 ft) above the bottom of the landfill. Water from the Kansas River caused the groundwater in the floodplain to change as the depth of flow in the river changed. Groundwater entered the bottom of the sanitary landfill the middle of May, reached the maximum level at the end of May, and dropped below the landfill the first part of August.

N. C. Burnett, D. A. Degner, and W. L. Mills collected and analyzed the water samples for the first year. N. C. Burnett was responsible for the microbiological analyses. D. A. Degner made all the chemical analyses for the first 8 months and W. L. Mills made the chemical analyses for the last 4 months. As expected, the microbiological data showed considerable variation. The 20°C Total Bacteria Count showed a median count of 8 x 10^4/ml for the water in front of the sanitary landfill. The water near the bottom of the sanitary landfill had a median count of 5.5 x 10^4/ml. The water just below the sanitary landfill had a median count of 2.5 x 10^6/ml. The water just downstream of the sanitary landfill had a median count of 5 x 10^4/ml. These data showed that the water leaching down out of the sanitary landfill had a higher bacteria count that the water in the landfill and the water downstream of the landfill. A more favorable environment just below the sanitary landfill allowed the better bacteria growth than inside the landfill, The water below the sanitary landfill had a median pH level of 6.8 in contrast to pH 5.8 near the bottom of the sanitary landfill. The median fungi colony count was 7/ml in the water ahead of the sanitary landfill, 5.5 x 10^3/ml near the bottom of the sanitary landfill, 4 x10^4/ml just under the sanitary landfill, and 6/ml downstream from the sanitary landfill. The use of the Most Probable Number test for coliform bacteria yielded little quantitative data. Only a few samples showed the presence of fecal coliform bacteria. It was readily apparent that the Lawrence sanitary landfill was not a source of fecal bacteria during this study.

The chemical analyses provided more data than the microbiological analyses. Data were collected on pH, temperature, specific conductance, alkalinity, chemical oxygen demand (COD), calcium, magnesium, iron, ammonia nitrogen, organic acids, chlorides, phosphates, and sulfates. The temperature within the sanitary landfill averaged 17°C in contrast to 15°C for the other three sample points. The COD data averaged 43,000 mg/L near the bottom of the sanitary landfill, 3,600 mg/L just below the sanitary landfill, 15 mg/L in ahead of the sanitary landfill, and 30 mg/L downstream from the sanitary landfill. There were 14,000 mg/L volatile acids near the bottom of the sanitary landfill, 1,500 mg/L just below the sanitary landfill, 9 mg/L ahead of the sanitary landfill, and 11 mg/L downstream of the sanitary landfill. Iron data showed 950 mg/L near the bottom of the sanitary landfill, 170 mg/L just below the sanitary landfill, 24 mg/L ahead of the sanitary landfill, and 19 mg/L downstream from the sanitary landfill. The NH_3-N averaged 350 mg/L near the bottom of the sanitary landfill, 35 mg/L just below the sanitary landfill, 2.7 mg/L ahead of the sanitary landfill, and 18 mg/L downstream of the sanitary landfill. Chlorides are quite soluble and make the best tracer of pollutant movement in the groundwater. The chlorides averaged 940 mg/L near the bottom of the sanitary landfill, 81 mg/L just below the sanitary landfill, 16 mg/L just ahead of the sanitary landfill, and 24 mg/L downstream of the sanitary landfill.

During this study the water rose into the bottom of the sanitary landfill and slowly dropped below the sanitary landfill. The water level inside the sanitary landfill dropped more slowly than the water table around the sanitary landfill. Considering the large paper fraction in the solid wastes, this is not unexpected. The rainfall during this period was 37.2", not sufficient to saturate the solid wastes in the sanitary landfill.

In March, 1973 the Kansas River level rose high enough to completely cover the sanitary landfill. It was April 19 before additional samples could be collected. D. A. Blackman, a graduate student at the University of Kansas, assisted in the sample collection and analyses. The water level had receded to about one foot below the top of the solid waste mass. As anticipated, there had been no scouring of the surface of the sanitary landfill. Silt carried by the river water was deposited on the surface of the sanitary landfill as the water level receded. The well ahead of the sanitary landfill could not be sampled at this time. All four wells within the sanitary landfill could be sampled. The chloride data were used as the primary measure of contamination. The chloride concentrations in the sanitary landfill were 68 mg/L in the middle of the 2 ft of soil cover, 540 mg/L one foot into the solid wastes, 1,850 mg/L three feet above the bottom of the sanitary landfill, 6,100 mg/L near the bottom of the sanitary landfill, and 1,320 mg/L just below the sanitary landfill. The downstream sample contained 110 mg/L chlorides. These data showed that flooding the sanitary landfill allowed additional chloride releases within the sanitary landfill and some release from the sanitary landfill. Samples collected in May, June, August, and September showed no significant change in the chloride concentrations except in the downstream samples. The chloride concentration in the downstream well was 120 mg/L in May, 240 mg/L in June, 350 mg/L in August, and 710 mg/L in September. As the water level dropped, some of the chlorides were leached from the sanitary landfill.

A second flooding of the sanitary landfill occurred the latter part of September, 1973. This time the Kansas River water rose at least 6 ft above the ground surface over the sanitary landfill. Again, there was no scouring of the landfill surface. When the water receded, samples were collected on October 19. The chloride concentrations in the sanitary landfill were 310 mg/L in the middle of the soil cover, 460 mg/L one foot into the solid wastes, 1,900 mg/L three feet above the bottom of the sanitary landfill, 1,250 mg/L near the bottom of the sanitary landfill, and 1,300 mg/L just below the sanitary landfill. The downstream sample contained 180 mg/L chlorides. These data showed that soluble contaminants were slowly leached from the sanitary landfill into the flowing groundwater adjacent to the Kansas River. Efforts by several research groups to find significant contamination of the groundwater downstream of the sanitary landfill over the next decade, as environmental awareness increased in the United States, failed to find any significant contamination from the sanitary landfill. The intake for the Lawrence municipal water supply on the Kansas River was located a few miles downstream of the sanitary landfill. At no time was the Lawrence water quality adversely affected by the sanitary landfill leachate. The federal EPA soon developed regulations prohibiting sanitary landfills being located in floodplains. This landfill was officially closed in 1981. The last samples were collected on July 25, 1984. The chloride concentrations were 29 mg/L in the well ahead of the sanitary landfill, 12 mg/L in the middle of the soil layer over the sanitary landfill, 12 mg/L one ft into the solid wastes, 88 mg/L three feet above the bottom of the sanitary landfill, 88 mg/L near the bottom of the sanitary landfill, 56 mg/L just below the sanitary landfill, and 17 mg/L downstream from the sanitary landfill.

It is important to realize the rise and fall of the Kansas River water in the Lawrence sanitary landfill provided the water needed to stimulate microbial degradation of the biodegradable materials in the Lawrence municipal solid wastes. When the river water covered the entire sanitary landfill, microbial metabolism was anaerobic. As the water level dropped, the upper layer of the solid wastes should have been exposed to some aerobic metabolism. It would make a good research study to determine the status of the solid waste materials in this test cell after 33 years of fluctuating water levels.

COMPOSTING

Composting of agricultural wastes has been practiced for centuries and is a satisfactory method for handling agricultural solid wastes. While some third world countries combine their organic municipal wastes with their organic agricultural solid wastes for composting, the United States has largely ignored composting as a method for handling the organic fraction of municipal solid wastes. As the municipal solid waste problems increased in the United States after WWII, efforts were directed towards using composting as a suitable treatment process for municipal solid wastes in small, rural communities where people recognized the value of agricultural compost. Composting research on municipal solid wastes proceeded along two separate lines, windrow composting and high rate, mechanical composting. It was not until the EPA Office of Solid Wastes set specific goals for the reduction in municipal solid wastes that composting became a part of solid waste processing in the United States. Most municipalities currently operate composting systems for yard wastes. Some municipalities also operate composting systems for processing wastewater sludge. The differences between yard wastes and wastewater sludge characteristics create two distinct treatment processes that utilize the same basic concepts.

Windrow Composting

Efforts to promote the use of special bacteria cultures to enhance composting of municipal solid wastes in California in the early 1950s stimulated the University of California at Berkeley to set up a research project to determine the fundamentals involved in composting municipal solid wastes. They used Berkeley, CA municipal solid waste in a windrow composting system. The solid waste underwent an initial separation process to remove all the non-biodegradable, inert materials, such as glass, metal objects, and plastics. About 66% of the municipal solid waste was considered suitable for composting. The solid wastes were ground to a particle size close to 1.5 inches across and then placed into elongated piles about 10 ft wide and 5 ft to 6 ft high. Samples were collected of the solid wastes for chemical analyses at the start of composting.

In a few days the temperature of the waste piles began to rise. When the temperature within the solid waste pile reached 75° C, a front loader was used to turn the pile of solid waste over to allow the heat to escape to the atmosphere and to allow fresh air to be trapped in the void spaces around the solid particles. Loss of moisture from the pile required adding water. It was quickly found that the moisture content played an important role in the composting process. Too much moisture in the pile absorbed the excess heat, keeping the temperature from rising to the desired level. It also allowed the solid particles to stick together, creating an anaerobic environment within the piles. Too little moisture prevented microbial growth. It was found that the proper range for moisture in the compost pile was between 40% and 65%. The 75° C temperature killed all of the pathogenic microorganisms in the solid wastes, allowing the finished compost to be handled without danger from pathogens.

After turning the compost pile, the temperature rose again to a maximum level. The compost pile was turned again and the moisture adjusted to the desired level. The temperature in the pile rose again but not as high as before. When the temperature reached its maximum level again, the compost pile was turned. After several turns, the temperature in the pile hardly rose at all, indicating the compost was relatively stable. The temperature rises in the compost piles indicated the rate of metabolism and the extent of metabolism by the microorganisms. The readily biodegradable organic compounds in the solid waste had been quickly metabolized; but the hard to metabolize organic compounds were barely touched. It was noted that bacteria were important in metabolizing the proteins in the solid wastes since the proteins provided sufficient nitrogen for the bacteria. Fungi and actinomycetes were found to metabolize the lignin-cellulose materials in the solid wastes. One of the important findings of the University of California was the fact that special microbial cultures were not needed for composting. The solid wastes already contained sufficient numbers and types of microorganisms that seeding had no effect on windrow composting.

High-Rate Composting

Windrow composting was a slow rate process and relatively labor intensive. European engineers had developed several different types of mechanical composting systems for municipal solid wastes. The mechanical composting systems were high-rate composting systems when compared with the windrow system. John R. Snell carried out a research study on high-rate, mechanical composting using a vertical reactor at Michigan State University between 1953 and 1955. His research showed that a finely ground solid waste gave the best composting results when the C/N ratio in the reactor was less than 50/1. The pH in the reactor was maintained between 5.5 and 8.0 with the moisture level between 50% and 60% by weight. The best microbial seed came from finished compost, between 2% and 10% of the solid waste load. The compost in the reactor was continuously mixed to allow the composting solids to have proper exposure to air. Air was continuously blown through the reactor to maintain adequate oxygen for the microorganisms. The temperature was controlled to maximize rapid microbial growth. Professor Snell found the compost was finished when there was no significant increase in temperature, no further loss of nitrogen, and the compost had no offensive odors.

With all the increased interest in high-rate composting John S. Wiley of the USPHS set up his own experiments to determine the primary factors affecting the high-rate systems. His research was carried out on a batch basis in 15 gallon composting units that were mechanically aerated and continuously mixed. John Wiley used solid wastes from the city of Savannah, GA. He ground the solid wastes in a hammermill to particle size less than 5/8 inches across. The results of this study showed that good mixing was essential to bring the microorganisms, the solid waste, and oxygen together for rapid metabolism. Under optimum conditions the temperature of the compost rose to between 60° C and 70° C before falling back to ambient temperature. Air was a critical parameter for the success of the high-rate composting. It was necessary to keep the airflow rate to between 10 and 30 cf/day/lb VS in the initial load of solid wastes. Too little airflow allowed anaerobic conditions to develop, slowing the overall composting process. Too much airflow dried out the compost and stopped metabolism. The optimum moisture level in the compost was between 55% and 69%. The putrescible materials in the solid waste were stabilized in 6 to 9 days. There was very little decomposition of paper products, wood, or rags in the compost after 9 days retention.

For the most part, interest in high-rate composting peaked with increased federal funding for demonstration plants. The operational problems that occurred in the demonstration plants together with the high cost of construction and operation led to the loss of interest in composting of municipal solid wastes. Municipalities were not equipped to produce, promote, and sell the solid waste compost. Municipal solid waste composting faded into the background except for isolated cases and specific applications.

Wastewater Sludge Composting

When Congress passed PL 92-500, all municipalities were required to establish secondary wastewater treatment. One of the consequences of constructing large numbers of secondary wastewater treatment plants was the production of increased quantities of wastewater sludge that had to be collected, treated, and returned back to the land environment. The initial concern was in finding a suitable method for return of the wastewater sludge to land without creating localized nuisance conditions. Equipment was developed to apply liquid sludge to agricultural land without creating nuisance conditions for the farmer or for adjacent residents. Overloading land areas with too much sludge led to regulations limiting the sludge application to specific land areas based on the nitrogen and phosphorus requirements of the crops being grown on the land area. As the sludge limitation regulations became accepted, concerns were raised about the survival of pathogenic microorganisms in the sludge and the potential effect on people in the vicinity of the sludge treated land. In an effort to eliminate the concerns over potential pathogenic microorganisms in the municipal wastewater sludge before it was added to agricultural land, attention was turned to composting as a possible pretreatment method. It was recognized that composting could produce high enough temperature to destroy the pathogenic microorganisms. The problems with composting wastewater sludge were the high moisture content, 80% to 90%, the density of the sludge particles, and a potential high oxygen demand. If composting of wastewater sludge was to be successful, these problems had to be solved.

Research at the USDA Research Laboratory in Beltsville, MD on composting wastewater sludge began in 1972. It quickly became apparent that a bulking agent, capable of adsorbing the excess moisture, was needed with wastewater sludge. Tests on a number of natural waste materials resulted in wood chips being the best bulking agent. Windrow composting with anaerobically digested wastewater sludge proved quite satisfactory. When raw wastewater sludge was composted, the demand for oxygen was too great and obnoxious odors resulted. Compressed air was supplied through perforated pipes located at the bottom of the windrow piles to supply sufficient oxygen. Blowing air through the compost pile resulted in the discharge of fungi spores into the surrounding environment, posing a potential health hazard to people around the windrow piles. The airflow was reversed with the air being pulled through the windrow pile. The inability to transfer sufficient oxygen to all parts of the compost pile resulted in some anaerobic metabolism and the production of obnoxious odors. The compressed air discharge was directed through a perforated pipe covered with a mound of finished compost to allow absorption and metabolism of the odorous materials from the windrow pile. Wastewater sludge finished the first stage of composting in 21 days. While the easily biodegradable organic materials were stabilized in 21 days, the slowly metabolized organic compounds required an additional 31 days curing before the composted material could be placed into bulk storage. The wood chips were recovered for reuse by screening. It was also found that placing a layer of compost over the pile before starting the compost cycle insulated the pile and helped retain the heat produced by microbial metabolism. By 1980 the Beltsville aerated-pile method for composting wastewater sludge was well established as a viable treatment process to destroy any pathogenic microorganisms that remained after secondary treatment.

The success of the Beltsville composting process stimulated equipment manufacturers to adapt their composting designs for municipal solid waste to wastewater sludge and to mixtures of municipal solid wastes and wastewater sludge. Mechanical compost systems appealed more to large wastewater treatment plants that lacked space and personnel for the windrow composting systems. Mechanical compost systems were built in vertical cylinders and in horizontal channels. The tendency of the mechanical systems was to provide for a rapid temperature rise to kill any pathogenic microorganisms, followed by a drying of the compost to prevent further microbial metabolism. The dry compost was stable until it became wet. Essentially, the composting process was completed when the compost was placed back into the natural environment and mixed with soil.

Yard Waste Composting

The large amount of yard wastes produced in the United States served as a potential for a significant reduction in municipal solid wastes sent to sanitary landfills. Initial efforts to develop a major household composting effort failed to generate the support desired by the EPA Office of Solid Wastes. Pressure on municipalities resulted in the development of special yard waste collections and yard waste composting. Yard wastes consist of grass clippings, branches from shrubs and trees, and leaves. Yard waste production is seasonal in the northern areas and continuous in the southern areas. Most communities use windrow composting because of its cost and simplicity. Tree branches are ground to produce wood chips that provide filler for grass clippings. The slow rate of wood decomposition compared with other plant materials allows the yard waste to compost slowly over time. Many communities allow the compost to accumulate for one to two years before using it.

Yard waste composting confirms the observation from other solid wastes that lignin-cellulose materials degrade slowly. Fungi and actinomycetes metabolize lignin-cellulose materials under aerobic conditions provided there are sufficient nitrogen, phosphorus, and trace metals for normal cell growth. These microbes require less moisture, less nitrogen, and less phosphorus for cell mass than bacteria. Low nutrient concentrations and low moisture allow the fungi and actinomycetes to grow normally while limiting the bacteria growth. Fungi also favor low pH environments. Bacteria grow best at neutral or alkaline pH levels and high moisture levels. While bacteria cannot prevent the growth of fungi and actinomycetes, bacteria grow at a faster rate and quickly overgrow the fungi and actinomycetes. Bacteria are responsible for the rapid temperature increase during the start of composting. Bacteria continue rapid metabolism under aerobic conditions, producing carbon dioxide, water, and new cell mass as their end products with the heat generation directly proportional to metabolism. Once the easily metabolized organic compounds have been metabolized, the bacteria shift their metabolism to endogenous respiration with a release of nitrogen, phosphorus and trace metals for other microorganisms to use. The complex chemical structure of wood limits the overall rate of metabolism. Metabolism by the fungi and actinomycetes is quite slow. Essentially, the composting process is controlled by the ability of the different microorganisms to metabolize the different organic materials.

Agricultural Composting

Agricultural composting is primarily anaerobic composting rather than aerobic composting. The limited labor available for composting solid wastes on most farms favors anaerobic composting. In the United States agricultural composting is limited to animal manure and specific crop residues. Manure from cattle feedlots is allowed to collect on the ground and undergo initial decomposition. Periodically, the manure is scraped from the feedlots and placed into large bunker cut into a hillside. The manure is added periodically to the bunker until it is full. As the bunker is filled, the top of the manure heap is covered with plastic film to minimize surface exposure. Old tires are often used to hold down the plastic film. Anaerobic decomposition proceeds very slowly since the composting materials are not mixed. Essentially time is exchanged for mixing. The manure remains in the bunker for 6 to 12 months before being removed and used as a natural fertilizer on agricultural fields.

Dairy farms normally have a mixture of straw bedding and manure that is removed at regular intervals and composted in windrow piles or earthen bunkers. Since fresh manure contains considerable water, the straw bedding helps to absorb the excess moisture. The straw also provides organic material high in carbohydrates. The windrow piles tend to be aerobic around the edges and anaerobic in the center of the piles. Periodic turning of the piles allows the actinomycetes and the fungi to metabolize the organic materials the bacteria have not metabolized. Use of earthen bunkers for composting does not allow turning of the composting materials. The normal farm cycle allows the composting materials to be used on an annual basis.

Cattle manure contains a very diverse group of bacteria that provide a proper seed for composting agricultural residues. Most of the bacteria from rumen animals are either facultative or strict anaerobes. The high protein content of the manure helps keep the pH high as NH_3-N is released. The organic acids formed as end products of anaerobic degradation are quickly neutralized by the excess NH_3-N, allowing the methane bacteria to metabolize the organic acids. The breakdown of lignin-cellulose compounds by the cattle allows cellulose bacteria to grow in the manure. Returning the composted manure to agricultural fields helps to supply organic matter that can hold more moisture, trace metals for good plant growth, and a diverse population of bacteria needed for a healthy soil.

SOIL STABILIZATION

Soil can provide an excellent medium for the stabilization of biodegradable solid wastes. Soil is an excellent medium for creating an aerobic environment and supplying a diverse supply of microorganisms necessary for waste stabilization. Soil stabilization of suitable solid wastes can improve the soil characteristics. It is important to recognize that every solid waste must ultimately be returned back to the soil at some stage in the treatment cycle. The microorganisms in the soil will complete the waste stabilization process.

All the microorganisms found in the complex waste treatment processes can be found in soil. Diverse bacteria, actinomycetes, fungi, protozoa, rotifers and worms are all found in soil. It is not surprising that soil is the source of the microorganisms predominating in all of the various biological waste treatment processes. The upper layer of soil provides an aerobic environment except for periods when the void spaces between soil particles are filled with water. When the void spaces are filled with water for an extended period of time, the soil environment will shift to being anaerobic. While the active microbial populations will reflect the existing environment in the soil, microorganisms suited to the opposite environment can also be found as spores or cysts. The spores and cysts can remain viable for extended periods of time. Thus, it is possible to isolate anaerobic bacteria in aerobic environments and to isolate aerobic microorganisms in anaerobic environments. Quantitative enumeration of the predominant microorganisms is essential to understand the significance of different microorganisms isolated from any waste treatment system.

Wastewater Treatment Plant Sludge

Wastewater treatment plant sludge disposal onto and into soil has been the most widely accepted soil stabilization process for solid wastes. Prior to the EPA being established in the United States, anaerobically digested wastewater sludge was often placed on the soil surface for final disposal. Concern over loss of sludge contaminants from the soil into adjacent watercourses led to the development of injection methods for adding liquid sludge under the soil surface and covering the sludge with soil. The injection and cover technique allowed raw wastewater sludge to be added to the soil and stabilized in the soil without the creation of nuisance conditions. Over time the microorganisms in the soil and in the sludge metabolized the biodegradable organic matter in sludge. The readily biodegradable organic compounds were rapidly metabolized while the complex organic compounds were slowly metabolized. The non-biodegradable organic compounds, the inorganic compounds, and the dead cell residue accumulated in the soil, improving its characteristics for agricultural crops. Only crops for consumption by domestic animals are grown on soil enriched with wastewater sludge in most countries. Some countries permit the growth of crops for human consumption where the harvested products are thoroughly cooked prior to consumption. Thorough cooking kills all of the pathogenic microorganisms that might be on the harvested products.

Agricultural Solid Wastes

Crop residues that have little economic value have long been allowed to decompose on the soil surface prior to being plowed back into the soil. There are differences of opinions on the “pros and cons” of leaving crop residues on agricultural soils. Soil conservationists favor leaving crop residues in the fields to reduce soil erosion. Some cattle growers favor leaving the crop residues in the fields and allowing cattle to feed on the crop residues during the winter months. Other cattle growers favor harvesting the crop residues for feeding cattle as part of their overall rations. No-till advocates definitely favor leaving crop residues on the land. The opposite view indicates that crop residues tend to accumulate on the soil surface, making planting of new crops more difficult. It has also been indicated that no-till cropping can lead to increased transmission of microbial plant diseases.

Crop residues have two major components, the above ground fraction and the below ground fraction. The above ground crop residues have the greater mass of material than the below ground crop residues. The value of the above ground crop residues in soil conservation is in absorbing much of the energy of the falling rain hitting the land surface. The below ground crop residues have extensive networks of large and small root structures that help hold the soil particles together. The major problems with crop residues are their very large lignin-cellulose complex and the lack of much nitrogen, phosphorus, and trace metals. Microbial degradation of above ground crop residues are related primarily to actinomycetes and a few fungi. The actinomycetes predominate over the fungi as a result of their smaller size and their ability to grow at a faster rate. The lack of nitrogen, phosphorus and trace metals limits the metabolism of crop residues by bacteria. Bacteria and actinomycetes combine to metabolize the below ground crop residues since there are more nutrients and trace metals available from the soil. A study reported by the USDA Agricultural Research Service in TEKTRAN indicated a long-term study of crop residue reduction in three no-till sites in eastern Colorado found a 43% yearly average residue loss. The residue loss also followed a first order relationship. It appears that limited mixing of the crop residues in the soil, together with limited nutrient elements and moisture, prevents rapid microbial metabolism of the crop residues.

Municipal Solid Wastes

Soil stabilization of municipal solid wastes has been rejected on the basis of the large manpower and energy requirements. Pretreatment of municipal solid wastes would require separation of the biodegradable materials from the metal, glass, and plastic materials. The biodegradable materials would need to be ground to a small particle size, about ¾ inches. The solid wastes would then be plowed into the surface of soil. Periodic mixing would be required to insure contact between the microorganisms, the nutrients and the solid wastes. The food wastes would be the readily biodegraded; but the lignin cellulose materials in the paper wastes would be slowly metabolized. The value of soil stabilization of municipal solid wastes would be the conversion of sandy soils to higher organic residue soils that could hold moisture and permit increased crop cultivation. Composting of municipal solid wastes and the addition of the finished compost to sandy soils would accomplish the same end results as soil stabilization with less effort.

The treatment of solid wastes is more difficult than the treatment of liquid wastes. Solid wastes have a limited surface area that must be increased for rapid metabolism. Grinding can reduce the particle size for solid wastes and increase the surface area. The solid wastes lack reactive sites as much as liquid wastes. Bacteria, actinomycetes and fungi must come into direct contact with the solid waste and hydrolyze the reactive site. In effect, the microorganisms must convert the solid waste to liquid wastes for metabolism. Adequate moisture must be available for the microorganisms to metabolize the organic materials. Each microorganism must obtain the building blocks and the energy to synthesize new cell protoplasm from the solid waste. If the solid waste is deficient in nutrient elements or trace metals, the microorganisms must obtain these materials from a nearby source in the environment. Coating the solid waste particle with microorganisms limits the growth since food becomes limiting for the outer layer of microorganisms. The lack of mixing allows end products of metabolism to accumulate and poison the local environment. Mixing can remove some of the surface microorganisms and the toxic end products, allowing more microorganisms to grow and speed the rate of stabilization of the solid wastes. As with the case of all waste treatment systems, creating the optimum environment allows for the best stabilization of the solid wastes.

THINGS TO REMEMBER

1. Solid wastes are more difficult to treat than liquid wastes.

2. Obtaining a true measure of solid waste characteristics requires much more effort than measuring liquid waste characteristics.

3. Municipal solid waste characteristics can be grouped into 9 major groups.

4. The data on municipal solid waste characteristics is still quite limited.

5. Sanitary landfills are engineered burial of solid wastes.

6. Considerable efforts have been made to eliminate sanitary landfills in the United States but no other suitable method exists for properly handling solid wastes.

7. Sanitary landfills are currently designed to collect gas and leachate produced in the landfill and to properly treat these materials before discharging them back into the environment.

8. Bacteria are the primary microorganisms in sanitary landfills.

9. Sanitary landfills operate in an anaerobic environment, limiting metabolism of organic compounds.

10. Many municipalities separate and compost yard wastes to minimize their placement in sanitary landfills.

11. Economics does not favor municipal solid waste composting in the

United States.

12. Some WWTP compost their wastewater sludge.

13. High rate composting systems only stabilize the readily biodegradable organic compounds.

14. Curing is required for at least 3 to 6 months or longer before most compost is stable enough to be used.

15. Soil stabilization of wastewater sludge has been quite successful at many WWTP.

16. Crop residues are decomposed at a slow rate, requiring years for stabilization.

17. Mixing, moisture, and nutrients control biodegradation of solid wastes in soil systems.

REFERENCES

Brown, S. M. (1970) *The Establishment of a Sanitary Landfill for Lawrence, Kansas*, PhD Thesis, University of Kansas, Lawrence, KS.

Burnett, N. C. (1972) *A Biological Evaluation of the Effect of a Flood Plain Sanitary Landfill Site on Water Quality*, PhD Thesis, University of Kansas, Lawrence, KS.

Degner, D. A. (1974) *A Chemical Investigation of the Effect of a Flood Plain Solid Waste Disposal Site on Ground Water Quality*, PhD Thesis, University of Kansas, Lawrence, KS.

Hart, S. A. (1968) *Solid Waste Management/Composting; European Activity and American Potential*, PHS Pub. No. 1826.

Magnuson, A. (1998) Leachate Recirculation, *MSW Management*, 24, March/April.

McGauhey, P. H. (1971) *American Composting Concepts,* **SW-2r**, USEPA.

Metcalf & Eddy (1991) *Wastewater Engineering*, 3rd Ed., McGraw-Hill.

Richman, M. (1966) Wrestling with In-Vessel Composting, *WEF Operations*

Forum, **13**, 11, November.

Snell, J. R. (1957) Some Engineering Aspects of High Rate Composting, *Jour. San. Engr. Div., ASCE*, **83,** SA1, Paper No. 1178.

Tietjen, C. and Hart, S. A. (1969) Compost for Agricultural Land, *Jour. San. Engr. Div., ASCE,* **95,** SA2, 269.

Tilsworth, T. (1970) *Aerobic Surface Stabilization of Refuse*, PhD Thesis, University of Kansas, Lawrence, KS.

USDA (1980) *Manual for Composting Sewage Sludge by the Beltsville Aerated-Pile Method*, **EPA-600/8-80-022**.

USDA (1998) *Long Term Decomposition of Crop Residues in Dry Land Agroecosystems*, **TEKTRAN**, Ag. Res. Service, March.

USEPA (2002) *Municipal Solid Waste – Basic Facts*, http://www.epa.gov, Web Site.

University of California (1953) *Reclaiming Municipal Refuse by Composting,* **Tech. Bull. No. 9, Series 37**, Sanitary Engineering Research Project, Berkeley , CA.

Wiley, J. S. (1955) Studies of High-Rate Composting of Garbage and Refuse, *Proc. 10th Ind. Waste Conf., Purdue Univ.,* 306, Lafayette, IN.

Chapter 14

HAZARDOUS WASTES

Hazardous wastes is a term that was introduced in the 1970s to replace the term, *toxic wastes*. For years engineers and scientists recognized that specific compounds in gaseous, liquid and solid wastes were toxic to biological life. It was also recognized that toxic wastes required special handling and treatment before the waste materials could be returned to the environment. In effect, the toxic wastes needed to be converted to non-toxic compounds. Concentrated inorganic acid wastes, primarily sulfuric acid, nitric acid, hydrochloric acid, and hydrofluoric acid, had to be neutralized and diluted prior to discharge in the environment. Concentrated inorganic alkaline wastes, containing sodium hydroxide, calcium hydroxide, and magnesium hydroxide, also needed to be neutralized and diluted prior to discharge. The same was true of strong oxidizing compounds, such as chlorine and ozone. Strongly reduced compounds, such as hydrogen sulfide, required oxidation and dilution. One of the major problems with toxic wastes lay with the multitude of organic compounds that had varying degrees of toxicity. There were no simple methods to treat the toxic organic wastes. Unfortunately, too many industrial plants ignored their toxic wastes and created serious pollution problems. Rachel Carson's book, *Silent Spring*, focused on the potential dangers of organic pesticides such as DDT, methoxychlor, chlordane, heptachlor, and benzene hexachloride. Rachel Carson's concerns with the dangers of pesticides struck a special chord with the rising environmental movement in the late 1960s. Widespread use of the herbicides 2,4-D and 2,4,5-T, during the unpopular war in Vietnam became a focal point of opposition to the organic chemical industry as a whole. As public opinion against toxic organic chemicals increased, efforts were soon directed towards federal legislation to prohibit their manufacture and use. Part of the public relations ploy against toxic wastes was the change from in terminology from toxic wastes to hazardous wastes. The term, hazardous wastes, conveyed a far greater danger to the public than toxic wastes and was used to convey the impression that hazardous wastes had been totally neglected by regulatory agencies and by industries producing these materials.

In 1976 Congress passed the Resource Conservation and Recovery Act (RCRA). One of the major provisions of RCRA was designed to control the handling and disposal of hazardous wastes in the environment. The net result was EPA's development of the "cradle to grave" concept for controlling hazardous waste production, storage, transportation and disposal. The ultimate objective of this legislation, in the eyes of the avid environmentalists, was the elimination of hazardous waste production. Unfortunately, it was not practical to eliminate all hazardous waste production. Instead of eliminating hazardous wastes, one of the most complex, bureaucratic systems ever devised to handle waste materials was created. In addition to RCRA, which dealt with hazardous wastes, Congress passed the Toxic Substances Control Act (TOSCA) to control potentially toxic materials before they were used. TOSCA required chemicals to be tested for health effects and evaluated for potential health risks. Efforts by industries to meet the requirements of these laws have created an entirely new group of technical specialists in the waste-handling field. It has also produced major changes in the way many industries and business are able to operate. Since all toxic chemicals cannot be made non-toxic during manufacturing processes, hazardous waste generation is inevitable. The basic problem for hazardous wastes is one of containment and processing for safe return back into the environment.

DEFINING HAZARDOUS WASTES

Hazardous wastes have been defined by RCRA into four basic categories.

1. *Ignitability*: poses a fire hazard during routine handling.

2. *Corrosivity*: liquid wastes having a pH equal to or less than 2.0 or equal to or greater than 12.5.

3. *Reactivity*: unstable material, reacts violently with water, produces toxic gases or is explosive.

4. *Toxicity*: extractable material in water that is biologically toxic.

While these definitions cover the range of hazardous wastes, it soon became evident that these four groups are much too broad for day-to-day use. Many commonly used chemicals could be classified as “hazardous” using this classification. In an effort to clarify what constitutes hazardous wastes, the EPA developed four lists of chemicals and materials that could be easily identified as being hazardous. The *F-list* contains hazardous wastes from non-specific sources and includes solvents used in degreasing, metal plating wastes and various chlorinated organics. The *K-list* contains hazardous wastes generated by specific industrial processes, such as, wood preservation, pigment production, chemical production, petroleum refining, iron and steel production, explosives manufacturing, and pesticide production. The *P-list* includes specific discarded commercial chemical products, container residues, and spillages that are acutely toxic and are accumulated in amounts greater than 1.0 kg/month. The *U-list* is similar to the P-list except they can be accumulated up to 25 kg/month without regulation. The four lists of hazardous waste materials are sufficiently specific that both industries and regulatory agencies know what materials need to be examined under EPA regulations. Needless to say, it requires a major educational effort to reach every hazardous waste producer across the United States and to make certain new users of chemicals are aware of the potential threat that hazardous wastes pose. The hazardous waste division of EPA will never have to worry about running out of work. The total workload is increasing with our expanding population.

SOURCES OF HAZARDOUS WASTES

Hazardous wastes are produced from specific industrial operations. Like most industrial wastes, hazardous wastes can be created from the raw materials used in the industrial operations, from intermediates formed during the processing of raw materials, and even from the final products packaged for sale to the public. Most industrial plants do not plan on producing hazardous wastes since they create serious internal management problems. Yet, some plants cannot help but produce hazardous wastes in view of the raw materials they use, the intermediates they produce, and their final products. Most hazardous wastes produced inside industrial plants come from spills, leaks, and cleaning equipment and workspaces. Most consumer hazardous wastes come from discarding containers containing products having hazardous characteristics. Considerable efforts are being made to reduce consumer hazardous wastes by changing the formulations of products from hazardous materials to non-hazardous materials wherever possible. Many industries are changing raw materials and the production of specific intermediates to minimize the production of hazardous wastes. Unfortunately, the leaks and spills of hazardous materials from old chemical processes have created major hazardous waste problems that have yet to be fully addressed. Many of the industrial plants that created considerable hazardous wastes have closed operations; but the hazardous wastes remain in the soil around the old plants.

Originally, hazardous wastes were believed to be largely solid wastes deposited in sanitary landfills. Many industries placed their hazardous wastes into steel barrels and buried them in either municipal landfills or industrial landfills. For this reason the early hazardous waste regulations were concerned with solid waste management. It did not take long before it was recognized that many of the hazardous wastes in the barrels were liquid wastes and semi-solid sludge. Over time the steel barrels began to leak and the apparent solid wastes became liquid wastes that moved out of the sanitary landfills into the ground water that was being used for local water supply. It is currently recognized that hazardous wastes can include gaseous, liquid, or solid.

TREATMENT CONCEPTS

It is important to understand how hazardous wastes can be treated if industries are to eliminate the production of hazardous wastes and if old hazardous waste sites are to be properly cleaned for reuse. Needless to say, understanding the chemical characteristics of the hazardous wastes is the first step of the treatment process. Next, it is necessary to recognize the required concentration of the hazardous wastes that can be safely discharged back into the environment. The difference between the initial concentration and the final concentration of the hazardous materials establishes the degree of treatment required. Treatment can be physical, chemical, or biological or a combination of these three.

PHYSICAL TREATMENT

Physical treatment is the simplest form of treatment and the least costly. It is also the least efficient form of hazardous waste treatment. With dilute hazardous wastes open storage ponds are often used in dry areas, if the wastes are not volatile. Solar evaporation removes water from the hazardous wastes, allowing the slow concentration of the hazardous materials at the bottom of the pond. The storage ponds can hold the hazardous wastes for long periods of time before the ponds must be cleaned out. In dry climates with limited water resources waste heat from the industrial processes can be used to evaporate the water from the hazardous wastes for reuse within the industrial plant. If the hazardous wastes have collected in the soil around the industrial plant, the simplest form of physical treatment is to collect the wastes by drilling wells and pumping the hazardous materials from the ground into waste storage ponds for further treatment. Some industrial plants have used deep well injection of liquid hazardous wastes in the past. The deep wells discharge the hazardous liquid into geological formations that retain the wastes indefinitely. The major disadvantage of deep well injection of hazardous wastes is the potential for leaks in the well piping, allowing the hazardous wastes to contaminate groundwater used for water supplies or leaks in porous geological strata. Deep well injection of hazardous wastes is no longer recommended for hazardous wastes. The risk of serious problems is too great. Ionic membranes can be used with some dilute hazardous wastes to separate the water from the hazardous materials, creating a concentrated stream of hazardous materials that could be reused in the industrial processes. Freezing the wastewaters also results in concentrating the hazardous materials in a smaller volume of water for easier handling and processing. Freezing forces the contaminants into the center of the frozen mass. Incineration is both a physical process and a chemical process. Organic wastes can be burned to oxidize the hazardous compounds to their basic components for discharge to the atmosphere. The heat produced from combustion can be captured and used for heating process units and work space, as well as for generating electrical power. It is also possible to treat organic wastes in a high temperature-high pressure reactor with an excess of oxygen to oxidize the organic waste materials in the liquid state. Heat is recovered to increase the temperature of the incoming wastes and to generate power for operating the pressure pumps.

CHEMICAL TREATMENT

With some hazardous wastes chemical treatment can be used to change the hazardous organics to a non-hazardous form. Inorganic acids and bases can be neutralized to a pH level between pH 6-9. Organic hazardous wastes can be oxidized with either hydrogen peroxide or ozone. Normally, the organic compounds are only partially oxidized to change their characteristics to a non-toxic form. Chlorine has been used to oxidize organic compounds in the past. Unfortunately, partial oxidation of the organic compounds with chlorine produces chlorinated organic compounds that are often more toxic than the original compounds. Chlorine is useful only for complete oxidation of hazardous organic compounds. Some organic compounds can be precipitated by the addition of polymers to increase the molecular size of the resulting compound. Strongly hydrophobic organic compounds can be separated from aqueous wastes with different solvents that have greater affinity for the organic compounds than water. Once the hazardous organics are extracted into the solvent, the solvent is removed by distillation for reuse. The concentrated organic compounds remain as the final residue in the distillation vessel. Activated carbon has been used to remove specific organic compounds. When the carbon is regenerated with heat, the concentrated organics are removed and often oxidized. There are many different chemical treatments available for organic hazardous wastes. Handbooks on organic hazardous waste treatment are a good source of detailed information on both chemical treatment and physical treatment systems that have been evaluated.

BIOLOGICAL TREATMENT

Biological treatment has been used for many years in the treatment of organic wastewaters containing hazardous materials. The chemical industry, petroleum refineries, and iron and steel mills have developed a number of biological treatment systems to handle hazardous waste materials. Biological treatment systems have been developed for above ground systems and in-situ systems. The above ground biological treatment systems are the easiest to design and to operate. The organic wastewaters from industrial plants are discharged into conventional or semi-conventional biological treatment systems. Activated sludge systems and high rate anaerobic systems have been used in the treatment of organic hazardous wastes. In plants where organic hazardous wastes have saturated the soil, the hazardous wastes have been pumped out of the ground for biological treatment. In a number of locations efforts have been directed towards in-situ biological treatment rather than pumping and treating. In-situ systems require the addition of nutrients and an oxygen source. Oxygen can come from diffused aeration or chemical oxygen, primarily hydrogen peroxide. Developing sufficient numbers of active, aerobic bacteria in the hazardous wastes located underground in the soil is a difficult process. The hazardous waste stream is normally pumped from the ground. The required chemicals are added and the wastes are pumped back into the ground. The major problem is proper distribution of the nutrients and the oxygen, if aerobic bacteria growth is desired. The injected liquid follows the path of least resistance in the soil. The path of least resistance may not be the desired path to reach the hazardous wastes. For many hazardous wastes it is necessary to develop a population of acclimated bacteria. Some in-situ systems have used the natural soil bacteria as a starter with the hazardous wastes determining the specific bacteria for growth. Where natural bacteria have failed to develop, acclimated bacteria have been injected with the return liquid into the soil environment. Once the bacteria begin to develop, there is concern that the bacteria mass will bridge across soil particles and retard normal fluid flow. In-situ treatment has had mixed results compared with the success of above ground treatment. Yet, the simplicity of in-situ treatment has attracted considerable attention.

Early Biological Treatment Systems

The early biological hazardous waste treatment systems were all above ground systems accepting industrial plant wastewaters, including the hazardous organics from the various process streams. Although there had been many studies on biological treatment of toxic industrial wastewater, one of the first large-scale treatment projects was the Dow Chemical Company study of the biological degradation of phenolic wastewater using trickling filter pilot plants in 1935. By 1937 they began the design of a full-scale biological treatment plant.

Phenol was an important industrial chemical and Dow Chemical was the world's largest producer of phenol. One problem that phenol created was tastes and odors in drinking water at concentrations below the toxic level. Phenol reacted with the residual chlorine in drinking water to produce a medicinal taste. Discharge of phenolic wastewater into streams and rivers used for drinking water required very high dilution rates to prevent tastes and odor complaints downstream. Biological treatment of the phenolic wastewater removed enough phenol that the downstream tastes and odors were eliminated. The Dow Chemical trickling filters were about 10 ft deep. Since phenol was a recognized biocide, the concentration of phenol in the feed wastewater was kept relatively low. Small volumes of strong phenolic wastewater were mixed with larger volumes of weaker phenolic wastewater and with the treated recycle flow. The trickling filters reduced the phenol concentration and the BOD5 concentration in the wastewaters about the same percentage, 77% to 78%. Best operations were obtained at elevated temperatures, as expected for trickling filters. The trickling filter plant was followed by 37 acres of storage ponds, providing 2 days retention at 10 mgd flow rates.

During the 1940s, R. Y. Stanier became interested in the biochemistry of aromatic organic compounds. He found that the common soil bacteria, *Ps. fluorescens*, easily adapted to the metabolism of aromatic organic compounds. In a short time Stanier isolated 22 strains of *Pseudomonas* capable of metabolizing aromatic compounds. The ease of metabolism of aromatic compounds led Stanier to examine the metabolic pathways for their metabolism. Essentially, Stanier confirmed that phenol metabolism was not as difficult as many engineers believed. Without knowing it, Stanier had begun to bridge the gap between basic microbiology and biological industrial wastewater treatment.

As Dow increased phenol production, the two trickling filters became four trickling filters. Activated sludge was added to polish the effluent instead of using large lagoons. Studies on the bacteria in the treatment systems indicated that *Bacillus, Pseudomonas, Alcaligenes, Achromobacter, Flavobacterium, Micrococcus,* and *Escherichia* were present. It appeared that a number of different soil bacteria had the ability to metabolize phenol. Expansion of petroleum refineries after World War II created many different sources of phenolic wastewater. In 1952 R. H. Coe examined the biological treatment of refinery wastewater in a small, laboratory activated sludge system. The system treated wastewater having an average of 100 mg/l phenol. He obtained 90% to 95% reduction in both BOD5 and phenol, confirming that activated sludge could be used to treat refinery wastewaters. W. W. Mathew also used activated sludge to treat ammonia still liquors generated by U.S. Steel in Gary, Indiana. He found that diluting ammonia still liquor 40:1 to 50:1 reduced the toxicity of these wastewaters and allowed the municipal activated sludge plant to reduce the influent phenol 99.94%. It was necessary to increase the oxygen supply and the concentration of MLSS to obtain good treatment. Even then, the WWTP did not show nitrification. Either the system was deficient in oxygen transfer or the phenolic compounds remaining were toxic to nitrifying bacteria. Sheets, Hamdy and Weiser examined a trickling filter pilot plant for treating catalytic cracker wastewater containing 100 to 400 mg/l phenol, 2,000 to 5,000 mg/l sulfides and 20 to 30 mg/l cyanide at a pH of 8.5. They found three major groups of bacteria in the trickling filter. *Pseudomonas, Bacillus* and *Micrococcus* were active in the treatment system. The high concentration of sulfides and the shallow filter depth, one foot, limited the phenol reduction to between 23% and 28% at 52°C to 54°C. The high temperature of the wastewater resulted in the development of aerobic, thermophilic bacteria. These studies confirmed that both mesophilic bacteria and thermophilic bacteria were able to metabolize phenolic compounds contained in industrial wastewaters.

The 1950s saw increased construction of trickling filters and activated sludge plants to treat petroleum wastewaters. The primary problems that developed in these petroleum wastewater treatment plants were related to toxicity created by highly variable wastewater characteristics. There were no controls over spills or discharges to the wastewater treatment plants. This lack of controls over the influent characteristics resulted in periodic overloading the treatment plants and the production of highly variable effluent quality. The variable effluent quality and the high cost for biological treatment stimulated refinery engineers to seek chemical and physical treatment systems to reduce the concentration of contaminants in the wastewater sent to the treatment plants. As the contaminant concentrations were reduced by pretreatment, wastewater lagoons became feasible where land was readily available. The bacteria reduced the organic contaminants, while the algae supplied the oxygen for aerobic metabolism. As the organic loads increased, floating, surface, mechanical aerators were added.

Research Studies

During this period considerable research was carried out on biological treatment of industrial wastes at MIT. Initially, the research under Professor C. N. Sawyer was directed towards determining the nutrient requirements for different industrial wastes. It was shown that a BOD_5/N ratio of 17/1 and a BOD_5/P ratio of 100/1 would provide good treatment of industrial wastewaters. Additional studies on the biological treatment of toxic organic wastes in the Sanitary Engineering Microbiology Laboratory at MIT indicated that complete mixing activated sludge (CMAS) provided the best system for maximum metabolism of both toxic and non-toxic organic wastewaters. It was also shown that acclimation was required for the proper development of bacteria populations capable of metabolizing toxic organic compounds. Most important was the demonstration that oxygen transfer, rather than toxicity, was the limiting factor in many industrial wastewater treatment plant designs. It was very important to keep the rate of organic addition below the oxygen transfer limit to prevent the buildup of toxic organics in the bioreactor.

While the results of university research were made available to practicing engineers in the field, there was a reluctance to use the research results in field scale designs. It was necessary for the MIT faculty to become involved as wastewater treatment consultants before the CMAS research was used in the field. In the 1950s a few industrial plants provided the opportunities needed to demonstrate the practical value of the new approaches to biological treatment of toxic organics. One of the more interesting demonstrations of the MIT research was at the Dominion Tar and Chemical Company Ltd. plant in Hamilton, Ontario. A small completely mixed activated sludge plant was constructed to treat the process wastewaters from coal tar distillation. The wastewaters contained about 1,000 mg/l NH_3-N and 5,000 mg/l COD as mixed phenols. A six month study of the treatment plant operations in 1960 indicated that the influent wastewater COD ranged from 3,000 to 12,490 mg/l with a median value of 7,500 mg/l. The treated effluent contained from 0.01 mg/l to 0.6 mg/l phenol with a median value of 0.06 mg/l. This treatment plant clearly demonstrated that the CMAS design could produce a very high degree of organic removal with proper operations. It was also demonstrated that laboratory studies on the biological degradation of toxic organics could be used as a basis for developing field scale treatment systems. One of the most important operating parameters for the Dominion Tar and Chemical Company wastewater treatment plant was the daily microscopic examination of the MLSS to determine protozoa activity. Since protozoa were more sensitive to toxic concentrations of pollutants than bacteria, a healthy group of protozoa indicated the activated sludge system was working normally. Dead protozoa indicated toxic conditions had developed in the activated sludge. Observation of large numbers of dead protozoa indicated to the operator that the incoming wastewaters should be turned off, allowing the system to aerate without additional wastes. Normally, the protozoa showed recovery within 24 hours, allowing the plant wastewaters to be turned back on. When chemical analyses of the incoming wastewater showed high concentrations of phenolic compounds, the feed rate was slowed to permit feeding the maximum rate of phenolic compounds that were not toxic to the bacteria. The rate of oxygen transfer was the ultimate controlling factor in determining the influent wastewater flow rate.

The key to metabolizing aromatic organic compounds is acclimation to increase the desired enzyme systems that exist normally in bacteria. The common soil bacteria have natural enzymes that are used to synthesize small quantities of aromatic amino acids in their cell protoplasm. By slowly increasing the concentration of aromatic compounds in the wastewaters, the aromatic enzymes in the bacteria are stimulated to permit both oxidation of the aromatic compounds and the synthesis of cellular aromatic compounds. Once the bacteria in the completely mixed activated sludge system have metabolized the toxic aromatic compounds, the protozoa are able to grow and help produce the high quality effluent. The greater sensitivity of the protozoa to toxic organics than bacteria permits the protozoa to be used as primary indicators of the metabolism of the toxic organics. As previously indicated, routine microscopic examination of the protozoa activity in the MLSS can assist the WWTP operator in recognizing potential problems from toxicity before the toxicity adversely affects the bacteria in the wastewater treatment system.

Toxic Nitrogen Compounds

While phenols and substituted phenols attracted considerable attention in the 1950s, the nitro- substituted aromatics also attracted attention. R. B. Cain reported on the isolation of two bacteria that could metabolize para- and ortho-substituted nitrobenzoic acids. *Pseudomonas fluorescens* and *Nocardia erythroplis* were both able to metabolize these nitrobenzoic acids. N. N. Durham showed that *Pseudomonas f*luorescens metabolized para-nitrobenzoic acid by reducing the nitro- group to an amino- group and then hydrolyzing the amino group to form para-hydroxybenzoic acid, which was metabolized as indicated by Stanier. The concern over 2,4,6-trinitrotoluene (TNT) in the environment has led to a number of studies over the years. It was found that the natural soil bacteria could be used to stimulate the metabolism of TNT with the addition of other organic compounds and phosphates. A combined anaerobic-aerobic metabolism was required for complete degradation. Research on nitro- aromatics has shown that the nitro- group can either be removed by oxidation or by reduction and

hydrolysis. The oxidation reaction results in the release of nitrite into the environment. The reduction reaction converts the nitro- group to an amino group before the hydrolysis reaction removes the amino group and produces ammonia. The differences in metabolic pathways have resulted in a number of publications using different bacteria and different nitro- aromatic compounds. The most important aspect of the research to date lies in the fact that many soil bacteria are able to metabolize the nitro- aromatic compounds under normal environmental conditions.

One of the more interesting groups of organic nitrogen compounds is the cyanide group, more commonly known as *nitriles*. The cyanide radical has the carbon atom triple bonded to the nitrogen atom, -C≡N. The cyanide radical becomes a nitrile radical when attached to another carbon atom. Cyanide and nitriles are widely used industrial chemicals and are quite toxic. Hydrogen cyanide exists as a highly toxic gas. Adding potassium hydroxide to hydrogen cyanide creates potassium cyanide, a water soluble compound. Although potassium cyanide undergoes ionization, the primary form is the unionized sodium cyanide at pH levels 7-8. Bacteria metabolism of potassium cyanide occurs aerobically and anaerobically. It appears that the initial metabolic reactions are the same aerobically or anaerobically. The carbon triple bond nitrogen reacts with water to replace the sodium with hydrogen to produce formamide. The formamide is hydrolyzed to ammonium formate. Under aerobic conditions the bacteria oxidize the ammonium formate to carbon dioxide, water, and new cell mass with DO. Anaerobically, the bacteria metabolize the formate to methane, carbon dioxide, and water with very little energy for cell formation. Research has shown that cyanide metabolism occurs more rapidly in the presence of other organic compounds that are easily metabolized. While it is possible to treat cyanide biologically, biological treatment is not used very often. Chemical treatment of cyanide is easier and less expensive than biological treatment.

The organic nitriles are more easily metabolized by bacteria than cyanide. The presence of additional carbon groups gives the bacteria more energy for growth. Biological metabolism of the nitrile radical is quite similar to the cyanide metabolism, adding water to produce the ammonium salt of the corresponding organic acid. The ammonium salt of the organic acid can be metabolized either aerobically or anaerobically under the proper environmental conditions. The additional energy content of the various nitriles makes biological treatment suitable for treating industrial wastewaters containing nitriles.

There is no doubt that cyanide and nitriles are toxic organic compounds that must be removed from wastewaters before being discharged back into the environment. With acclimated bacteria in an aerobic CMAS system, these toxic compounds can be completely metabolized to carbon dioxide, water, ammonia, and new cell mass. With proper design and operation of the biotreatment plant, the ammonia nitrogen can be oxidized to nitrates and the nitrates can be reduced to nitrogen gas, if desired. In effect, the toxic nitrogen organic compounds are converted to non-toxic compounds by bacteria metabolism.

Chlorinated Organic Compounds

Chlorinated organic compounds had been used to a limited extent prior to World War II. During World War II chlorinated organic solvents and pesticides were used extensively, creating large production facilities that shifted to peacetime production after the war without regard to their potential impact on environmental systems. Adding chlorine to organic molecules increases their hydrophobic properties and their resistance to biodegradation. The increased hydrophobic characteristics make these compounds less soluble in water and more soluble in fats and oils. Their low concentrations in water and their chemical structures make them more difficult to be metabolized by bacteria and fungi. DDT, dichloro-diphenyl-trichloroethane, was a very effective pesticide and helped save millions of lives during World War II. DDT also accumulated in the fatty tissues of fish and higher animals that ate fish for food. Two herbicides, 2,4-D, dichlorophenoxyacetic acid, and 2,4,5-T, trichlorophenoxyacetic acid, were widely used as defoliants in the war with Vietnam. Large amounts of these herbicides were used in Vietnam in an effort to reduce the foliage in battle zones. On the home front these herbicides helped control broadleaf weeds in lawns and in agriculture. As these chlorinated materials entered the environment in large quantities, concern was raised that their resistance to biodegradability posed a threat to major biological systems. Over the years there have been extensive studies on the biodegradation of these herbicides. As expected, 2,4,5-T is more difficult for bacteria to metabolize than 2,4-D. Research has shown that there are numerous groups of soil bacteria which can utilize these herbicides as their soul source of organic carbon and derive adequate energy from metabolism to generate new cells.

Another group of chlorinated organic compounds, poly-chlorinated biphenyls (PCB), found a perfect use as a transformer electrolyte. PCB had all the right properties as a commercial transformer electrolyte. Unfortunately, when PCB electrolytes were accidentally spilled into the environment, the PCB remained unchanged. Like the other complex chlorinated aromatics, PCB was relatively insoluble in water and accumulated like DDT in the fatty tissues of animals. Carbon tetrachloride, CCl_4, was widely used as a solvent for removing grease from different products. Carbon tetrachloride also remained unchanged in the environment. A chlorinated phenol, pentachlorophenol, was used to protect wooden posts placed into the ground, extending their normal life. The success of pentachlorophenol in protecting wooden posts in the soil was related to the inability of microorganisms to metabolize this chlorinated organic compound. It is not surprising that the accumulation of chlorinated organic compounds in the environment has been used by environmental activists as prime examples of the dangers created by modern American industry.

Because of the issues raised about chlorinated organic compounds and the potential environmental damage, considerable research has been published in the literature on various studies related to bacteria metabolism of various chlorinated organic compounds. It is not possible to adequately cover all the published research studies in a few pages. Detailed information can be found in the major technical journals. It suffices to say that research studies have found that both pure cultures and mixed cultures of bacteria can metabolize many different chlorinated organic compounds under the proper environmental conditions. Some chlorinated organic compounds are more easily metabolized under aerobic conditions than under anaerobic conditions; while others are more easily metabolized under anaerobic conditions. The key to metabolism lies in the chemical structure of the chlorinated organics and the ability of the bacteria to adapt their enzyme systems to removing the chloro- group. The common chlorinated organics include both aliphatic and aromatic compounds. The aliphatic compounds include saturated and unsaturated hydrocarbons with carbon lengths from 1 to 3 carbon atoms. The aromatic compounds are all unsaturated with one or two benzene ring structures. The enzymes used for removing chloro- groups appear to be hydrolytic enzymes that result in the production of HCl and the addition of the hydroxyl ion to the organic compound where the chloro- group was previously located. Dechlorination can be measured by determining the increase in chloride ions in the aqueous environment and the decrease in pH. It is not surprising that monochlorinated organic compounds are more easily metabolized than dichlorinated organic compounds. The dichlorinated organic compounds are more easily metabolized than trichlorinated organic compounds. The trichlorinated organic compounds are more easily metabolized than polychlorinated organic compounds. Under aerobic conditions the chlorinated organics are metabolized to carbon dioxide, water, hydrochloric acid, and cell mass. Under anaerobic conditions the end products of metabolism will include methane, carbon dioxide, hydrochloric acid, and cell mass. Nutrients and trace metals are required for metabolism. Adequate alkalinity is required to neutralize the hydrochloric acid produced by the dechlorination reactions. Some bacteria find it easier to metabolize the chlorinated organics with other, non-chlorinated, organic compounds. The low solubility of the chlorinated organic compounds requires a non-chlorinated organic compound having a similar chemical structure to stimulate bacterial growth in sufficient numbers to permit metabolism of the chlorinated compounds. Research publications, dealing with bacteria metabolism of chlorinated organic compounds, have been a prime source of information in the technical literature over the past 30 years. Unfortunately, very few research papers have dealt with the quantitative metabolism of chlorinated organic compounds, compared with their non-chlorinated organic counterpart. Recently, a few investigators have begun to make quantitative balances, demonstrating the energy relationships in the metabolism of chlorinated organic compounds. More research is definitely needed into the energy relationships for metabolism of the different chlorinated organic compounds compared with their non-chlorinated analog.

DEVELOPING BIOTREATMENT SYSTEMS FOR HAZARDOUS WASTEWATERS

Finding bacteria that can metabolize hazardous organic compounds is the first step in developing biological treatment systems to treat these materials in hazardous waste streams. Normally, the best source of bacteria is from soil around where the hazardous materials have been used in manufacturing. Most manufacturing plants have had spills and leaks from time to time. Equipment containing the hazardous materials is often washed outside of the normal manufacturing areas with some of the materials being absorbed by the soil. Over the years some hazardous materials have accumulated in the surrounding soil where they have stimulated limited growth of bacteria that can tolerate the toxic materials and can partially metabolize the toxic materials as a source of energy. Since soil often lacks sufficient nutrients for large populations of bacteria or fungi, metabolism of the toxic materials has been limited, allowing the toxic materials to accumulate with time. It is not surprising that large amounts of hazardous organic chemicals have been found in soil environment under a number of chemical plants. The concentration of toxic materials may have reached the level where all biological life is unable to survive in the contaminated soil. If this is the case, the desired bacteria can normally be found further from the high concentrations of toxic materials where the concentration of toxic chemicals begin to drop below the acutely toxic level and some acclimation has occurred in the soil. Examination of the bacteria populations in the surface soil along a line radiating from the manufacturing area can be a useful indicator of where to look for acclimated bacteria.

Simple bacteria plate counts can be used as a rough indicator of the bacteria populations in the soil. As the total number of bacteria per g soil rises, it indicates that the level of toxicity has diminished. When the bacteria population reaches its maximum level and remains steady as you move further from the source of contamination, you have reached the outer limit of the toxic material at the surface. If it is desired to isolate specific metabolic types of bacteria from the contaminated soil, the bacteria growth media should contain a non-toxic organic compound of similar chemical structure as the toxic material being examined. Sodium benzoate can be used as a nutrient when evaluating aromatic, hazardous organic compounds. Sodium benzoate is readily metabolized while stimulating the enzymes for metabolizing benzene ring structures. Sodium butyrate can be used to stimulate bacteria capable of metabolizing aliphatic toxic organics chemicals. Aerobic bacteria can be stimulated by growth in a shallow layer in an Erlenmeyer flask on a mechanical shaker. It is important that the media contains adequate nitrogen, phosphorus, trace metals and alkalinity for good metabolism. If the organic compounds are volatile, the flask should be tightly stopped to prevent loss of the volatile organic compounds from the total system. The volatile organic compounds will come to equilibrium between the aqueous phase and the gaseous phase. As the bacteria metabolize the volatile organic compounds in the aqueous phase, the volatile compounds in the gaseous phase will move back into the aqueous phase and will be metabolized. Care should be taken to insure that the total COD of the organic compounds being treated does not exceed the available oxygen in the gas phase for aerobic metabolism. Anaerobic bacteria will be found in greater numbers in soil at least one to two feet below the surface than in surface soils. Care must be taken to provide a good anaerobic environment if anaerobic bacteria are to be isolated. It is much more difficult to isolate anaerobic bacteria than aerobic bacteria. Special anaerobic jars or anaerobic hoods must be used for isolation and growth of anaerobic bacteria.

Often, the desired bacteria can be found downstream of the industrial plant discharge sewers in streams and rivers. Long-term discharge of industrial wastewaters allows acclimated bacteria to accumulate in the bottom mud and in quiescent zones along the stream banks. If the wastewaters are quite toxic, it may be necessary to travel several miles downstream to where the hazardous organic compounds have been diluted below their toxic concentration and bacteria metabolism has occurred. Dense organic compounds may be covered with settled silt, making it necessary to look below the bottom surface of adjacent streams to find the desired bacteria. Natural development of bacteria will occur when the environment reaches suitable conditions. With some synthetic organic compounds it may not be possible to find bacteria capable of metabolizing the synthetic organic compounds.

LABORATORY TREATMENT UNITS

If the initial isolation techniques yield some bacteria capable of metabolizing the hazardous organics, laboratory treatment units are normally used as the second step to demonstrate that biological treatment of the hazardous organic waste is feasible. Laboratory biotreatment reactors range in size from one liter to 20 liters. Biological treatment units are usually started on non-toxic organic compounds to buildup a large population of microorganisms before starting to feed the hazardous organic compounds and adding the previously isolated bacteria. The closer the non-toxic organic chemical structure is to the chemical structure of the hazardous organic compounds, the easier it will be to change the biological population from the non-toxic organic substrate to the hazardous organic compounds. The simplest laboratory treatment unit is an aerobic system employing diffused aeration to create a batch fed, activated sludge system, operating on a 24-hour cycle. The initial feed should have about 1,000 mg/L to 1,500 mg/L BCOD with sufficient nitrogen, phosphorus, and trace metals to produce a good microbial population. If the hazardous materials are aromatic compounds, sodium benzoate can be used to create a good bacteria population. If the hazardous wastes are non-aromatic compounds, sodium butyrate can be used as the substrate. These two substrates will produce some alkalinity during metabolism without raising the bioreactor pH too high. Ammonium chloride or ammonium sulfate can be used as the nitrogen source. Dibasic potassium phosphate is a good source of phosphorus. The bacteria need a BCOD/N ratio of about 30/1 and a BCOD/P ratio of 150/1 for good growth.

Many bacteriologists like to use phosphate buffers to control the pH in biological reactors. Unfortunately, phosphate buffers will create an artificial environment with very high phosphorus concentrations and should not be used. Bicarbonate is the natural buffer system in biological wastewater treatment systems. Metabolism of the sodium salts of the organic acids will produce some sodium bicarbonate as alkalinity and help hold the pH at the desired level. If the metabolism of the organic compounds produces too much acid, additional sodium bicarbonate should be added to keep the pH between 6.5 and 9. Settled municipal wastewater can be used to provide the trace metals that the bacteria require, additional bacteria seeding, and the liquid for diluting the organic substrate to the desired concentration. Tap water can also be used for making the substrate if municipal wastewaters are difficult to obtain or if the laboratory personnel have concerns about using municipal wastewater. Porous air stone diffusers used in home aquariums and laboratory air systems are normally adequate for both oxygen transfer and mixing. Care should be taken to provide just enough aeration to keep the bioreactor well mixed. Excessive aeration can strip more carbon dioxide from the system than desired and push the alkalinity from bicarbonates to the carbonates, raising the pH above the desired level.

Once the non-toxic substrate has been mixed and placed in the bioreactor, it should be seeded with contaminated soil that showed reasonable bacteria growth, or with some of the bacteria isolated from the initial study on the soil, or just with settled municipal wastewater. Aeration is started; and the system is aerated for about 23 hours. Next, the aeration is turned off for about 1/2 hour. If the settled supernatant is turbid with a few suspended solids settling out, 2/3 of the supernatant should be carefully poured off, trying to retain any settled solids, and replaced with the organic substrate prepared with settled domestic sewage or tap water as the liquid. Aeration should be started and the cycle repeated every 24 hours. After about 3 or 4 days a good activated sludge should form that settles quickly, leaving a relatively clear supernatant. Microscopic examination of the mixed liquor should show good bacteria floc and an active protozoa population. If the protozoa are missing, the bioreactor should be reseeded with settled municipal wastewaters and the feeding continued until a good population of ciliated protozoa occurs. Once a balanced population of bacteria and protozoa are observed in the bioreactor, the acclimation process can be started.

The complexity of the acclimation process depends upon the toxicity of the hazardous organic compound being evaluated. If the hazardous compound has a low toxicity, acclimation can usually be obtained in about 5 days by starting with 20% hazardous material and 80% non-toxic organics in the feed. Each day the hazardous material fraction of the feed is increased another 20% until only the hazardous material is being fed. It is important to make certain the feed has adequate N, P, and trace metals for good bacteria production. If the hazardous material is too toxic, the addition of hazardous material should be reduced to 10%/day or even to 5%/day. Complex organic compounds may require long aeration periods before a suitable bacteria population develops. Failure to develop an acclimated bacteria population in the batch fed system should not be taken as evidence that the hazardous material is non-biodegradable. Some systems have been known to take several months to develop an acclimated bacteria population of adequate size. Microscopic examination of the mixed liquor can be useful in evaluating the growth of the bacteria and the protozoa. Changes in the COD of the mixed liquor over time can be used to indicate loss of organics from the system by metabolism or evaporation. Metabolism of the substrate will create a corresponding increase in MLSS. Evaporation of water from the laboratory unit will require the addition of distilled water to make up for the water loss.

If the batch fed system has difficulty in acclimating to the hazardous wastes, continuously fed systems may be more successful. Small CMAS systems can be constructed from plastic pipe or polyethylene plastic sheets. Small porous air stones are used to supply oxygen for metabolism and mixing the same as in the batch fed bioreactors. The continuously fed bioreactors can employ a small sedimentation section within the aeration section or separate sedimentation units with the settled sludge returned to the aeration unit by a small peristaltic pump or an airlift pump. Separate sedimentation units tend to have sludge accumulation on the wall that must be cleaned at regular intervals. The combination aeration-sedimentation bioreactors tend to provide the best results in small laboratory units. The continuously fed CMAS pilot plants utilize slow speed, peristaltic pumps to move the desired substrate through the aeration tanks at the rate of 1.0 to 2.0 g BCOD/L aeration volume/d. The MLSS are allowed to accumulate to at least 1,000 mg/L VSS before starting the acclimation process. A slow shift from the non-toxic organic substrate to the toxic organic substrate should allow the buildup of the desired microorganisms in the activated sludge. Acclimation can be determined by the effluent COD and by daily microscopic observations of the MLSS to determine both bacteria and protozoa activity. Once the bioreactor has acclimated to the toxic substrate, it will be necessary to waste the daily accumulation of TSS to keep the system in equilibrium. The acclimated activated sludge can also be useful as a source of acclimated microorganisms for use in the BOD_5 test. The laboratory pilot plants are useful in demonstrating that the wastewaters are treatable in an aerobic system. Since it is possible to transfer oxygen at a higher rate in the laboratory unit than in the full scale treatment plant, care should be taken not to use the laboratory parameters for plant design.

Anaerobic bioreactors can also be used to examine the biodegradability of concentrated organic hazardous materials. Unfortunately, anaerobic bioreactors are much more complex than aerobic systems and more difficult to establish and operate. Studies with aerobic systems normally last only 4 to 6 weeks to obtain the desired results. Anaerobic studies take at least 4 to 6 weeks to get started and 4 to 6 months to collect the minimum amount of data for evaluation. Normal anaerobic bioreactors will require at least 10 to 12 months operation to obtain sufficient data to insure a full understanding of the microbial reactions with the hazardous materials. Anaerobic bioreactors are far more sensitive to operational upsets and can be frustrating to everyone concerned. It is essential to understand that there is no such thing as a "short term anaerobic study". Data obtained from all short term studies have been shown to have no real value when it comes to the design and operation of field systems. Anyone interested in anaerobic studies should be familiar with R. E. Speece's book, *Anaerobic Biotechnology*, as well as, the latest technical literature, especially technical papers in the *Purdue Industrial Waste Conference Proceedings*.

FULL SIZE TREATMENT PLANT DESIGN

Laboratory bioreactor studies demonstrate the treatability of the hazardous organic materials and the type of system that can be used for field scale treatment plants. Laboratory studies do not provide design data for full size plants. Recognition of this fact is very important. Too often, management expects that the pilot plant studies will yield important design criteria for the full-scale treatment plant. Laboratory systems provide better mixing and greater biological metabolism than full-scale treatment systems. On the other hand, laboratory systems do not produce as good solids separation as full-scale treatment plants. Treatment plant designs must be based on combining the laboratory treatment results with experience from field plants. The wastewater characteristics, the aeration equipment, and the hydraulic retention times of both the wastewater being treated and the microbial suspended solids in the aeration tanks are the major design factors affecting aeration tank design. Oxygen transfer and mixing are two important parameters that determine the upper limits for the design of aeration tanks. Conventional activated sludge plants with coarse bubble diffusers along one wall show oxygen transfer rates from 20 to 30 mg/L/hr. Placing the same type of coarse bubble diffusers along both side walls of the aeration tank can increase the oxygen transfer to 40 mg/L/hr with the same overall air flow rate. Placing more air diffusers over the entire tank floor and adding more air overall can produce oxygen transfer rates of 60 mg/L/hr or more. Fine bubble diffusers have increased the oxygen transfer rates to 100 mg/l/hr. The usefulness of high oxygen transfer aerators is limited by the oxygen demand rate created by the microorganisms metabolizing the organic matter in the wastewater being added to the aeration tank. Unfortunately, a high rate of metabolism is created only at high organic feed rates and results in more organics in the effluent. High quality effluents can only be produced at relatively low organic loading rates, negating the benefits of high efficiency aeration equipment. Operating aeration tanks in series in large treatment plants allows the aeration systems to be adjusted to the specific oxygen demands. While oxygen demand rates will control aeration requirements in the first aeration tank, mixing will control aeration requirements in the second aeration tank.

Because of the toxic nature of hazardous organic wastewaters, CMAS treatment systems are preferred over other modifications of activated sludge. CMAS plants minimize the toxic effect of the incoming organic compounds, while allowing the bacteria to respond most efficiently to the organic compounds. When organic materials are discharged intermittently from industrial processes, a well-mixed surge tank ahead of the biological reactor will produce a relatively uniform organic loading rate together with a more uniform hydraulic loading rate. While the more uniform organic loading rate will minimize oxygen demand variations in the aeration tanks, the more uniform hydraulic loading rate will minimize suspended solids separation problems in the secondary sedimentation tanks. The secondary sedimentation tanks must capture 98% to 99% or more of the MLSS entering from the aeration tank, if effluent criteria are to be met without tertiary treatment units. The MLSS are separated from the treated wastewaters by entrainment in the dense layer of suspended solids at the bottom of the sedimentation tanks and mechanically recycled back to the aeration tank on a continuous basis to maintain the desired MLSS concentration in the aeration tanks. Fundamental concepts of activated sludge systems have shown that living microbes, dead microbes, and non-biodegradable suspended solids, contained in the incoming wastewater, accumulate in the activated sludge on a continuous basis. A small fraction of the excess suspended solids are carried out in the treated effluent. The majority of the excess suspended solids are removed by wasting from either the secondary sedimentation tanks or the return sludge pipes on a daily basis. The active bacteria in the WAS requires further treatment before the WAS can be returned back into the land environment.

Large anaerobic treatment plants have been more difficult to design than aerobic treatment plants simply because of the lack of field scale plant experience to go with the laboratory experience. Anaerobic systems have worked best when hazardous organic materials are a small part of the total organic load. The non-toxic organic compounds help create the bacteria necessary to metabolize the hazardous organic compounds. Some successes have been obtained with upflow anaerobic sludge blanket (UASB) systems and with fixed media systems, using plastic trickling filter media. The complexity of the microbial populations in anaerobic systems makes anaerobic systems more sensitive to toxic organic materials than aerobic systems. Care must be made to use well-mixed surge tanks to level out variations in both organic concentrations and hydraulic loading rates. Even with 99% organic stabilization in anaerobic systems, the effluent will contain from 10 to 1000 times more organics in the treated effluent than an activated sludge effluent. Higher organic loading rates in the anaerobic bioreactors, as compared to activated sludge systems, account for this difference in effluent quality. Normally, aerobic treatment units are placed after anaerobic treatment units to produce a high quality effluent for discharge back to the environment.

IN-SITU TREATMENT

Accumulation of spilled hazardous organic materials in the soil has resulted in efforts toward in-situ treatment of these materials. The basic problem with in-situ treatment is the inability to see and evaluate accurately the complete treatment process underground. Limited sampling and analyses are inadequate for proper evaluation of any in-situ treatment system. The failure of microorganisms to metabolize the hazardous organic materials in the soil over time is a clear indication that the microbial environment is currently unfavorable and must be adjusted if in-situ treatment is to become successful. The concentration of toxic organic compounds could be too high, exceeding the toxic threshold. There could be a deficiency of N, P, or any of the trace elements required for microbial growth. If aerobic microbial metabolism is desired, there could be a lack of an oxygen source. If anaerobic metabolism is desired, there could be a lack of sulfides to depress the oxidation-reduction potential low enough for growth of methane bacteria. Lastly, there could be a lack of sufficient numbers of active bacteria acclimated to the hazardous materials.

If the organic materials are located near the soil surface, 10 to 15 ft, surface treatment is possible. Surface treatment consists of digging up the contaminated soil, adding the necessary nutrients, and allowing the normal soil bacteria to develop aerobically. As the bacteria metabolize the organic compounds in the upper layer of soil, the contaminated soil is mixed and the moisture adjusted to allow complete stabilization of the all of the organic materials, toxic and non-toxic. Over time the lower levels of contaminated soil are mixed with the treated soil to continue the stabilization process. Because soil mixing is difficult compared with hydraulic mixing in tanks, it normally takes several months to several years to completely stabilize the hazardous organic materials. If the hazardous organic compounds have migrated deeper than 15 to 20 ft below the soil surface, treatment is much more difficult. It is necessary to pump nutrients into the soil in an effort to stimulate microbial growth. It has even been proposed to pump acclimated bacteria into the contaminated soil to provide a suitable population of bacteria for rapid stabilization of the organic compounds. Pushing actively growing bacteria through soil is not an easy task, since the fluid flow takes the path of least resistance rather than flowing past the hazardous organics. The bacteria tend to produce films of cells that grow outward from the soil particles into the void spaces, creating resistance to fluid flow through the void spaces. It may also be necessary to pump water into the soil to provide a suitable fluid environment for the bacteria. If water is pumped into the soil, it must be pumped back out of the soil if potential contamination of groundwater is to be avoided

Pump-and-treat of hazardous wastes is also complicated by the hydraulic characteristics of the soil itself. The simplest pump-and-treat system requires that the hazardous organics are pumped to treatment systems located on the land surface. It is often necessary to pump water into the contaminated soil to push the contaminants towards the recovery wells. Detergents have also been added to the injection water where complex, hydrophobic organics are involved. The detergents remove the hydrophobic organics from the soil particles, permitting their removal in the water phase. Since removal of complex hydrophobic organics by detergents requires partitioning of the organics between the soil and the detergents, considerable effort is required to remove the organics to very low levels. Surface treatment of the contaminated water requires that the detergent be biodegradable. The hazardous organics may or may not be biodegradable. If the hazardous organics are biodegradable, the treatment system must be designed for both the detergents and the hazardous organics. If the hazardous wastes are not biodegradable or are slowly biodegraded, metabolism of the detergent fraction will allow the organics to be released as insoluble organics which can be removed by physical separation.

There is no doubt that biological treatment of hazardous waste is an important part of environmental pollution control. Hazardous wastes have been ignored for years and have created serious environmental pollution problems in many areas of the world. It will take years to clean up all of the land contamination that has occurred from improper waste disposal practices. Proper treatment of hazardous wastewaters that are currently being generated on a daily basis is essential to prevent the hazardous waste problem from increasing. Industrial plants must recognize that continued operations depend on proper waste processing for the return of all residual materials not incorporated into the finished products.

THINGS TO REMEMBER

1. The term, "hazardous waste", is a relatively new term to designate what was previously known as "toxic waste".

2. Hazardous wastes in the United States are designated by federal legislation and controlled by the EPA.

3. Hazardous wastes are created by specific chemical industries and people using the products of those industries.

4. Hazardous wastes can be treated by physical, chemical, and biological systems.

5. Although chlorinated hazardous wastes are the most persistent in nature, biological treatment systems can destroy the toxicity of these chlorinated organic compounds.

6. Laboratory studies can be used to determine if biological systems can be used to treat hazardous wastes.

7. Laboratory data can be used to develop full-scale treatment plant design concepts.

8. Full-scale treatment plant design must be adapted to fit available, mechanical equipment.

9. In-situ treatment of hazardous waste is difficult because of the lack of mixing in the soil system and the difficulty in creating the desired environment under the soil surface.

10. Bio-augmentation with special cultures have not been as successful as naturally developed acclimated cultures from soil around the specific hazardous waste site.

REFERENCES

Cain, R. B. (1958) The Microbial Metabolism of Nitroaromatic Compounds, *J. Gen. Microbiol.*, **19**, 1.

Coe, R. H. (1952) Bench Scale Biological Oxidation of Refinery Wastes With Activated Sludge, *Sew. & Ind. Wastes*, **24**, 731.

94th Congress (1976) *The Toxic Substances Control Act of 1976*, Public Law 94-469.

94th Congress (1976) *Resource Conservation and Recovery Act of 1976,* Public Law 94-580.

98th Congress (1984) *The Hazardous and Solid Waste Amendments of 1984*, Public Law 98-616.

Durham, N. N. (1959) Studies on the Metabolism of p-Nitrobenzoic Acid, *J. Microbiol.*, **4**, 141.

Environmental Protection Agency (1980) Hazardous Waste and Consolidated Permit Regulations, *Federal Register*, **45,** 33063.

Fallon, R. D. (1992) Evidence of a Hydrolytic Route for Anaerobic Cyanide

Degradation, *Appl. Environ. Microbiol.*, **58**, 3163.

Harlow, I. F., Powers, T. J. and Ehlers, R. B. (1938) The Phenolic Waste Treatment Plant of the Dow Chemical Company, *Sew. Works Jour.*, **10**, 1043.

Helmers, E. N., Frame, J. D., Greenberg, A. E. and Sawyer, C. N. (1951) Nutritional Requirements in the Biological Stabilization of Industrial Wastes, II. Treatment with Domestic Sewage, *Sew. & Ind. Wastes*, **23**, 884.

Ka, J. O., Holben, W. E., and Tiedje, J. M. (1994) Genetic and Phenotypic Diversity of 2,4-Dichlorophenoxyacetic Acid (2,4-D)-Degrading Bacteria Isolated from 2,4-D-Treated Field Soils, *Appl. Environ. Microbiol.*, **60**, 1106.

Kansas Dept. of Health and Environment (1992) *Hazardous Waste Generator Handbook*, Topeka, Kansas

Kilbane, J. J., Chatterjee, D. K., and Chakrabarty, A. M. (1983) Detoxification of 2,4,5-Trichorophenoxyacetic Acid from Contaminated Soil by *Pseudomonas cepacia*, *Appl. Environ. Microbiol.*, **45**, 1697.

Lynn, G. E. and Powers, T. J. (1955) Bacterial Studies In Oxidation of Phenolic Wastes, *Sew. & Ind. Wastes*, **27**, 61.

McKinney, R. E., Symons, J. M., Shifrin, W. G. and Vezina, M. (1958) Design and Operation of a Complete Mixing Activated Sludge System, *Sew. & Ind. Wastes*, **30**, 287.

Mathews, W. W. (1952) Treatment of Ammonia Still Wastes by the Activated Sludge Process, *Sew. & Ind. Wastes*, **24**, 164.

Nishino, S. F. and Spain, J. C. (1993) Degradation of Nitrobenzene by a *Pseudomonas pseudoalcaligenes*, *Appl. Environ. Microbiol.*, **59**, 2520.

Sheets, W. D., Hamdy, M. K. and Weiser, H. H. (1954) Microbiological Studies on the Treatment of Petroleum Refinery Phenolic Wastes, *Sew. & Ind. Wastes*, **26**, 862.

Stanier, R. Y. (1947) Simultaneous Adaptation: A New Technique for the Study of Metabolic Pathways, *J. Bacteriol.*, **54**, 339.

Stanier, R. Y. (1948) The Oxidation of Aromatic Compounds by Fluorescent Pseudomonads, *J. Bacteriol.,* **55**, 477.

INDEX